DEUTSCHE AKADEMIE DER WISSENSCHAFTEN ZU BERLIN

Zentralinstitut Physik der Erde (ZIPE)

Seismologischer Dienst Jena

Seismological Bulletin 1967
Station Moxa (MOX)

By

Peter Bormann and Johannes Stelzner

with 15 Figures

AKADEMIE-VERLAG · BERLIN

1972

Erschienen im Akademie-Verlag GmbH, DDR-108 Berlin, Leipziger Str. 3—4

Lizenznummer: 202 · 100/582/72
Gesamtherstellung: VEB Druckerei „Thomas Müntzer", DDR-582 Bad Langensalza
Bestellnummer: 2004/B/1967. ES 18 E 2
EDV 761 678 5
Printed in German Democratic Republic

PREFACE

With the beginning of the volume 1967, the "Seismological Bulletin of the Station Moxa" is published as an independent publication.

The former series of „Veröffentlichungen des Instituts für Geodynamik Jena" was stopped as a result of the constitution of the „Zentralinstitut Physik der Erde" 1969, February, in which new institution the „Institut für Geodynamik Jena" is incorporated.

During 1967, the equipment of station Moxa has been little changed. Since 1967, 20 January a new short-period vertical seismometer from type VSJ-II with the magnification of 100000 at the period of 1 s is working. The records of seismometers from type Krumbach ($V_{max} = 2000$, $T_0 = 2$ s) are stopped in August 1967.

The Seismological Bulletin has been developed from provisional evaluations of the Seismological Service under the direction of JOHANNES STELZNER, who was assisted by DOROTHEA GÜTH.

Final interpretations of the registrations were done by PETER BORMANN. Some changes in the scheme of the evaluation are noticed in the "Preliminary Notes for the Interpretation of Seismograms".

Control of the instruments of the station Moxa was carried out under the direction of CHRISTIAN TEUPSER.

There is added to the Bulletin a scientific contribution concerning "Relative Frequencies of Body Wave Onsets in Seismic Registrations of the Station Moxa".

TABLE OF CONTENTS

Preliminary notes for the interpretation of seismograms

In the Bulletin the international code is used:

1. Phase interpretation

Pg — direct longitudinal wave in near epicentral distances ($D < 10°$)

Pb, Pn — guided longitudinal head waves along the CONRAD- or MOHOROVIČIĆ-discontinuity ($D < 10°$)

P — direct longitudinal wave travelled through the earth mantle

PKIKP — direct longitudinal wave travelled through the inner core (travel-time branch DF)

PKHKP — direct longitudinal wave refracted in the intermediary zone between inner and outer core. Phase symbol according to BOLT [1] (travel-time branch GH)

PKP2 — direct longitudinal wave travelled through the outer core only (travel-time branch AB)

PKP — first noticeable onset of longitudinal core phase, not identified

PP, PPP — waves reflected on the earth surface with permanent longitudinal character

PKKP — core phase reflected once within the core at the outer core boundary

PKPPKP — longitudinal core phase reflected at the earth surface

Sg — direct transversal wave in near epicentral distances ($D < 10°$)

Sb, Sn — guided transversal head waves along the CONRAD- or MOHOROVIČIĆ-discontinuity ($D < 10°$)

S — direct transversal wave travelled through the earth mantle

SKS — direct wave travelled transversal through the mantle and longitudinal through the core

SS, SSS — waves reflected on the earth surface with permanent transversal character

SKKS — wave travelled transversal through the mantle, longitudinal through the core and reflected within the core at the outer core boundary

PcP, ScS, PcS, ScP — longitudinal and transversal waves with steady or changing character reflected at the outer core

PS, SP, PPS — longitudinal and transversal waves with changing character reflected at the surface of the earth

pP, sP, pPP, sPP,
pPKIKP, sPKP2, pS — phases of deep-focus earthquake of longitudinal or transversal waves with steady or changing character. p;s — reflected near the epicentre

pPKP, sPKP — phases of deep focus earthquakes of longitudinal core waves not exactly to be coordinated

SKP, PKS — core phases with different character before and after the direct transit of the core

SKSP — SKS wave with longitudinal character after the reflection at the surface of the earth

P_1, P_2, P_3, . . . ,
PP_1, PP_2, . . . ,
S_1, S_2, . . . ,
$PKKP_1$, $PKKP_2$, . . . ,
$PKPPKP_1$, $PKPPKP_2$, . . . — multiple onsets of body waves

Pn, Sn — teleseismic Pn and Sn waves in the epicentral distances $23° < D < 40°$ after BÅTH [2]

Pa, Sa — waves probably guided in the asthenosphere channel or higher modes of surface waves

PL — leaking modes, normal dispersed train of waves of periods greater than about 10 s, beginning at or near the time of initial P-wave

X, Y, Z — remarkable phases of body waves, not to be identified

Li, Lg1, Lg2, Rg — guided waves in the continental crust, probably higher modes of surface waves

L(3.20), R(2.90) — waves guided in the continental crust, probably higher modes of surface waves with the apparent horizontal velocity of 3.20 km · s^{-1} and 2.90 km · s^{-1} respectively

LQ — beginning of LOVE waves

LR — beginning of RAYLEIGH waves

LmV, LmH — maximum of the vertical and horizontal component respectively of longperiodical surface waves. If there are several maxima with comparable proportions in A/T, the numeration was carried out in a temporal sequence e. g. Lm1H, Lm2H

The phase symbol is followed by the designation of the type of seismometer from which the time of onsets is taken.

A — seismograph with amplitude characteristic of type A (short-period)

B — seismograph with amplitude characteristic of type B (middle-period)

C — seismograph with amplitude characteristic of type C (long-period)

2. Measurement of amplitudes and calculation of magnitudes

All data of amplitudes and periods printed in the column "remarks" are always taken from the records of the same instruments, from which are taken the onset-times of the corresponding phases. In case of doubt the symbol of phase and component is followed by the symbol of the type of instruments in parenthesis e. g.: PV(A), PV(B), SH(B)

Data of amplitudes obtained from records of instruments of type A are given in units of lenght of nm (1 nm = 1 nanometre = 10^{-6} millimetre). Data of amplitudes obtained from instruments of type B and such obtained from instruments of type C are given in units of length µm (1 µm = 1 mikrometre = 10^{-3} millimetre) e. g.: PV 1,25 s 38,6 nm, SH: 10 s 3,2 µm, LmH: 22 s 15 µm.

Magnitudes are determined from all those phases, for which calibrating functions are known and internationally used, i. e.

for maxima of body waves (h < 60 km) P(PH, PV), PP(PPH, PPV), and S(SH)-Q-functions from GUTENBERG and RICHTER [3] — and

for maxima of surface waves (h < 100 km) LmH, LmV — calibrating functions from Prague σ [4] —.

The station correction S was not yet taken into consideration.

MPV, MP1V, MP2V, MPPV — magnitude of vertical component V of the adequate body waves

MPH, MPPH, MSH — magnitude of horizontal component H of the adequate body waves

MLV, ML1V, ML2V — magnitude of the vertical component V of the maximum surface waves

MLH, ML1H, ML2H — magnitude of the horizontal component H of the maximum surface waves

If there are several evaluations of amplitudes from different types of seismographs for the same wave, the symbol of magnitudes is followed by the symbol of instruments e. g.: MPV(A), MPV(B).

3. Direction of body-wave onsets

If the direction of motion at the beginning of a wave onset is clearly to be recognized, the sign + or — is placed before the phase symbol. It means:

in the Z component	+ ground motion upwards, compression
	— ground motion downwards, dilatation
in the N component	+ ground motion to the north
	— ground motion to the south
in the E component	+ ground motion to the east
	— ground motion to the west

4. Further abbreviations

i — sharp beginning of phase motion (impetus)

e — gradual beginning of phase motion (emersio)

D — epicentral distances in degree (°), calculated according to geocentric coordinates, the maximum error of the own calculations amounts to ± 0,1°

Az — azimuth: clockwise measured angle between north direction in epicentre and the connecting line from epicentre to station Moxa

h — depth of focus in km, our data for depth of focus are based on travel-time curves for deep focus earthquakes after GUTENBERG and RICHTER [5]

H — origin time in GMT (Greenwich Mean Time)

USCGS — United States Coast and Geodetic Survey, Washington

BCIS — Bureau Central International de Séismologie, Strasbourg

ANUSSR — Academia Nauk USSR, Moscow

UPP — Seismological Institute Uppsala, Sweden

Round brackets indicate uncertainties in interpretation of phase, time, depth of focus or epicentral distances respectively.

[1] BOLT, A., The velocity of seismic waves near the earths center. Bull. Seism. Soc. Am. 54 (1964) 1, 191—208.

[2] BÅTH, M., Propagation of Sn and Pn teleseismic distances. Pure and Applied Geophysics 64 (1966/II) 19—30.

[3] GUTENBERG, B. and RICHTER, C. F., Magnitude and energy of earthquakes. Annali di Geofisica 9 (1956) 1, 1—15.

[4] KÁRNÍK, V., KONDORSKAJA, N. V. u. a., Standardization of the earthquake magnitude scale. Stud. Geophys. et Geodet., Prague 6 (1962) 41—48.

[5] GUTENBERG, B. and RICHTER, C. F., Materials for the study of deep-focus earthquakes. Bull. Seism. Soc. Am. 26 (1936) 4, 341—390.

Seismological Station Moxa (MOX)
of the Institute of Geodynamics, Jena

Elevation above mean sea level:	**455 m**
Bedrock:	**clay slate of the lower carboniferous formation**
Geographic coordinates:	$\varphi = 50°38'46''$ **N** $\lambda = 11°36'58''$ **E**
Address:	**Zentralinstitut Physik der Erde (since February 1969) Institutsteil Jena, Seismologischer Dienst** **DDR-69 Jena, Burgweg 11** **German Democratic Republic** **Telex: 058 8668 seis dd**

Seismographs and their parameters 1967

T_0 — seismometer free period
T_g — galvanometer free period
D_s — seismograph damping
D_g — galvanometer damping
V_0 — magnification factor
N — north-south component
E — east-west component
Z — vertical component

Type	Comp.	T_s [s]	T_g [s]	D_s	D_g	V_0	σ^2
Krumbach (modif.)	Z	2.0	0.21	0.54	3.68	23500	0.221
SKM-III	N	1.53	0.34	0.5	2.23	19000	0.10
	E	1.45	0.35	0.5	2.08	23000	0.12
	Z	1.49	0.35	0.5	2.12	19000	0.13
VSJ-II	Z	1.0	1.0	0.50	0.50	43700	0.56
SSJ-I	N	20	1.13	0.48	8.79	220	0.074*)
		20	1.13	0.50	8.87	112	0.074**)
		20	1.14	0.50	8.79	1120	0.074
	E	20	1.13	0.51	8.83	210	0.071*)
		20	1.13	0.49	8.85	107	0.071**)
		20	1.16	0.49	8.61	1090	0.071
	Z	20	1.13	0.51	8.83	210	0.048*)
		20	1.13	0.48	8.82	107	0.048**)
		20	1.24	0.48	8.05	1070	0.048
SSJ-I/L	N	30	70.7	1.24	0.5	1500	0.15
	E	30	79.1	1.3	0.5	1200	0.087
	Z	30	77.2	1.3	0.5	1300	0.066
Strain/L	N		48.6		0.60	55***)	
	E		53.4		0.69	74***)	
Wiechert	N	6.2		0.27		230	
1200 kp	E	7.2		0.34		205	
Mainka 150 kp	N	19.3		0.31		50	
200 kp	E	21.4		0.47		47	

*) 1. 1.—30. 8.
**) 1. 9.—31. 12.
***) for wave velocity 5 km s^{-1}

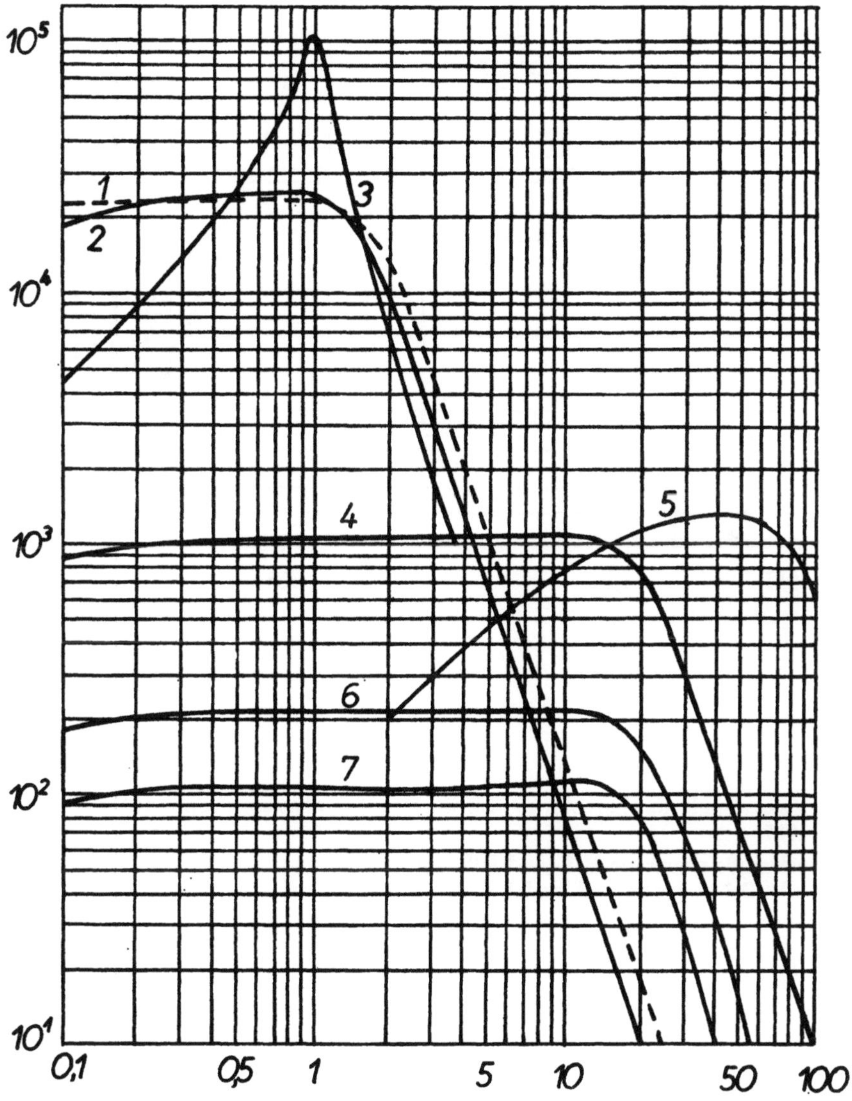

Fig. 1. Mean amplitude characteristics of the electromagnetic seismographs of the station Moxa 1967

1 — Modified Krumbach Seismograph (Z-component)
2 — Seismograph Kirnos Modernised-III (SKM-III) (NS-, EW- and Z-component)
3 — Seismograph Type Jena II (Z-component)
4 — Seismic Station Apparatus Type Jena I/1000 (SSJ-I/1000) (NS-, EW- and Z-component)
5 — Seismic Station Apparatus Type Jena I/L (SSJ-I/L) (NS-, EW- and Z-component)
6 — Seismic Station Apparatus Type Jena I/200 (SSJ-I/200) (NS-, EW- and Z-component)
7 — Seismic Station Apparatus Type Jena I/100 (SSJ-I/100) (NS-, EW- and Z-component)

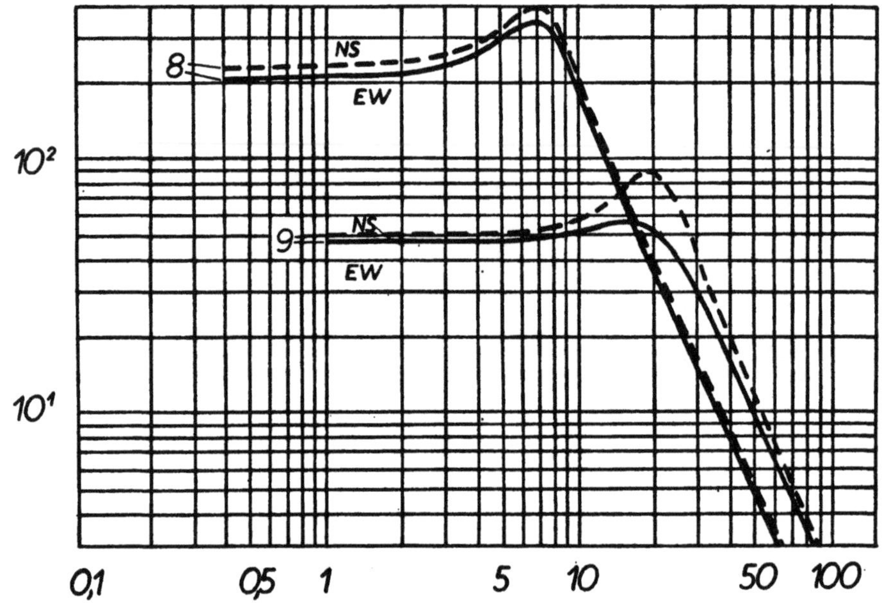

Fig. 2. Mean amplitude characteristics of the mechanical seismographs of the station Moxa 1967

8 — Wiechert Seismograph (NS- and EW-component)
9 — Mainka Seismograph (NS- and EW-component)

Seismological Recordings at Station Moxa 1967

Day	Phase		h m s	Remarks
1.	ePKP	A	00 40 28	Santa Cruz Islands 12.07°S 166.22°E
	ePP	A	43 07	H = 00 21 06.6 h = normal MAG=4.9
				D = 136.3° Az = 336.7° (USCGS)
1.	eP	A	03 11 18	Andaman Islands Region 10.71°N 92.78°E
	e	A	11 37	H = 02 59 33.8 h = 60 km MAG=5.2 (USCGS)
				D = 76.2°
1.	ePKP	A	07 25 22.5	Tonga Islands 15.27°S 173.65°W
	+i	A	25 24.2	H = 07 05 48.6 h = normal MAG=6.0 (USCGS)
	-i	B	25 32.5	D = 144.5°
	ePP	B	28 38	PV1:1.2s 46.0nm PV2:1.3s 117nm
	eSKKS	C	35 31	PV3:11s 3.1 /um
	iSKSP	B	38 44	LmH:18.5s 4.7 /um LmV:17.5s 4.4 /um
	ePSKS	C	38 52	MLH=6.2 MLV=6.2
	ePPS	C	41 10	i 25 28.0 e 25 35
	eSS	C	47.8	
	LmH	B	08 30.5	
	LmV	B	36.6	
1.	ePKIKP	A	09 04 19	Fiji Islands 20.51°S 178.41°W
	ePKHKP	A	04 23	H = 08 45 42.2 h = 627 km MAG=5.6
	ePKP2	A	04 29	D = 148.9° Az = 347.6° (USCGS);
	e	A	06 24	(h = ca. 610 km)
	e(pPKHKP)	A	06 40	PV2:1.4s 48.5nm
1.	e	A	09 17 13	Santa Cruz Islands 12.26°S 165.98°E
	e(PP)	A	17 30	H = 08 55 16.8 h = normal MAG=4.7 (USCGS)
	LmH	B	10 21.2	D = 136.5°
	LmV	B	21.2	LmH:17s 1.1 /um LmV:16s 1.3 /um
				MLH=5.6 MLV=5.7
1.	iP	A	10 06 03.3	
1.	LmH	B	15 44.6	Santa Cruz Islands 12.39°S 165.78°E
	LmV	B	44.6	H = 14 18 51.4 h = normal MAG=5.0 (USCGS)
				D = 136.5°
				LmH:ca.20s ca.0.7 /um LmV:ca.20s ca.0.9 /um
				MLH=ca.5.3 MLV=ca.5.4

Day	Phase		h m s	Remarks
1.	ePKP	A	22 18 22	Santa Cruz Islands 11.13°S 165.46°E
	e	A	18 29	H = 21 58 57.8 h = normal MAG=5.4 (USCGS)
	e	A	18 40	D = 135.0°
	e	A	21 38	LmH:16s 1.2 µm LmV:18s 1.5 µm
	LmH	B	23 22.5	MLH=5.7 MLV=5.7
	LmV	B	25.6	e 23 29
1.	LmV	B	23 49.2	LmH:16s 1.1 µm LmV:16s 1.5 µm
	LmH	B	49.5	
2.	e	A	04 16 38	
	i	A	16 50.0	
2.	LmH	B	07 57.5	Near Coast of Northern Chile
	LmV	B	57.5	25.21°S 70.98°W
				H = 06 53 17.5 h = 38 km MAG=5.0 (USCGS)
				D = 104.6°
				LmH:20s 0.97 µm LmV:20s 0.9 µm
				MLH=5.3 MLV=5.3
2.	eP	A	08 24 08	Near Coast of Libya 32.21°N 22.75°E
	i	A	24 10.5	H = 08 19 32.8 h = 20 km MAG=4.5
	i	A	24 15.5	D = 20.2° Az = 339.1° (USCGS)
	ei	A	24 22	PV1:1.4s 24.2nm PV2:1.5s 60.0nm
	ePP	A	24 29	PV3:1.5s 80.0nm PV4:1.8s 122nm
	eS	B	27 54	LmH:17s 2.0 µm
	LmH	B	33.0	MPV1=4.2 MPV2=4.6 MLH=4.6
				Successive P-onsets with increasing amplitude.
2.	-iP	A	09 58 14.3	Republic of the Congo 10.18°S 28.54°E
	epP	A	58 24	H = 09 47 53.3 h = normal MAG=5.6
	isP	A	58 29.0	D = 62.3° Az = 347.9° (USCGS); h = 40 km
	ei	A	58 37.5	PV:1.2s 41.0nm pPV:1.4s 24.2nm
	e	A	58 58	MPV=5.4
2.	eP	A	13 56 57.5	Iran 30.63°N 50.43°E
	e	A	57 18	H = 13 50 06.2 h = 40 km MAG=5.2
				D = 35.1° Az = 316.0° (USCGS)

Day	Phase		h m s	Remarks
2.	eP	A	14 03 12.5	
2.	iPg	A	14 10 04.8	Explosion/GDR 51.37°N 12.89°E
	iSg	A	10 20.5	D = 1.1°
2.	ePKIKP	A	20 19 17	Santa Cruz Islands 12.33°S 166.41°E
	e	A	19 21	H = 19 59 58.2 h = normal MAG=5.2
	e	A	19 29	D = 136.6° Az = 336.8° (USCGS)
	ePP	B	22 00	PV2:1.2s 17.9nm PV3:1.6s 41.6nm
	eSKP	A	22 53	SKPV:1.8s 51.0nm
	eSS	C	40.2	LmH:20s 1.5 µm LmV:16s 1.5 µm
	LmH	B	21 24.2	MLH=5.7 MLV=5.8
	LmV	B	25.1	Successive PKP-onsets with increasing amplitude.
3.	eP	B	05 39(21)	Kenai Peninsula/Alaska 60.86°N 151.53°W
	ipP	B	39 43	H = 05 28 29.3 h = 92 km MAG=4.6 (USCGS)
	e	B	40 03.5	D = 68.0° h = 90 km
				pP is greater than P.
3.	eSS	C	06 15.6	Santa Cruz Islands 10.91°S 165.50°E
	LmH	B	59.4	H = 05 35 46.6 h = normal MAG=5.2 (USCGS)
	LmV	B	07 02.6	D = 135.0°
				LmH:18s 4.1 µm LmV:16s 3.2 µm
				MLH=6.2 MLV=6.1
3.	LmH	B	07 07.2	LmH:16s 5.9 µm
3.	LmH	B	07 16.5	Santa Cruz Islands 11.16°S 165.46°E
	LmV	B	19.5	H = 05 52 51.8 h = normal MAG=5.3 (USCGS)
				D = 135.2°
				LmH:17.5s 4.7 µm LmV:17s 3.5 µm
				MLH=6.2 MLV=6.1
				AN USSR gives: 10.4°S 165.0°E
				H = 05 53 06 h = 134 km
3.	LmV	B	07 43.2	LmH:15.5s 2.9 µm LmV:16s 3.5 µm
	LmH	B	43.7	

Day	Phase		h m s	Remarks
3.	LmH	B	12 19.2	Probably North Atlantic Ocean
	LmV	B	(20)	
3.	LmH	C	12 55	Probably Santa Cruz Islands (USCGS)
3.	iPKP	A	20 32 20.5	New Hebrides Islands 20.50°S 169.27°E
	epPKP	A	32 32	H = 20 12 48.4 h = 46 km MAG=4.4
				D = 145.1° Az = 335.0° (USCGS); h = 42 km
				PV:1.2s 25.6nm
3.	ePKP	A	21 42 43	Santa Cruz Islands 12.38°S 166.36°E
	ePP	A	45 25	H = 21 23 21.8 h = normal MAG=5.0
	eSKP	A	46 13	D = 136.6° Az = 336.7° (USCGS)
	eSS	C	22 03 38	LmH:19s 0.8 µm LmV:18s 0.76 µm
	eSSS	C	08 48	MLH= 5.4 MLV=5.4
	LmH	B	49	e 42 52 e 45 30 e 45 37 e 46 35
	LmV	B	(49)	
4.	-iP	A	03 54 13.0	Philippine Islands Region 20.32°N 120.02°E
	+ipP	A	54 23.0	H = 03 41 36.4 h = normal MAG=5.6
	e	A	57 24	D = 85.5° Az = 322.7° (USCGS); h = 37 km
	e(PP)	A	57 33	PV:1.6s 75.5nm pPV:1.5s 56.7nm
	LmV	B	04 36.8.	LmH:13s 0.97 µm LmV:12s 1.26 µm
	LmH	B	37.3	MPV=5.6 MLH=5.4 MLV=5.5
4.	LmH	C	04 05	Bismarck Sea 3.58°S 148.95°E
	LmV	C	05	H = 02 53 28.1 h = 9 km MAG=4.9 (USCGS)
				D = 121.0°
				LmH:23s 0.66 µm LmV:24s 0.6 µm
				MLH=5.2 MLV=5.2
4.	i	A	04 44 33.0	Taunus/GFR 50.1°N 8.2°E
	eiSg	A	45 06	H = 04 43 52 (BCIS)
	ei	A	45 10	D = 2.3°
4.	e(P)	A	06 02 18.5	Greece 38.64°N 22.14°E
	e	A	02 21	H = 05 58 54.1 h = 7 km MAG=5.2
	e	A	02 27	D = 14.1° Az = 331.6° (USCGS)
	iPP	A	02 30.3	Successive P-onsets with increasing amplitude.

Day	Phase		h m s	Remarks
cont.				
4.	eSS	C	06 05 16	
	LmH	C	08.2	
4.	eSg	A	08 02 44	Taunus/GFR 50.1°N 8.2°E H = 08 01 32 (BCIS) D = 2.3°
4.	epP	A	11 37 56	Burma-India Border Region
	e	A	38 06	23.43°N 93.91°E
	e	A	38 22	H = 11 26 45.4 h = 58 km MAG=5.4 (USCGS) D = 67.5° (h = ca. 70 km) The P-onset must be 18 s earlier than pP. An alternative interpretation would be that the phase given as pP is the P-onset of a second stronger shock in the same place.
4.	e	A	14 36 18	Probably explosion.
	e	A	36 22	
4.	eP	A	18 12 38	Near S. Coast of Southern Honshu 33.64°N 135.82°E H = 18 00 58.7 h = 416 km MAG=4.3 (USCGS) D = 82.7°
4.	eP	A	20 27 11.5	Near Coast of Venezuela 10.74°N 62.50°W
	epP	A	27 31	H = 20 15 55.8 h = 74 km MAG=5.5
	esP	A	27 40.5	D = 71.7° Az = 40.2° (USCGS); h = 80 km AN USSR gives: h = normal
4.	e	A	20 43 28	Explosion?
4.	LmH	C	23 47	Probably South Sandwich Islands (USCGS)
5.	+eiP1	A	00 24 15.5	Mongolia 48.08°N 102.80°E
	iP2	A	24 21.5	H = 00 14 40.4 h = normal MAG=6.4
	iP3	AB	24 28.0	D = 55.7° Az = 309.6° (USCGS)
	ePP2	B	26 27.5	PV1:1.2s 71.8nm PV2:1.8s 1120nm

Day	Phase		h m s	Remarks
cont.				
5.	ePP3	B	00 26 34	PV3(A):1.6s 1575nm PV3(B):8s 16.3 μm
	eS2	B	32 04	SH3:18s 60 μm
	iS3	B	32 11	LmV:17s (610 μm)
	eiSS	B	35 56	MPV1=5.6 MPV2=6.6 MPV3(A)=6.8
	iSSS	B	37 44	MPV3(B)=7.1 MSH3=7.3 MLV=ca.7.8
	LmV	B	48.0	Multiple body wave onsets with successively increasing amplitudes. It seems to be possible to interpret this as successive shocks with increasing magnitude in the same location. Compare Jan. 20 at 02 06. P has a period of about 23 s in the long-period seismographs of the SSJ-1/L type (type C).
5.	eP	A	00 51(50)	Mongolia 48.38°N 103.05°E
				H = 00 42 13.3 h = normal MAG=5.6 (USCGS)
				D = 55.6°
5.	eP	A	06 26 18.5	Mindoro/Philippine Islands
	e	A	26 22	13.82°N 120.68°E
	e	A	27 06	H = 06 13 31.6 h = 166 km MAG=5.4
	i	A	27 14.8	D = 91.1° Az = 323.0° (USCGS)
	e	A	27 27	PV:1.4s 30.3nm
	LmH	C	07(04)	MPV=5.2
				AN USSR gives: h = normal
5.	+iP	A	10 16 03.4	Kirgiz SSR 39.41°N 72.93°E
	ei(pP)	A	16 09	H = 10 07 58.3 h = 11 km MAG=5.3
	ePP	A	17 45	D = 43.6° Az = 305.9° (USCGS);
	eS	C	22 30	(h = 25 km)
	eSS	C	25(48)	PV:1.2s 28.2nm
	LmH	B	34.3	LmH:18s 1.9 μm LmV:14s 2.3 μm
	LmV	B	38.3	MPV=5.1 MLH=5.1 MLV=5.3
				e 16 14 i 16 17.0 e 17 55
5.	e	A	16 38(00)	Savoy/France 46.2°N 6.4°E
	eSg	A	38 06	H = 16 35 01 (BCIS)
	iLg2	A	38 11.0	D = 5.7°

Day	Phase		h m s	Remarks
5.	eSg	A	20 09 30	Austria 46.8°N 13.7°E H = 20 07 20 (BCIS) D = 4.1°
5.	eSg	A	20 11 43	Austria 46.8°N 13.7°E H = 20 09 36 (BCIS) D = 4.1°
6.	+iP	A	00 07 57.0	Mongolia 48.10°N 102.90°E
	eS	C	15 40	H = 23 58 21.4 h = normal MAG=5.4
	LmH	B	33	D = 55.8° Az = 309.6° (USCGS)
	LmV	B	34.2	PV:1.3s 72.2nm LmH:15s 3.7 μm LmV:16s 5.2 μm MPV=5.6 MLH=5.6 MLV=5.7
6.	+eiP	A	00 16 03	Hokkaido/Japan Region 41.80°N 143.34°E
	e	A	16 10	H = 00 04 02.7 h = 35 km MAG=5.5
	epP	A	16 13.5	D = 78.7° Az = 331.0° (USCGS); h = 40 km
	eS	C	25 55	PV:2.0s 214nm
	e(ScS)	C	26 11.5	LmH:17s 4.6 μm LmV:17s 5.6 μm
	LmH	B	53.8	MPV=5.9 MLH=5.9 MLV=6.0
	LmV	B	58.1	
6.	ePn	A	04 45 10	Northeastern Italy 45.9°N 11.5°E
	e(Pg)	A	45 29	H = 04 43 58 (BCIS)
	eSn	A	46 05.5	D = 4.8°
	eSg	A	46 30.5	e 46 37
7.	ePKIKP	A	00 46 34	Southeast Indian Rise 48.79°S 112.75°E
	eSKP	B	49 59	H = 00 27 25.2 h = normal MAG=5.8
	eSKKS	C	55 48	D = 131.3° Az = 303.8° (USCGS)
	eSS	C	01 06 28	PV1:1.8s 61.2nm SKPV:8s 1.2 μm
	eSSS	C	11 15	e 49 47.5 e 07 18
	LmH	C	43.0	
	LmV	B	48.8	

Day	Phase		h m s	Remarks
7.	ePKP	A	11 52(20)	Santa Cruz Islands 12.28°S 166.05°E
	LmH	B	12 59	H = 11 33 00.5 h = normal MAG=4.7 (USCGS)
	LmV	B	59	D = 136.4°
				LmH:18s 0.7 µm LmV:18s 0.8 µm
				MLH=5.4 MLV=5.4
7.	LmH	B	14 40	South of Mariana Islands 11.80°N 142.67°E
	LmV	B	40	H = 13 34 48.3 h = 36 km MAG=5.6 (USCGS)
				D = 104.7°
				LmH:20s 0.8 µm LmV:20s 1.2 µm
				MLH=5.3 MLV=5.4
7.	ePP	B	17 03 02	Santa Cruz Islands 11.87°S 166.06°E
	e	B	03 30	H = 16 41 03.0 h = normal MAG=5.1 (USCGS)
	eSKP	B	03 48	D = 136.1°
	eSS	C	21.0	LmH:18s 1.2 µm LmV:18s 1.0 µm
	LmH	B	18 04.5	MLH=5.6 MLV=5.6
	LmV	B	04.5	
8.	eP	B	05 14 08	Near East Coast of Kamchatka
	i	A	14 09.5	55.97°N 162.87°E
	ePP	B	16 48	H = 05 02 52.1 h = normal MAG=5.1 (USCGS)
	eS	C	23 26	D = 71.5°
	ePPS	C	24 04	PV:8s 0.85 µm PPV:10s 0.4 µm
	eSS	C	28.0	LmH:15.5s 3.1 µm
	LmH	C	45.5	MPV=5.9 MPPV=5.6 MLH=5.7
8.	eP	A	05 18 04.5	Near East Coast of Kamchatka
				59.09°N 162.91°E
				H = 05 06 47.7 h = 23 km MAG=4.9
				D = 71.0° Az = 341.1° (USCGS)
				PV:1.5s 30.0nm
				MPV=5.2
8.	eP	A	06 54 47.5	Near East Coast of Kamchatka
	LmH	B	07 26	56.05°N 162.79°E
	LmV	B	30	H = 06 43 32.3 h = 44 km MAG=4.9 (USCGS)
				D = 71.6°
				PV:1.4s 54.5nm
				LmH:16s 0.8 µm MPV=5.5 MLH=5.1

Day	Phase		h m s	Remarks
8.	eP	A	08 43 17	<u>Near East Coast of Kamchatka</u>
	LmH	B	09 14.5	56.18°N 162.68°E
	LmV	B	(20)	H = 08 31 59.7 h = 24 km MAG=4.9
				D = 70.9° Az = 341.0° (USCGS)
				PV:1.5s 40nm
				LmH:16s 0.77 µm
				MPV=5.3 MLH=5.1
8.	LmH	B	16 48.5	<u>Santa Cruz Islands</u> 12.19°S 166.50°E
	LmV	B	49	H = 15 27 16.9 h = 40 km MAG=5.1 (USCGS)
				D = 136.6°
				LmH:20s 0.6 µm LmV:20s 0.7 µm
				MLH=5.3 MLV=5.4
9.	iP	A	02 02 47.0	<u>Southern Iran</u> 27.71°N 54.47°E
	e	A	03 07	H = 01 55 13.6 h = 17 km MAG=5.3
	e	A	03 41	D = 39.6° Az = 317.2° (USCGS)
	ePP	A	04 24	PV:1.9s 58.8nm PPV:1.9s 35.3nm
	eS	C	08(48)	LmV:12s 0.6 µm
	eSS	C	11(45)	MPV=5.2 MPPV=4.7 MLV=4.7
	LmH	B	(22)	
	LmV	B	24.2	
9.	e(P)	A	14 07 05	
9.	eP	A	18 21 01	<u>Near West Coast of Columbia</u> 5.15°N 77.63°W
	eSKS	C	31 27	H = 18 08 23.9 h = 40 km MAG=5.2
	eS	C	31 36	D = 85.6° Az = 39.7° (USCGS)
	ePS	B	32 40	PV:9s 0.5 µm
	eSS	C	37(10)	LmH:16s 0.6 µm LmV:16s 0.6 µm
	LmH	B	19 06	MPV=5.7 MLH=5.1 MLV=5.1
	LmV	B	06	i 21 05.0 e 21 11.5 e 21 18 e 21 26
				e 24 31
9.	ePKP	A	21 36 34	<u>New Hebrides Islands</u> 19.75°S 169.46°E
				H = 21 17 23.1 h = 204 km MAG=4.3 (USCGS)
				D = 144.5°
9.	e	A	22 00 50	

Day	Phase		h m s	Remarks
9.	e	A	23 22 21	
10.	LmH	C	09 19	Honshu/Japan
10.	ePKP	A	13 53 53	Tonga Islands 19.62°S 175.84°W
	e	A	54 04	H = 13 34 05.8 h = normal MAG=5.0
	e	A	54 15	D = 148.5° Az = 350.9° (USCGS)
	LmH	C	14 56	
10.	e	A	14 00 03	
	i	A	00 13.0	
	i	A	00 26.0	
10.	LmH	C	18 29	Near East Coast of Kamchatka
	LmV	C	(33)	55.98°N 162.84°E
				H = 17 42 38.5 h = normal MAG=4.4 (USCGS)
				D = 71.3°
11.	eP	A	06 07 50	Northern Celebes 0.13°S 120.10°E
	ePP	A	12 03	H = 05 54 00.1 h = 23 km MAG=5.6 (USCGS)
	LmH	B	07 05.8	D = 101.8°
	LmV	B	05.8	LmH:17s 0.4 µm LmV:16s 0.5 µm
				MLH=4.8 MLV=5.1
11.	e	A	10 33 02	Explosion?
	e	A	33 06	
	e	A	33 29	
11.	e	A	11 19(10)	Explosion?
11.	+eP	A	11 26 50	Iran-Iraq Border Region 34.13°N 45.71°E
	+epP	A	26 56.5	H = 11 20 45.7 h = 34 km MAG=5.6
	ePPP	A	27 59.5	D = 29.8° Az = 314.1° (USCGS); h = 33 km
	ePcP	A	29 52	PV:2.0s 79.0nm SH:10s 1.0 µm
	eS	B	31 48	LmH:15s 2.7 µm LmV:14s 3.8 µm
	iSS	C	32 36	MPV=5.3 MSH=5.5 MLH=5.0 MLV=5.2
	LmH	B	41.8	e 27 02 e 27 37 ei 28 13
	LmV	B	42.8	

Day	Phase		h m s	Remarks
11.	i(P)	A	12 39 16.0	
11.	+eP	A	16 21(00)	<u>South of Panama</u> 5.34°N 82.50°W
	eSKS	C	31 32	H = 16 08 06.1 h = 22 km MAG=5.3
	eS	C	31 48	D = 88.5° Az = 39.4° (USCGS)
	eSS	C	37(36)	LmH:22s 1.9 µm LmV:24s 2.3 µm
	LmH	B	53.5	MLH=5.5 MLV=5.5
	LmV	B	53.5	e 21 23 e 21 29
12.	e	A	03 36 31	
	e	A	36 47	
	e	A	37 04	
13.	e	A	06 49 23	
13.	e(P)	A	10 33 05.5	
13.	eSn	A	10 35 15.5	<u>Apennines/Italy</u> 44.7°N 9.1°E
	eSg	A	35 59	H = 10 32 37 (BCIS)
	eLg2	A	36 07	D = 6.2°
13.	iPg	A	14 06 51.5	<u>Explosion/GDR</u>
	iSg	A	07 07.6	D = 1.2°
13.	e	A	14 10 16	<u>Solomon Islands</u> 10.63°S 161.36°E
	eSKP	B	10 54	H = 13 48 11.7 h = 32 km MAG=5.7 (USCGS)
	ePPS	C	21.7	D = 133.0°
	eSS	C	27 20	LmH:17s 3.4 µm LmV:17s 3.5 µm
	LmH	B	15 13	MLH=6.1 MLV=6.1
	LmV	B	13	e 10 39 e 29 20
13.	e(P)	A	19 52 27	
	e	A	52 41	
	e	A	52 55	
13.	e(P)	A	23 27 06.5	
13.	e(P)	A	23 58 30	

Day	Phase		h m s	Remarks
14.	e	A	11 08 17.5	
14.	eP	A	12 16 39	<u>Rat Islands/Aleutian Is.</u> 52.08°N 175.45°E
	epP	A	16 50	H = 12 04 50.7 h = 41 km MAG=5.1
	LmH	C	48	D = 76.7° Az = 349.5° (USCGS); h = 43 km
				PV:1.3s 19.5nm pPV:1.1s 12.2nm
14.	e	A	13 38 22	<u>Mindoro/Philippine Islands</u>
	LmH	C	14(26)	13.59°N 120.65°E
	LmV	C	(26)	H = 13 24 53.2 h = 38 km MAG=4.7 (USCGS)
				D = 91.2°
				The first onset of P is not detectable.
14.	ePS	C	14 33 08	<u>Prince Edward Islands Region</u>
	e	C	33 50	43.44°S 39.13°E
	eSS	C	38 20	H = 14 06 48.3 h = normal MAG=5.3 (USCGS)
	LmH	B	15 08.2	D = 96.9°
	LmV	B	08.2	LmH:16s 1.4 µm LmV:16s 1.6 µm
				MLH=5.5 MLV=5.6
14.	eP	A	14 54 56	PV1:1.4s 15.2nm PV3:1.3s 19.5nm
	e	A	54 59.5	
	e	A	55 05	
14.	+iP	A	15 37 35.8	<u>Northern Sinkiang Prov./China</u>
	e	A	37 46.0	44.57°N 81.49°E
	+i(pP)	A	37 50.7	H = 15 29 15.0 h = normal MAG=5.0
	e	A	37 56	D = 45.9° Az = 303.6° (USCGS);
				(h = 67 km)
				PV:1.0s 21.4nm
				MPV=5.2
14.	iPg	A	19 09 17.0	<u>Italy-Austria Border Region</u> 46.5°N 13.5°E
	e	A	10 10	H = 19 07 50 (BCIS)
	eSg	A	10 13	D = 4.4°
15.	eP	A	09 27 44.5	<u>Kodiak Island Region</u> 56.49°N 153.17°W
	epP	A	27 51	H = 09 16 16.5 h = 8 km MAG=4.4
				D = 72.5° Az = 10.1° (USCGS); h = 25 km

Day	Phase		h m s	Remarks
15.	iPg	A	14 35 11.0	<u>Northeastern Italy</u> 46 1/4°N 12 3/4°E
	iSn	A	35 37.7	H = 14 33 42 (BCIS)
	iSb	A	35 46.5	D = 4.4°
	iSg	A	35 59	e 35 54.5
15.	eP	A	20 08 13	<u>Lake Baikal Region</u> 55.72°N 110.72°E
	e	A	08 18	H = 19 58 45.6 h = 32 km MAG=5.1
	i	A	08 25.2	D = 54.6° Az = 309.6° (USCGS)
	eS	C	15 52	PV:2.1s 91.0nm
	eSS	C	19(36)	LmH:15s 1.4 μm LmV:15.5s 1.5 μm
	LmH	B	33.5	MPV=5.5 MLH=5.2 MLV=5.2
	LmV	B	33.6	ei 08 35.5 e 20 14
16.	e	A	03 37 35	Explosion?
	e	A	37 51	
	e	A	37 56	
16.	e(P)	A	03 44 30	<u>Japan</u> 36.2°N 138.2°E
	e	A	44 50	H = 03 32 12.3 h = 38 km MAG=4.6 (USCGS)
	LmH	B	04 21.4	D = 81.4°
				LmH:12s 0.65 μm
				MLH=5.2
				AN USSR gives:
				Sichote Alin Mountains 44.9°N 135.0°E
				H = 03 33 24 h = 215 km
16.	ePP	C	05 06(30)	<u>Santa Cruz Islands</u> 11.30°S 165.68°E
	ePKS	C	07(20)	H = 04 44 27.3 h = normal MAG=5.3 (USCGS)
	ePPS	C	18.5	D = 135.4°
	eSS	C	24.3	LmH:18s 0.75 μm LmV:18s 0.64 μm
	LmH	B	06 07.3	MLH=5.4 MLV=5.3
	LmV	B	07.8	
16.	LmH	B	12 34.2	<u>Solomon Islands</u> 10.66°S 161.33°E
	LmV	B	34.2	H = 11 09 08.4 h = 40 km MAG=5.1 (USCGS)
				D = 133.1°
				LmH:18s 0.64 μm LmV:18s 0.64 μm
				MLH=5.4 MLV=5.4

Day	Phase		h m s	Remarks
16.	+ePKIKP	A	14 45 47	Santa Cruz Islands 11.22°S 165.71°E
	ePP	B	48 24	H = 14 26 22.9 h = 6 km MAG=5.3
	ePKS	C	49 18	D = 135.3° Az = 336.7° (USCGS)
	eSS	C	15 06.4	PV:1.4s 21.2nm
	LmH	B	44.2	LmH:16s 1.9 µm LmV:18s 2.2 µm
	LmV	B	49.8	MLH=5.9 MLV=5.9
16.	eP	A	20 04 12	Dodecanese Islands 36.63°N 26.91°E
				H = 20 00 12.4 h = 158 km (USCGS)
				D = 17.8°
17.	e	A	00 32 42	Explosion
	e	A	32 43.5	
17.	-eP	A	01 20 46	Santiago del Estero Prov./Argentina
	epP	A	22 59	27.41°S 63.28°W
	iPP	A	25 03.0	H = 01 07 54.3 h = 590 km MAG=5.5
	epPP	A	27 02	D = 101.9° Az = 38.9° (USCGS);
	eSKS	C	30 32	h = ca. 610 km
	iPS	C	34 35	PV:2.3s 100nm PPV:2.5s 350nm
	esS	C	35 30	PSH:20s 1.9 µm sSH:24s 1.4 µm
	iPPS	C	37 02	PPSH:26s 3.2 µm (SKKP)V:2.0s 66.6nm
	e(SKKP)	A	39 20.5	MPV=5.8 MPPV=6.1
	eSa(4.57)	C	49(10)	e 20 48.5 e 23 03 ei 23 06.5 e 24 09
				e 27 13.5 e 33 09 e 39.1 e 42.1
				The phase interpreted as (SKKP) is about 10 s earlier than the SKKP-onset must be after the travel time table.
				AN USSR gives: h = normal
17.	ePKIKP	A	01 36 35	New Hebrides Islands 14.73°S 167.25°E
	e(PP)	A	39 20.5	H = 01 17 19.4 h = 90 km MAG=4.9 (USCGS)
				D = 139.1°
				PV1:1.2s 10.2nm PV2:2.0s 66.8nm
				The stronger second onset is about 15 s earlier than the PP-onset must be for this quake after the travel time table. An alternative interpretation would be to attribute this phase to the SKKP-phase of the Argentina deep-earthquake.

Day	Phase		h m s	Remarks
17.	ePKP	A	03 40 09	Fiji Islands Region 20.18°S 177.77°W
				H = 03 21 18.3 h = 500 km MAG=4.6
				D = 148.7° Az = 348.5° (USCGS)
17.	+iP	A	12 11 44.5	Near East Coast of Honshu/Japan
	epP	A	11 57	38.29°N 142.07°E
				H = 11 59 31.5 h = 44 km MAG=5.9
				D = 81.3° Az = 330.7° (USCGS); h = 46 km
17.	e	A	16 18 23	
18.	e(Sg)	A	04 19 58	Explosion?
	e	A	20 12.5	
	e	A	20 13	
18.	eP	A	04 32 37	Kurile Islands 48.88°N 154.88°E
	e(pP)	A	32 43	H = 04 20 52.9 h = 40 km MAG=5.4
				D = 76.0° Az = 336.9° (USCGS);
				(h = 23 km)
				PV:1.6s 30.2nm
				MPV=5.2
18.	+eP	A	05 44 28.5	Eastern Russia 56.64°N 120.84°E
	-iP	B	44 28.5	H = 05 34 32.6 h = 11 km MAG=6.1 (USCGS)
	i	A	44 31.5	D = 58.0°
	ePP	B	46 34	PV(A):2.0s 719nm PV(B):5s 3.8 μm
	ePa(8.26)	C	47 34	PPV:6s 1.5 μm SH(A):3.4s 3700nm
	i(PPP)	B	47 50	LmH:17s 103 μm LmV:14s 52.8 μm
	eS	C	52 23	MPV(A)=6.5 MPV(B)=6.8 MPPV=6.4
	iS	C	52 32	MSH(A)=6.8 MLH=7.0 MLV=6.8
	iS	A	52 32.1	i 44 33.0 i 44 38.5 i 52 34 ei 03 50
	LmH	B	06 07.9	ei 05 20
	LmV	B	11.2	Clear higher mode surface waves.
18.	eP	A	06 33 25	Kurile Islands 45.31°N 150.69°E
				H = 06 21 27.0 h = normal MAG=4.3
				D = 78.1° Az = 334.8° (USCGS)

Day	Phase		h m s	Remarks
18.	-iP	A	08 30 14.4	Fox Islands/Aleutian Is. 52.51°N 168.30°W
	+iP	B	30 15	H = 08 18 22.0 h = 37 km MAG=5.7
	i	A	30 21.5	D = 77.2° Az = 0.0° (USCGS)
	eS	BC	40 00	PV(A):1.3s 161nm SH(B):16s 0.7 µm
	iScS	C	40 27	MPV(A)=6.0 MSH(B)=5.6
	eLQ	C	50.7	i 30 26.5 i 30 30.5
18.	iP	A	08 40 58.0	Hokkaido/Japan Region 42.04°N 142.43°E
				H = 08 29 03.4 h = 65 km MAG=4.8 (USCGS)
				D = 78.0°
18.	ePKP	A	14 49 08	Tonga Islands 17.94°S 175.38°W
				H = 14 29 51.5 h = 226 km MAG=4.6 (USCGS)
				D = 146.6°
				PV:1.2s 20.6nm
18.	eP	A	15 39 38.5	Kurile Islands 47.35°N 152.06°E
				H = 15 28 02.7 h = 140 km MAG=4.8
				D = 76.6° Az = 335.4° (USCGS)
18.	eP	A	21 59 01	Mongolia 48.13°N 102.92°E
	i	A	59 04.7	H = 21 49 25.8 h = normal MAG=5.2 (USCGS)
	e	A	59 35	D = 56.0°
	ePP	A	22 01 06	LmH:16s 2 µm LmV:14s 2.2 µm
	LmH	B	20.5	MLH=5.3 MLV=5.4
	LmV	B	25.3	
19.	eSg	A	04 37(18)	France-GFR Border Region 48.7°N 7.7°E
	e	A	37 26	H = 04 35 33 (BCIS)
				D = 3.2°
19.	ePKIKP	A	12 57 35	Santa Cruz Islands 11.83°S 166.44°E
	e	A	58 11.5	H = 12 38 31.3 h = 156 km MAG=5.5
	e(sPKIKP)	A	58 35	D = 136.2° Az = 337.0° (USCGS);
				(h = 165 km)
19.	ePKP	A	12 59(45)	Fiji Islands Region 14.84°S 178.76°W
	i	B	59 52.5	H = 12 40 12.6 h = 18 km MAG=6.6 (USCGS)
	e	B	13 03 54	D = 143.5°

Day	Phase		h m s	Remarks
cont.				
19.	iSS	B	13 21 36	LmH:22s 28.7 μm LmV:22s 38.8 μm
	iSS	C	21 38.5	MLH=6.9 MLV=7.1
	LmH	B	58.8	e 07 23 e 22(10)
	LmV	B	59.1	
19.	eP	A	14 50 05.5	<u>Off East Coast of Kamchatka</u> 55.32°N 163.15°E H = 14 38 44.6 h = normal MAG=4.5 (USCGS) D = 71.8°
19.	+eP	A	14 53 27.5	<u>Fox Islands/Aleutian Is.</u> 52.40°N 169.60°W
	e	A	53 33.5	H = 14 41 36.7 h = 55 km MAG=5.2
	e(PcP)	A	53 38.0	D = 77.3° Az = 359.2° (USCGS) PV1:1.0s 47.6nm PV2:1.1s 14.6nm PV3:1.0s 21.4nm MPV=5.6 If (PcP) was interpreted as pP then the focal depth would be h = 40 km.
19.	eP	A	16 57 18	<u>Nevada</u> H = 16 45 00 (UPP) D = ca. 81.2° PV:1.1s 17.8nm MPV=5.1 Probably underground explosion.
19.	e	A	17 56 05.5	
19.	e	A	19 55 48	<u>Probably Caribbean Sea</u> 16.99°N 85.74°W H = 19 43 24.3 h = normal MAG=4.5 (USCGS) D = 81.7° The first onset of P must be about 8 s earlier.
20.	+iP1	A	02 06 59.0	<u>Mongolia</u> 48.00°N 102.94°E
	+iP2	AB	07(00)	H = 01 57 23.1 h = normal MAG=6.1 (USCGS)
	-i	A	07 05.0	D = 55.8°
	-iPP	A	09 02.5	PV1:1.0s 42.8nm PV2(A):(2.3s 2200nm)

Day	Phase		h m s	Remarks
cont.				
20.	+iPP	B	02 09 05	PV2(B):6s 12.5 μm PH2(B):6.5s 7.5 μm
	−ePPP	B	10 14	PPV(B):5.5s 3.7 μm PPPV(B):5.5s 4.7 μm
	+iPPP	A	10 18.3	LmH:13s 75.2 μm LmV:13s 109 μm
	eS	C	14(40)	MPV1=5.5 MPV2(A)=6.9 MPV2(B)=7.2
	iSS	C	18 40	MPH2(B)=7.3 MPPV(B)=6.7 MLH=7.0
	LmH	B	32.8	MLV=7.2
	LmV	B	33.8	Multiple P. P1 is much smaller than P2. It seems possible to interpret this as two shocks in the same focus. Compare Jan. 5 at 00 24.
20.	eP	A	03 36 50	<u>Mongolia</u> 48.01°N 103.00°E
				H = 03 27 13.9 h = normal MAG=5.0
				D = 55.9° Az = 309.7° (USCGS)
20.	eP	A	03 38 22	<u>Mongolia</u> 47.78°N 102.77°E
				H = 03 28 44.8 h = normal MAG=5.1 (USCGS)
				D = 55.8°
20.	iP	A	05 24 58.5	<u>West Pakistan</u> 32.25°N 69.83°E
	e	A	25 06.5	H = 05 16 39.8 h = 70 km MAG=5.1 (USCGS)
	e	A	25 11.5	D = 46.1°
				AN USSR gives: h = normal
20.	eP	A	06 32 52.5	<u>Mongolia</u> 47.94°N 103.09°E
	i	A	32 54	H = 06 23 16.3 h = normal MAG=5.0
	e	A	34 31	D = 56.0° Az = 309.8° (USCGS)
	e	A	34 52	PV2:1.1s 27.0nm
	ePP	A	34 57.5	MPV2=5.3
20.	e	A	11 32 05	Explosion?
	e	A	32 17	
	e	A	32 20	
20.	ePg	A	11 43 17	<u>Explosion/GDR</u>
	iSg	A	43 33.5	D = 1.2°

Day	Phase		h m s	Remarks
20.	e	A	12 02 37	Explosion?
	e	A	02 43	
20.	e	A	12 18 54	
20.	ePg	A	12 53 05	Explosion/GDR
	i	A	53 10	D = 1.2°
	i	A	53 18	
	iSg	A	53 21.0	
20.	iPg	A	14 21 54.0	Explosion
	iSg	A	22 12.0	D = 1.4°
20.	e	A	14 48 45	
20.	e	A	17 41 05	Explosion?
	e	A	41 10	
	e	A	41 19	
20.	eP	A	17 52 21	Nevada
				H = 17 40 00 (UPP)
				D = ca. 81°
				Probably underground explosion.
21.	eP	A	00 51 08	Mongolia 48.09°N 102.92°E
				H = 00 41 32.2 h = normal MAG=4.9
				D = 55.8° Az = 309.7° (USCGS)
21.	ePKP	A	03 13 39	Easter Islands Cordillera 49.79°S 114.83°W
	ei	A	13 44.5	H = 02 54 00.8 h = normal MAG=5.3 (USCGS)
	ei	A	13 50	D = 146.3°
	eSS	C	36.2	PV1:1.1s 14.6nm PV2:1.3s 38.9nm
	eSSS	C	41.2	PV3:2.2s 290nm
	LmH	B	04 28.9	LmH:18s 1.9 μm LmV:17s 2.4 μm
	LmV	B	30.1	MLH=5.8 MLV=6.0
				Successive PKP-onsets with increasing amplitude.

Day	Phase		h m s	Remarks
21.	ePKP2	A	14 08 44	Kermadec Islands Region 30.74°S 178.22°W H = 13 48 14.1 h = 65 km MAG=4.9 (USCGS) D = 158.8°
22.	e	A	04 36 45	Near earthquake?
	e	A	36 52	
	e	A	37 11.5	
22.	e	A	10 41(51)	Fox Islands/Aleutian Is. 53.52°N 165.31°W H = 10 30 03.0 h = 69 km MAG=5.0 D = 76.2° Az = 2.0° (USCGS) The first onset of P, which must be at 10 41 45 is not detectable.
	e	A	42 20	
22.	e	A	11 27 57	Near earthquake?
	e	A	28 12	
	e	A	28 43	
22.	-eP	A	12 11 25	Mongolia 48.09°N 102.86°E
	e	A	11 27.5	H = 12 01 49.0 h = normal MAG=5.1
	LmH	C	35.5	D = 55.7° Az = 309.6° (USCGS)
22.	eP	A	12 21 49.5	Nicobar Islands Region 8.82°N 93.73°E
	e	A	21 55	H = 12 09 52.3 h = 36 km MAG=4.9
	e	A	22 02.5	D = 78.2° Az = 319.9° (USCGS)
	e	A	22 21.5	AN USSR gives: Andaman Islands 13.8°N 96.0°E H = 12 10 09 h = normal MAG=5.0 D = 75.9°
22.	eP	A	12 25 38	Mongolia 48.01°N 102.90°E H = 12 16 02.0 h = normal MAG=5.0 D = 55.8° Az = 309.7° (USCGS)
22.	iP	A	19 31 41.5	North of Ascension Island 0.82°S 15.97°W H = 19 22 01.4 h = normal MAG=4.6 D = 56.4° Az = 20.7° (USCGS)
22.	e	A	21 18 25.5	

Day	Phase		h m s	Remarks
22.	iPKP ei	A A	21 54 11.8 54 45	<u>Tonga Islands</u> 16.53°S 173.83°W H = 21 34 35.0 h = normal MAG=4.2 (USCGS) D = 145.7°
22.	ePKP epPKP	A A	22 54 26 56 40	<u>Fiji Islands Region</u> 18.04°S 178.52°W H = 22 35 50.6 h = 600 km MAG=4.5 (USCGS) D = 146.5° h = 600 km
22.	eP i	A A	23 20 52 20 54.3	<u>Taiwan Region</u> 21.83°N 121.76°E H = 23 08 17.4 h = 63 km MAG=4.9 (USCGS) D = 85.4°
23.	e	A	02 00 39.5	
23.	e e(Sg)	A A	11 24 32 24 40	Explosion?
23.	ePKP	A	11 29 45.5	<u>Kermadec Islands</u> 27.72°S 176.91°W H = 11 09 51.8 h = 60 km MAG=5.1 (USCGS) D = 156.1°
23.	e e e	A A A	14 40 27 40 32 40 43	Explosion?
23.	e(Sg)	A	15 38 09	
23.	eP i i	A A A	20 57 40 57 52.2 58 04.5	<u>North of Ascension Island</u> 1.64°S 15.64°W H = 20 47 56.7 h = normal MAG=5.1 D = 57.1° Az = 20.3° (USCGS)
23.	eLQ LmH LmV	C C B	21 03.3 (18) 20.2	<u>Revilla Gigedo Islands Region</u> 19.92°N 109.34°W H = 20 25 38.3 h = 56 km MAG=5.3 (USCGS) D = 92.7° LmV:17s 0.5 μm MLV=5.1

Day	Phase		h m s	Remarks
24.	+iP	A	03 17 35.0	Hokkaido/Japan Region 41.45°N 141.94°E
	epP	A	17 55	H = 03 05 39.0 h = 69 km MAG=5.7
	e	A	18 01	D = 78.5° Az = 330.3° (USCGS); h = 77 km
	e	A	18 51	PV:1.3s 77.7nm
	LmH	C	52	MPV=5.5
	LmV	B	55.2	e 20 05
24.	e	A	05 27 41	Probably near earthquake.
	e	A	28 35	
	e	A	28 47	
	e	A	29 16	
	e	A	29 35	
24.	eP1	A	09 39 04.5	Central Mid-Atlantic Ridge 0.58°S 21.04°W
	+iP2	A	39 17.5	H = 09 29 12.3 h = normal MAG=4.9
	ePP1	C	41(17)	D = 58.1° Az = 23.9° (USCGS)
	ePP2	C	41 28	PV1:1.2s 41.0nm PV2:1.3s 350nm
	e(Pa)	C	42 40	SH2:22s 33.4 µm
	eS1	B	47 10	LmH:15s 34 µm LmV:14s 44 µm
	iS2	B	47 22	MPV1=5.4 MPV2=6.3 MSH2=7.0 MLH=6.6
	eSS	B	51 08	MLV=6.8
	i(SS2)	C	51 16	e 42 45 e 42 58 e 54 20
	LmH	B	10 07.2	Multiple body wave onsets. Probably two
	LmV	B	10.5	shocks in the same area, the second larger
				than the first one.
24.	iP	A	14 56 20.8	Szechwan Province/China 30.12°N 104.12°E
	i	A	56 22	H = 14 45 16.0 h = normal MAG=5.4
	e	A	56 50	D = 68.8° Az = 317.0° (USCGS)
				PV2:1.8s 51nm
				MPV2=5.4
24.	e	A	15 57 02	
25.	+iP	A	01 58 04.5	Afghanistan-USSR Border Region
	+iP	B	58 05	36.64°N 71.63°E
	-isP	AB	59 31.5	H = 01 50 19.4 h = 281 km MAG=5.7
	ei(PcP)	A	59 48	D = 44.4° Az = 308.0° (USCGS); h = 275 km
	iPP	B	59 53	PV(A):1.2s 656nm PH(A):1.2s 300nm

Day	Phase		h m s	Remarks
cont.				
25.	isPP	A	02 01 12.5	PV(B):4s 3.1 /um sPV(A):1.6s 425nm
	isPP	B	01 15.5	SPV(B):4s 2.4 /um (PcP)V3:1.7s 278nm
	iS	B	04 16	PPV:10s 2.0 /um sPPV(B):10.5s 2.3 /um
	isS	B	05 58	MPV(A)=5.8 MPH(A)=5.8 MPV(B)=6.0
	eScS	C	07 28	MPPV=6.0
	eSS	C	07(40)	e 01 29 e 01(53) e 04 00
				AN USSR gives: h = 270 km
25.	iPKP	A	04 31 42.0	Fiji Islands Region 20.16°S 178.52°W
				H = 04 12 58.8 h = 576 km MAG=4.0 (USCGS)
				D = 148.5°
25.	ePKIKP	A	07 50 19	Fiji Islands Region 20.6°S 178.4°W
	iPKHKP	A	50 23.7	H = 07 31 35.3 h = 517 km MAG=4.8 (USCGS)
	ePKP2	A	50 30.5	D = 149.0°
25.	e	A	08 03 05	Probably explosion.
	e(Sg)	A	03 23	
25.	ePKHKP	A	11 03 07.5	Fiji Islands Region 19.87°S 178.12°W
	iPKP2	A	03 12.0	H = 10 44 27.2 h = 599 km MAG=4.2
				D = 148.3° Az = 348.2° (USCGS)
25.	e	A	20 16 25	Explosion (KHC)
	e	A	16 29	
	e	A	16 42.5	
25.	LmH	C	07 00	Revilla Gigedo Islands 21.38°N 108.91°W
	LmV	C	00	H = 06 04 33.9 h = normal MAG=5.3 (USCGS)
				D = 91.2°
26.	iSg	A	10 46 18.5	Explosion/GDR 51.0°N 14.4°E
	e	A	46 33	D = 1.8°
26.	e(P)	A	10 49 54	
26.	iPg	A	14 07 28.8	Explosion/GDR 51.37°N 12.89°E
	iSg	A	07 44.8	D = 1.2°

Day	Phase		h m s	Remarks
26.	e	A	14 29 57.5	
26.	eP	A	16 15 31.5	Tunisia 34.62°N 10.39°E
	e	A	15 50	H = 16 11 42.8 h = normal MAG=5.0
	LmH	B	21.9	D = 16.0° Az = 2.8° (USCGS)
	LmV	B	(23)	LmH:16s 2.0 μm
				MLH=4.4
				AN USSR gives: Africa 32.8°N 7.9°E
				H = 16 11 21 h = normal
				D = 18.0°
26.	eP	A	16 23 16	Mexico-Guatemala Border Region
	e	A	23 21	15.04°N 92.76°W
	e	A	23 47	H = 16 10 34.3 h = 56 km MAG=5.3 (USCGS)
	eS	C	33 48	D = 87.3°
	LmH	B	17 04.3	LmH:17s 1.5 μm LmV:18s 1.9 μm
	LmV	B	04.6	MLH=5.4 MLV=5.5
				e 24 06.5 e 24 22
27.	eP	A	08 21 42	Central Mid-Atlantic Ridge 0.89°N 28.14°W
	e	A	22 05	H = 08 11 36.5 h = normal MAG=4.7 (USCGS)
	e	A	22 30.5	D = 59.9°
	eS	C	29(55)	i 22 32.5 e 22 44
27.	e	A	10 27 14	
27.	ePKP	A	12 56 12	Fiji Islands Region 20.52°S 177.76°W
	i	A	57 27.0	H = 12 37 13.5 h = 440 km MAG=4.6 (USCGS)
				D = 149.0°
				PV:1.6s 22.8nm
28.	epP	A	01 53 07.5	Taiwan 24.79°N 121.83°E
	e	A	53 09	H = 01 40 26.9 h = 90 km MAG=5.2
				D = 83.0° Az = 323.0° (USCGS)
				pP is a clear onset in our short-period records. P seems to be much smaller and is not detectable.
				AN USSR gives: h = normal.

Day	Phase		h m s	Remarks
28.	eP	A	03 07 05	West Pakistan 30.23°N 69.53°E
				H = 02 58 33.7 h = 39 km MAG=4.5
				D = 47.2° Az = 312.7° (USCGS)
28.	+eiP	A	14 04 49.5	Fox Islands/Aleutian Is. 52.40°N 169.54°W
	+iP	B	04 50.5	H = 13 52 58.3 h = 47 km MAG=5.9
	e	A	04 53.5	D = 77.3° Az = 359.2° (USCGS)
	iS	C	14 39	PV(A):1.0s 234nm PV(B):11s 14 µm
	iSS	C	19 40	PH(B):12s 4.7 µm
	ePKPPKP	A	32 20	LmH:20s 38.5 µm LmV:16s 40 µm
	LmH	B	43.0	MPV(A)=6.3 MPV(B)=7.0 MPH(B)=6.9
	LmV	B	48.4	MLH=6.7 MLV=6.8
				e 04 56.5 i 05 00 i 05 10.0 i 05 26.2
				e 09 20
28.	+eP	A	14 17 49.2	Fox Islands/Aleutian Is. 52.33°N 169.43°W
	e	A	17 54	H = 14 05 58.1 h = 54 km MAG=5.0
				D = 77.4° Az = 359.3° (USCGS)
				PV:1s 21.4nm
				MPV=5.2
28.	eP	A	14 19 05	Aleutian Islands
				H = 14 07 14 (UPP)
28.	+eP	A	14 35 18.5	Fox Islands/Aleutian Is. 52.43°N 169.41°W
	e	A	35 21	H = 14 23 26.7 h = 47 km MAG=5.2
	e	A	35 45	D = 77.3° Az = 359.3° (USCGS)
	e	A	36 13	PV1:1.1s 24.4nm PV2:1.1s 24.4nm
	e	A	38 35	MPV=5.2
28.	eP	A	14 42 17.5	Fox Islands/Aleutian Is. 52.49°N 169.42°W
	e	A	42 20.5	H = 14 30 24.2 h = normal MAG=4.9
	e	A	42 22.5	D = 77.2° Az = 359.3° (USCGS)
				PV:1.1s 24.4nm
				MPV=5.2
28.	eP	A	14 53 14	Fox Islands/Aleutian Is. 52.38°N 169.47°W
				H = 14 41 23.6 h = 59 km MAG=4.5 (USCGS)
				D = 77.5°

Day	Phase		h m s	Remarks
28.	eP	A	15 01 28	Fox Islands/Aleutian Is. 52.30°N 169.33°W
				H = 14 49 35.7 h = 47 km MAG=4.1 (USCGS)
				D = 77.5°
28.	eP	A	16 43 15	Fox Islands/Aleutian Is. 52.34°N 169.35°W
	epP	A	43 25	H = 16 31 21.1 h = 32 km MAG=5.6
				D = 77.4° Az = 359.4° (USCGS)
				PV:1.4s 91.0nm
				MPV=5.7
28.	+eP	A	17 31 25	Fox Islands/Aleutian Is. 52.28°N 169.52°W
	epP	A	31 37.5	H = 17 19 32.7 h = 41 km MAG=4.7
	esP	A	31 42.5	D = 77.4° Az = 359.3° (USCGS); h = 47 km
	e	A	31 48	PV:1.3s 27.8nm pPV:1.2s 25.6nm
				sPV:1.5s 23.4nm
				MPV=5.2
28.	eP	A	17 38 27	Fox Islands/Aleutian Is. 52.32°N 169.43°W
				H = 17 26 32.8 h = 33 km MAG=4.3
				D = 77.4° Az = 359.3° (USCGS)
				AN USSR gives: 53.9°N 170.7°W
				H = 17 26 43 h = normal
				D = 75.8°
28.	eP	A	17 53 52.5	Fox Islands/Aleutian Is. 52.40°N 169.41°W
	+iP	B	53 53	H = 17 42 01.5 h = 50 km MAG=5.6
	-i	A	53 54.0	D = 77.3° Az = 359.3° (USCGS); h = 47 km
	ei	A	53 55	PV4:1.3s 122nm
	epP	A	54 05	LmH:19s 2.3 μm LmV:17s 2.9 μm
	esP	A	54 09.5	MPV4=5.9 MLH=5.5 MLV=5.6
	eSS	C	18 08(45)	e 53 59
	LmH	B	32	
	LmV	B	32.5	
28.	LmH	B	18 39.8	LmH:19s 2.8 μm LmV:15s 2.4 μm
	LmV	B	42.4	
28.	iP	A	21 00 25	Fox Islands/Aleutian Is. 52.47°N 169.40°W
				H = 20 48 34.0 h = 47 km MAG=4.4
				D = 77.3° Az = 359.3° (USCGS)

Day	Phase		h m s	Remarks
28.	+iP	A	22 39 12.0	Kamchatka 55.03°N 160.21°E
	ePcP	A	39 29.5	H = 22 28 01.2 h = 113 km MAG=5.1
	-ipP	A	39 43.5	D = 71.5° Az = 339.5° (USCGS); h = 130 km
				PV:1.0s 14.3nm
				MPV=4.7
				AN USSR gives: East of Kamchatka
				53.7°N 161.6°E
				H = 22 27 42 h = normal
				D = 73.0°
29.	+iPn	A	00 13 04.6	Austria 47.89°N 14.29°E
	iPg	A	13(14.5)	H = 00 12 13.6 h = 25 km MAG=4.6
	iSn	A	13 43.0	D = 3.3° Az = 328.6° (USCGS)
	eSg	A	13 55.5	
29.	iP	A	04 01 41.5	Southern Iran 26.54°N 55.32°E
				H = 03 53 58.8 h = 42 km MAG=5.1
				D = 41.0° Az = 317.9° (USCGS)
29.	eP	A	07 11 11	Mongolia 47.97°N 103.07°E
				H = 07 01 34.7 h = normal MAG=4.8
				D = 55.9° Az = 309.8° (USCGS)
29.	eP	A	07 20 54	Iran
				H = 07 13 11 (UPP)
				D = ca. 41.2°
29.	eP	A	07 21 21	Iran
	e	A	21 23	H = 07 13 38 (UPP)
	e	A	21 34	D = ca. 41.2°
	e	A	21 40	LmV:15s 1.1 μm
	LmH	C	39.3	MLV=4.8
	LmV	B	43	
29.	e(P)	A	07 23 13	Iran?
	e	A	23 16	Probably a separated shock.
29.	eP	A	08 04 20	Southern Iran 26.53°N 55.16°E
	e	A	04 25	H = 07 56 39.2 h = 38 km MAG=5.2
	e	A	04 36	D = 40.9° Az = 317.9° (USCGS)

Day	Phase		h m s	Remarks
cont.				
29.	e	A	08 04 55	SH:13s 1.75/um SSH:14s 1.9/um
	ePP	A	06(00)	LmH:16s 3.6/um LmV:15s 4.9/um
	eS	B	10 28	MSH=5.9 MLH=5.3 MLV=5.5
	eSS	C	13 38	e 06 30 e 06 37.5 e 06 46 e 07 11
	LmV	B	26.6	e 07 29 e 07 53
	LmH	B	27.3	Multiple P.
29.	e	A	08 10 04	
	e	A	10 13	
29.	e	A	08 11 34	
	e	A	12 31	
29.	eP	A	09 35 33	<u>Fox Islands/Aleutian Is.</u> 52.4°N 169.6°W
				H = 09 23 40.2 h = normal MAG=4.3 (USCGS)
				D = 77.4°
29.	ePKP	A	18 05 57	<u>Tonga Islands</u> 16.2°S 173.8°W
	e	A	06 05	H = 17 46 33.3 h = 134 km MAG=4.5 (USCGS)
	ipPKP	A	06 33.0	D = 145.4° h = 134 km
	esPKP	A	06 47	i 07 02
29.	eP	A	19 51 45	PV:1.8s 40.8nm
29.	e(Pg)	A	20 07 08	<u>Apennines/Italy</u> 44.6°N 10.1°E
	e	A	07 15	H = 20 05 10 (BCIS)
	eSn	A	07 44.5	D = 6.2°
	e(Sg)	A	08 30.5	e 07 12 e 07 46 e 07 48 e 07 50
				e 08 25 e 08 34
29.	e(Sn)	A	20 47 27.5	<u>Northern Yugoslavia</u> 45.5°N 14.25°E
	e	A	47 30.5	H = 20 45 09 (BCIS)
	e(Sb)	A	47 50	D = 5.5°
	iSg	A	48 04.5	e 48 07
30.	iP	A	01 25 50.5	<u>Western Caucasus</u> 41.00°N 44.23°E
	i	A	25 51.5	H = 01 20 31.7 h = normal MAG=5.0
	e	A	25 58.5	D = 24.5° Az = 304.2° (USCGS)

Day	Phase		h m s	Remarks
cont.				
30.	iPP	A	01 26 20.5	LmH:15.5s 3.7/um LmV:13s 3.8/um
	eS	C	30(12)	MLH=5.0 MLV=5.1
	eLQ	C	31.0	AN USSR gives: Caucasus 40.1°N 43.7°E
	LmH	B	35.9	H = 01 20 26 h = normal MAG=5¼-5½
	LmV	B	37.2	D = 24.6°
30.	e(P)	A	01 30 10	
30.	eiP	A	01 30 45	<u>Armenia</u> 41.0°N 44.2°E
				H = 01 25 27 (BCIS)
				D = 24.5°
30.	e(P)	A	01 36 16	
30.	iP	A	04 09 46	<u>Kazakh SSR</u> 50.0°N 78.0°E
				H = 04 02 00 (BCIS)
				D = 41.2°
				Underground explosion.
30.	eSg	A	05 08 23	<u>Austria</u> 47.9°N 14.2°E
	e	A	08 31.5	H = 05 06 39 (BCIS)
				D = 3.2°
30.	e(P)	A	06 28 28	
30.	e(P)	A	06 51 17	
	e	A	51 19	
30.	e(P)	A	07 20 00	
	e	A	20 35	
30.	eP	A	12 30 09.5	<u>Turkey</u> 39.26°N 41.26°E
	e	A	30 28	H = 12 25 00.6 h = normal MAG=4.8
	e	A	31 45	D = 23.7° Az = 308.4° (USCGS)
	eSS	C	35.1	
30.	e(P)	A	16 49 37	

Day	Phase		h m s	Remarks
30.	iP	A	21 16 20.0	Burma 26.16°N 96.16°E
	i	A	16 22.5	H = 21 05 30.4 h = 44 km MAG=5.5
	i	A	16 25.0	D = 66.9° Az = 316.4° (USCGS)
	e	A	16 40.5	PV1:1.2s 20.5nm PV3:1.6s 45.5nm
	LmH	B	45.5	LmH:20s 0.45 /um
	LmV	B	(49)	MPV=5.2 MLH=4.7
				Multiple P.
31.	eP	A	03 45 12.5	Mongolia 47.93°N 102.81°E
	e	A	45 30.5	H = 03 35 36.3 h = normal MAG=4.9
	LmH	C	04(09)	D = 55.8° Az = 309.7° (USCGS)
	LmV	B	11.3	PV:1.0s 16.7nm
				MPV=5.1
31.	e(P)	A	04 37 06	
31.	e(Sg)	A	10 24 50	Probably explosion.
31.	e	A	10 47 15	
	e	A	47 21	
31.	e	A	11 56 11	Explosion?
	e	A	56 29	
31.	eP	A	13 50 39	Off Coast of Central America
	epP	A	50 47	2.79°N 84.40°W
	eS	C	14 01.7	H = 13 37 34.3 h = normal MAG=5.3
	e(PS)	C	02.8	D = 91.7° Az = 39.3° (USCGS); h = 30 km
	eSS	C	08.0	PV:1.7s 30.4nm pPV:1.7s 43.5nm
	LmH	C	24.5	LmH:25s 0.8 /um LmV:24s 0.7 /um
	LmV	C	(25)	MPV=5.3 MLH=5.1 MLV=5.0
				e 51 07 e 51 18
31.	+iP	A	17 55 54.0	Hokkaido/Japan Region 42.84°N 145.35°E
	e(PcP)	A	56 04	H = 17 43 56.2 h = 44 km MAG=5.1
	LmV	B	18 31.8	D = 78.5° Az = 332.0° (USCGS)
	LmH	C	33.3	e 56 14.5 e 56 19

Day	Phase		h m s	Remarks
31.	eP	A	19 08 02	Iran 26.5°N 55.3°E
	e	A	08 09	H = 19 00 22.7 h = 16 km MAG=5.2 (USCGS)
	ei	A	08 50	D = 41.0°
	i	A	09 52.0	LmH:16s 0.4 µm LmV:15s 0.55 µm
	LmV	B	30.3	MLH=4.4 MLV=4.6
	LmH	B	31	
31.	LmH	C	21 52.5	Taiwan Region 24.0°N 121.6°E
	LmV	B	53	H = 20 58 13.9 h = 22 km MAG=4.9 (USCGS)
				D = 83.5°
				LmH:18s 0.4 µm LmV:16s 0.45 µm
				MLH=4.9 MLV=5.0

Day	Phase		h m s	Remarks
1.	eP	A	01 15 03	Southern Iran 26.65°N 55.28°E
	e	A	15 24	H = 01 07 19.2 h = 19 km MAG=5.0
	eS	C	21 12	D = 40.9° Az = 317.8° (USCGS)
	eSS	C	24.2	PV:1.4s 21nm
	eScS	C	25 04	LmH:19s 1.2 µm LmV:13s 0.9 µm
	LmH	B	34.2	MPV=4.9 MLH=4.8 MLV=4.8
	LmV	B	37.1	e 15 28 e 24 24
1.	ei	A	03 50 18.5	
1.	e(P)	A	09 06 34	
1.	eP	A	09 29 54	Kamchatka 55.8°N 160.7°E
	epP	A	29 29	H = 09 18 50.5 h = 140 km MAG=4.4 (USCGS)
				D = 70.8° h = 145 km
1.	eP	A	15 33 16	Southern Sumatra 4.84°S 103.16°E
				H = 15 19 56.8 h = 33 km MAG=5.3 (USCGS)
				D = 94.8°
1.	e(P)	A	17 43 44	
2.	ePKIKP	A	06 44 17.5	South Sandwich Islands Region
	epPKIKP	A	44 40	57.94°S 25.65°W
	ePP	A	45 03	H = 06 25 49.8 h = 81 km MAG=5.8 (USCGS)
	eS	C	52 54	D = 112.3° h = 83 km
	ePS	B	54 36	LmH:19s 3.2 µm LmV:18s 3.7 µm
	eSS	C	07 00 44	MLH=5.9 MLV=6.0
	eSSS	C	04.6	e 55 10 e 09(00)
	eLQ	C	11.3	
	LmH	B	30.8	
	LmV	B	31.1	
2.	eP	A	07 46 08	Southern Sinkiang Prov./China
	+ipP	A	46 14.8	39.71°N 75.51°E
	e(PcP)	A	47 49	H = 07 37 54.9 h = 39 km MAG=5.3
	e	B	08 02 10	D = 45.0° Az = 306.1° (USCGS); h = 30 km
	eLg2	B	02 55	PV:1.6s 53nm pPV:1.6s 53nm
	e	B	04 35	LmH:12.5s 2.3 µm LmV:14s 2.6 µm

Day	Phase		h m s	Remarks
cont.				
2.	LmH	B	08 06.4	MPV=5.3 MLH=5.3 MLV=5.4
	LmV	B	06.4	e 46 40 e 47 08.5 e 48 05 e 48 37
				e 02 25
				Well developed higher mode surface waves.
2.	e(P)	A	09 52 16	
2.	e	A	10 02 25	
2.	e	A	12 26 47	
2.	eP	A	16 36 17	<u>Hokkaido/Japan Region</u> 41.56°N 139.75°E
	ePcP	A	36 28	H = 16 24 39.1 h = 176 km MAG=5.4 (USCGS)
	eipP	A	37 02	D = 77.5° h = 186 km
	iS	B	45 52.5	PV:1.4s 30.2nm PcPV:1.0s 14.3nm
	eSS	B	50 44	pPV:1.2s 18nm SH:(8s) 1.3 µm
	LmH	B	17(08)	MPV=4.8 MSH=6.0
	LmV	B	(12)	e 36 41.5 e 36 47 e 37 17 e 38 57
2.	ePKIKP	A	18 36 47	<u>New Ireland Region</u> 4.34°S 153.73°E
				H = 18 18 17.4 h = 247 km MAG=5.0 (USCGS)
				D = 124.0°
				AN USSR gives: Solomon Islands
				5.4°S 154.1°E
				H = 18 17 48 h = normal
3.	e	A	03 53 36	
	e	A	53 40.5	
	e	A	53 58	
3.	eP	A	08 29 20	<u>Honshu/Japan</u> 36.54°N 138.03°E
	e(pP)	A	29 31	H = 08 17 05.4 h = 26 km MAG=4.7
	e	A	29 51	D = 81.2° Az = 328.8° (USCGS);
	LmH	B	09 05.5	(h = 40 km)
	LmV	B	06	LmH:14s 1.1 µm
				MLH=5.4
3.	e	A	09 38 24	

Day	Phase		h m s	Remarks
3.	e(Sg)	A	11 01 02	Explosion?
3.	e	A	11 32 24.5	
3.	e	A	12 38 44	
3.	e	A	12 54 15	
3.	eP	A	13 00 57	Java Sea 5.62°S 110.54°E
	ePP	A	05 03.5	H = 12 48 09.2 h = 560 km MAG=5.4
	eS	C	11 40	D = 100.0° Az = 320.3° (USCGS)
	eSP	C	13.2	e 04 54 e 22 08
	e(sSP)	C	16 44	AN USSR gives: h = 512 km
	eSS	C	18 40	
	e(sSS)	C	21(52)	
4.	iP	A	14 27 13.5	
4.	ePP	A	18 06(00)	Volcano Islands 25.55°N 142.72°E
	e	A	06 28	H = 17 49 02.4 h = normal MAG=4.9 (USCGS)
	LmH	C	47.7	D = 92.8°
5.	eP	A	19 05 48	Ascension Island Region 5.45°S 11.36°W
	ipP	A	05 54.2	H = 18 55 45.1 h = 19 km MAG=5.2
	esP	A	05(58)	D = 59.3° Az = 16.8° (USCGS); h = 24 km
	e	A	06 06	PV:1.5s 30nm pPV:1.5s 33.4nm
	e	A	06 16	LmH:16s 0.6 µm LmV:18s 1 µm
	LmH	B	33	MPV=5.2 MLH=4.8 MLV=5.0
	LmV	B	33	e 06 24.5
6.	e(P)	A	00 31 07	
6.	iPKHKP	A	03 30 18.3	South of Fiji Islands 22.82°S 176.06°W
	e(pPKHKP)	A	30 54	H = 03 10 33.4 h = 90 km MAG=5.1
				D = 151.6° Az = 349.7° (USCGS);
				(h = 130 km)
6.	eP	A	03 37 29	Southern Alaska 60.14°N 152.79°W
	e	A	37 38	H = 03 26 35.4 h = 110 km MAG=4.9

Day	Phase		h m s	Remarks
cont.				
6.	i	A	03 37 49.0	D = 68.9° Az = 10.6° (USCGS); h = 102 km
	epP	A	37 54	PV:1.0s 19nm
	i	A	37 58.5	MPV=4.9
				AN USSR gives: h = normal
6.	e	A	10 34 18.5	
6.	e(P)	A	14 38 47	
6.	eP	A	14 48 19.5	
7.	ePP	A	08 47 12.5	Mariana Islands 13.86°N 144.78°E
	e	A	47 28.5	H = 08 28 57.9 h = 138 km MAG=5.4 (USCGS)
	epPP	A	47 39	D = 103.9° h = ca. 110 km
				AN USSR gives: h = normal
7.	e	A	10 00 06	
	e	A	00 10	
7.	-eiP	A	15 04 36	Alaska Peninsula 56.69°N 157.18°W
	i	A	04 38.5	H = 14 53 13.9 h = 67 km MAG=5.6 (USCGS)
	e	A	04 45	D = 72.6° h = 67 km
	epP	A	04 53	PV:1.0s 52.5nm pPV:1.1s 39.0nm
	esP	A	05 00.5	MPV=5.6
7.	eP	A	19 51 41	Off East Coast of Kamchatka
	e	A	51 46	55.20°N 163.03°E
	e	A	51 57.5	H = 19 40 14.8 h = normal MAG=4.9 (USCGS)
				D = 71.8°
8.	ePKP	A	00 07 59.5	Fiji Islands Region 17.78°S 178.48°W
				H = 23 49 21.8 h = 571 km MAG=3.9 (USCGS)
				D = 146.2°
8.	e(Sg)	A	12 08 06	Upper Silesia
	e	A	08 10	
	e	A	08 15	

Day	Phase		h m s	Remarks
8.	ePP	A	15 54(52)	Southwestern Atlantic Ocean
	e	A	55 07.5	58.27°S 12.97°W
	ePS	C	16 04.4	H = 15 35 42.8 h = normal MAG=5.1 (USCGS)
	eSS	C	10.2	D = 110.1°
	eLQ	C	20.2	LmH:20s 0.65 /um LmV:20s 1.0 /um
	LmH	B	41	MLH=5.2 MLV=5.4
	LmV	B	41	
8.	eiP	A	17 28 41	Burma-India Border Region 23.22°N 93.92°E
	i	A	28 48	H = 17 17 45.7 h = normal MAG=5.1
	e	A	29 02	D = 67.6° Az = 317.0° (USCGS)
	e	A	29 23	LmH:24s 0.45 /um
	LmH	C	57	MLH=4.7
9.	e(Sg)	A	03 57 17	Near earthquake?
	e	A	57 29.5	
9.	e(Sg)	A	12 00 20	
9.	iPg	A	12 00 36.5	Explosion/CSSR 50.29°N 13.57°E
	iSg	A	00 56.5	D = 1.3° Yield: 5.8 t
9.	e(Sg)	A	12 02 08	Rock burst/Upper Silesia
	e	A	02 14	
	i	A	02 17.2	
9.	-eP1	A	14 11 14.5	Greece-Albania Border Region
	iP2	A	11 18.0	39.98°N 20.26°E
	eSS	C	13(45)	H = 14 08 18.7 h = 3 km MAG=5.6
	ei	C	14 15	D = 12.3° Az. = 333.2° (USCGS)
	eLg1	B	14 40	PV1:1.2s 118nm PV2:0.9s 354nm
	eLg2	B	15 09	LmH:12s 25 /um LmV:12s 21.4 /um
	LmH	B	17.0	MLH=5.4 MLV=5.5
	LmV	B	17.3	i 11 29.2 ei 13 11 e 13(58) e 14 11
				P is multiple. Clear higher mode surface waves.
9.	+iP	B	15 37 20	Colombia 2.85°N 74.89°W
	+eiP	A	37 20.2	H = 15 24 47.2 h = 58 km MAG=6.3

Day	Phase		h m s	Remarks
cont.				
9.	eipP	A	15 37 37.3	D = 85.6° Az = 39.6° (USCGS); h = 65 km
	ipP	A	37 38.8	PV(B):12s 18 /um PH(B):12s 10.8 /um
	-iPP	B	40 33	PV(A):2.4s 910nm pPV2:2.7s 2220nm
	iS	B	47 46	PPV:14.5s 8.5 /um SH:17.5s 57 /um
	isS	B	48 15	SSH:15s 18.7 /um
	iPS	B	48 54	LmH:20s 64 /um LmV:20.5s 100 /um
	iSPP	B	49 18	MPV(B)=7.1 MPH(B)=7.4 MPV(A)=6.5
	iSS	B	53 16	MPPV=7.0 MSH=7.5 MLH=7.0 MLV=7.2
	ePKPPKP	A	16 03 30	i 37 31.0 i 37 49.0 i 37 56.5 (PV:2.1s
	LmH	B	12.5	682nm) ei 38 12 (PV:2.1s 835nm) i 48 42
	LmV	B	12.5	i 49 00 i 49 10
10.	ePg	A	05 04 59	Apennines/Italy 44 1/4°N 10 1/4°E
	iSn	A	05 34.0	H = 05 02 52 (BCIS)
	eiSg	A	06 18.5	D = 6.5°
	eLg2	A	06 28	e 06 05 e 06 24 e 06 31
10.	eP	A	06 00 00	Southern Sinkiang Prov./China
	epP	A	00 09	41.56°N 86.16°E
				H = 05 51 01.9 h = 23 km MAG=5.1
				D = 50.4° Az = 307.2° (USCGS); h = 38 km
				PV:1.0s 9.5nm pPV:1.0s 9.5nm
				MPV=4.8
10.	e	A	06 50 01.5	Austria 47.7°N 16.0°E
	eiSg	A	50 04.5	H = 06 47 56 (BCIS)
	i	A	50 10.0	D = 4.1°
	iLg2	A	50 11.7	
10.	eP	A	06 54 34	West of Svalbard (UPP)
	e	A	54 38.5	
10.	iPg	A	07 59 44.5	Explosion
	iSg	A	59 58.5	D = 1.1°
10.	e	A	09 01 50	
	e	A	01 55	
	i	A	02 01.5	
	ei	A	02 06.5	

Day	Phase		h m s	Remarks
10.	e	A	09 02 43.5	
10.	ePKIKP	A	11 48 15	Banda Sea 7.29°S 128.54°E
	e(PP)	A	49 17	H = 11 29 46.4 h = 27 km MAG=5.3 (USCGS)
	e	A	49 26	D = 112.5°
	e	C	12 13 40	
	eLQ	C	16.0	
10.	iPg	A	14 15 16.0	Explosion/GDR 51.37°N 12.89°E
	iSg	A	15 31.5	D = 1.2°
10.	e	A	16 26 47	Explosion?
	e	A	27 03	
11.	e	A	00 44 43	
	i	A	44 45.2	
11.	eP	A	02 51 25	Off East Coast of Kamchatka
	e	A	51 28.5	51.72°N 159.49°E
	e(pP)	A	51 36.5	H = 02 39 47.1 h = 21 km MAG=4.6
	e	A	51 43	D = 74.4° Az = 339.4° (USCGS);
				(h = 43 km)
				PV:1.2s 20.5nm
				MPV=5.1
11.	eP	A	09 37 03.5	Lake Baikal Region 52.01°N 106.23°E
	e	A	37 10	H = 09 27 29.6 h = 5 km MAG=5.4 (USCGS)
	ePcP	A	38 06.5	D = 54.8°
	e	A	38 10	PV1:1.2s 12.8nm PV2:1.1s 26.9nm
	LmH	B	10 03.6	LmH:13s 4.6 µm LmV:13s 7.2 µm
	LmV	B	03.7	MPV1=4.9 MPV2=5.3 MLH=5.8 MLV=6.0
11.	e(Sg)	A	13 01 07	
11.	eP	A	14 44 55.5	Kurile Islands 48.25°N 154.76°E
	LmH	C	15 19.8	H = 14 33 06.3 h = 26 km MAG=4.7
				D = 76.5° Az = 336.9° (USCGS)

Day	Phase		h m s	Remarks
11.	eiP	A	15 25 01	Iran 30.46°N 50.72°E
	LmH	C	45	H = 15 18 06.3 h = 42 km MAG=5.0 (USCGS)
				D = 35.3°
11.	iP	A	15 37 26.7	Greenland Sea 79.58°N 3.43°E
	i	A	37 29.8	H = 15 31 27.1 h = normal MAG=4.9
	ePP	A	38 23	D = 29.2° Az = 169.3° (USCGS)
12.	eP	A	06 30 24	
12.	e(S)	C	10 33.7	North of Ascension Island 2.10°S 12.30°W
	LmH	C	51	H = 10 16 13.5 h = normal MAG=4.8 (USCGS)
	LmV	B	51.5	D = 56.3°
				LmH:18s 0.6/um
				MLH=4.7
12.	eP	A	14 22 04.5	Near Coast of Northern Chile
	LmH	C	15 03.5	21.69°S 70.12°W
				H = 14 08 12.5 h = 18 km MAG=5.5
				D = 101.4° Az = 40.0° (USCGS)
				LmH:24s 0.4/um
				MLH=4.9
12.	eP	A	16 14 51	Hindu Kush Region 35.76°N 70.99°E
	e	A	16 23	H = 16 06 47.8 h = 100 km MAG=5.2 (USCGS)
	ePP	A	16 38	D = 44.5° (h = ca. 120 km)
	e(pPP)	A	17(03)	AN USSR gives: h = 160 km
	e	A	18 11	
12.	e(P)	A	21 24 50.5	PV:1.4s 21.2nm
	e	A	24 58	
	e	A	25 04.5	
13.	eiP	A	10 19 25	Fox Islands/Aleutian Is. 52.49°N 169.61°W
				H = 10 07 34.5 h = 51 km MAG=4.5 (USCGS)
				D = 77.2°
13.	eiP	A	11 33 24	Honshu/Japan 36.25°N 139.72°E
				H = 11 21 08.4 h = 58 km MAG=4.9 (USCGS)
				D = 82.2°

Day	Phase		h m s	Remarks
13.	eP	A	11 42 44	Kurile Islands Region 43.95°N 148.36°E
	e	A	43 05.5	H = 11 30 45.0 h = normal MAG=4.6 (USCGS)
				D = 78.6°
				PV:1.0s 19nm
				MPV=5.2
13.	ePP	A	15 49 46	New Ireland Region 5.09°S 153.11°E
				H = 15 29 05.1 h = 57 km MAG=5.0 (USCGS)
				D = 124.3°
13.	eP	A	17 17 12	Kurile Islands 43.56°N 147.42°E
				H = 17 05 10.3 h = 30 km MAG=4.7 (USCGS)
				D = 78.5°
13.	+iP1	A	23 20 13.0	North Atlantic Ocean 52.69°N 34.08°W
	+iP2	A	20 16.2	H = 23 14 19.6 h = 10 km MAG=5.5
	+iP2	C	20 16.4	D = 28.1° Az = 75.6° (USCGS)
	+iP3	B	20 19.5	PV1:1.0s 23.8nm PV2(A):2.4s 522nm
	iPP	B	20 57.5	PV2(C):16s 11.3 μm PV3:5s 7 μm
	iS	B	25 00	PLV:15s 14 μm PPV:12.5s 27 μm
	i	B	26 10	PPH:12.5s 26.7 μm SH:15s 38.4 μm
	i(PcS)	B	27 08	LmH:17s 388 μm LmV:13s 308 μm
	i	B	28 15	MPV1=4.9 MPV2(A)=5.9 MPV3=7.0 MPLV=6.5
	i	B	28 20	MPPV=7.0 MPPH=7.1 MSH=6.8 MLH=7.1
	LmH	B	29.8	MLV=7.2
	LmV	B	31.6	Successive P-onsets with increasing amplitude. In the registrations of the long-period seismographs long-period PL-waves (T = 12 - 16 s) are recorded, superimposed by shortperiod P-waves (T = 3 - 5 s).
14.	iP1	A	01 47 53.0	Andaman Islands Region 13.75°N 96.55°E
	iP2	A	47 57.5	H = 01 36 04.7 h = 27 km MAG=5.6
	iP3	A	48 00.2	D = 76.3° Az = 319.3° (USCGS)
	i(S)	B	57 54	PV1:1.5s 53.4nm PV2:1.7s 152nm
	eSPP	B	58 32	PV3:2.3s 1500nm (S)H:14.5s 7.6 μm
	eSSS	B	02 06 20	LmH:18s 34.8 μm LmV:16s 36.0 μm
	LmH	B	26.0	MPV1=5.4 MPV2=5.8 MPV3=6.7 MLH=6.7
	LmV	B	30.2	MLV=6.8 Successive P-onsets with increasing amplitude.

Day	Phase		h m s	Remarks
14.	iP	A	01 51 11.8	<u>Andaman Islands</u>
				H = 01 39 25.1 (UPP)
14.	e(Sn)	A	02 16 12	<u>Yugoslavia</u> 44.2°N 19.2°E
	eSg	A	17 09.5	H = 02 12 37 (BCIS)
	e	A	17 29	D = 8.3°
14.	ePKIKP	A	05 20 55.5	<u>New Hebrides Islands Region</u>
	e(pPKIKP)	A	23 27	13.25°S 171.32°E
	eSKP1	A	23 35.5	H = 05 02 38.4 h = 635 km MAG=5.6
	eSKP2	A	23 48	D = 139.2° Az = 340.2° (USCGS);
	ePP	A	24 00	(h = 668 km)
	e(PKS)	A	24 38	PV:1.8s 51nm SKP2V:1.6s 26.5nm
				e 22 37 e 25 28
				SKP1 and SKP2 belong to different branches of the SKP travel time curve.
14.	eP	A	14 45 15	
	e	A	45 20	
	e	A	45 24	
	e	A	45 42	
14.	ePKP	A	18 33 01	<u>Tonga Islands</u> 19.42°S 172.75°W
	epPKP	A	33 10	H = 18 13 14.4 h = normal MAG=4.9 (USCGS)
	e	A	33 34	D = 148.7° h = 32 km
15.	iP	A	01 53 38.8	<u>Western Iran</u> 34.46°N 47.57°E
	i(PP)	A	54 32.0	H = 01 47 27.7 h = 38 km MAG=5.0 (USCGS)
				D = 30.6°
15.	+eiP	A	06 08 37.5	<u>Burma</u> 20.37°N 94.08°E
	-ipP	A	08 40.8	H = 05 57 24.6 h = 10 km MAG=5.5
	-eiX	A	08 52	D = 69.8° Az = 317.7° (USCGS); h = 12 km
	i	A	08 57.5	PV:1.6s 68.1nm pPV:1.2s 30.6nm
	LmH	C	37	XV:2.2s 158nm
				MPV=5.6
				X is a clear phase in all records of our short-period vertical components. Another shock in the same focus? If this phase

Day	Phase		h m s	Remarks
cont. 15.				was interpreted as pP or sP, than the focal depth would be 56 km or 40 km respectively.
15.	e(P)	A	06 49 06	
15.	e(P)	A	07 03 30	
15.	ei(Sg)	A	11 36 49	Explosion?
15.	i	A	14 38 50.0	
15.	eiPg	A	15 05 30.5	Explosion/GFR 50.96°N 9.22°E
	eSg	A	05 51	H = 15 05 01.2
				D = 1.6°
15.	-iP	AB	16 23 20.5	Peru-Brazil Border Region 9.01°S 71.28°W
	i	A	23 22.5	H = 16 11 11.8 h = 597 km MAG=6.2
	+epP	B	25 31	D = 92.4° Az = 39.2° (USCGS); h = 600 km
	ePP	C	27 16	PV(B):9s 5.3 µm PV1(A):2.4s 364nm
	iSKS	B	33 00	PV2(A):1.8s 694nm pPV:(9s) 6.8 µm
	eS	A	33 32	SKSH:10s 13 µm SH(B):8s 6.3 µm
	iS	B	33 34	SPV:14.5s 11.5 µm SPH:17s 12.6 µm
	eiSP	C	34 44	MPV(B)=6.6 MPV1(A)=6.0 MPV2(A)=6.4
	isSP	C	38 24	MSH(B)=6.6
	isSS	C	43 24	AN USSR gives: 8.4°S 76.3°W
	e	B	47 12	H = 16 10 19 h = 174 km
	ePKPPKP	A	48 38	D = 95.2°
	eSa(4.48)	C	49.4	
15.	i(PKIKP)	A	19 52 06.0	South of Fiji Islands 24.58°S 177.63°W
	e	A	52 19	H = 19 32 21.0 h = 70 km MAG=4.9 (USCGS)
	e(PKP2)	A	52 26	D = 152.9°
	e	A	52 37	
15.	iP	A	23 48 50.0	North Atlantic Ocean 52.65°N 33.94°W
				H = 23 42 55.9 h = normal MAG=4.4 (USCGS)
				D = 28.0°

Day	Phase		h m s	Remarks
16.	i	A	10 50 17.5	
17.	eP	A	00 51 30	Talaud Islands 4.44°N 125.58°E
	e	A	51 39	H = 00 37 42.5 h = 66 km MAG=5.5 (USCGS)
	e	A	52 17.5	D = 101.4°
	LmH	C	01 39	
17.	-ePKIKP	A	10 30 41	Tonga Islands Region 23.69°S 175.24°W
	e(pPKIKP)	A	30 44.5	H = 10 10 51.5 h = 19 km MAG=6.4 (USCGS)
	ePKHKP	A	30 48.5	D = 152.6° (h = 12 km)
	ei(pPKHKP)	A	30 52.0	PV1:2.2s 324nm PV2:2.3s 600nm
	eiPKP2	A	31 00	PV3:1.7s 465nm PV4:1.9s 775nm
	ePP	B	34 30	PV5:2.0s 613nm PPV:9s 2.3 µm
	eSKKS	C	41 26	LmH:19s 12.2 µm LmV:19s 13.7 µm
	ePS	C	44 47	MPPV=6.4 MLH=6.8 MLV=6.8
	ePPS	C	47 40	i 31 05.0 i 31 10.5 i 31 27.0 i 31 44.5
	ePPPS	C	49 25	i 35 15 e 38 48 i 55 40
	eSS	C	54.0	
	eLQ	C	11 11.5	
	LmH	B	42.4	
	LmV	B	42.4	
17.	iPg	A	11 37 15.5	Explosion/GDR
	iSg	A	37 31.5	D = 1.2°
17.	iPg	A	13 00 23.5	Explosion/GDR 51.29°N 12.75°E
	iSg	A	00 38	H = 13 00 00
	e	A	00 52	D = 1.1°
17.	e	A	17 38 35	
18.	i	A	00 40 22	North Atlantic Ridge 28.95°N 43.42°W
	i	A	41 26.0	H = 00 31 48.2 h = normal MAG=5.2 (USCGS)
				D = 46.3°
				P must be 10 s earlier than the first given onset.
18.	iPKIKP	A	02 58 15.5	New Ireland Region 5.90°S 153.17°E
	ei	A	58 18	H = 02 39 19.4 h = 41 km MAG=5.4

Day	Phase		h m s	Remarks
cont.				
18.	ei(pPKP)	A	02 58 30.5	D = 125.1° Az = 331.1° (USCGS);
	e	A	59 44	(h = ca. 45 km)
	ePP	A	03 00 07	e 00 33
				The second onset is bigger than the first.
18.	iSg	A	09 22 09.5	Explosion/ČSSR 50.18°N 14.40°E
				D = 1.8°
18.	e	A	19 33 12	
18.	e	A	20 32 26	
18.	e	A	22 49 36.5	Explosion
	eiSg	A	49 43	
19.	e	A	01 58 44	
19.	eiP	A	06 44 33	Kurile Islands 49.54°N 154.10°E
	ei	A	46 54	H = 06 32 51.8 h = normal MAG=4.6 (USCGS)
				D = 75.1°
19.	i(P)	A	12 54 08.0	
19.	ePKHKP	A	14 41 43.5	Tonga Islands 21.70°S 174.93°W
	ePKP2	A	41 52	H = 14 21 52.8 h = normal MAG=4.4
	e	A	42 11.5	D = 150.7° Az = 351.5° (USCGS)
19.	ePKP	A	19 45 13	Tonga Islands 18.89°S 174.02°W
	i	A	45 22.0	H = 19 25 26.7 h = normal MAG=4.5 (USCGS)
				D = 148.1°
19.	-iP	A	20 17 03.5	Northern Sinkiang Prov./China
	e	A	17 10	42.02°N 83.55°E
				H = 20 08 20.9 h = normal MAG=5.1
				D = 48.6° Az = 306.2° (USCGS)
				PV:1.4s 22.4nm
				MPV=5.0

Day	Phase		h m s	Remarks
19.	iP	A	21 41 34.5	Fox Islands/Aleutian Is. 52.37°N 169.49°W
	i(pP)	A	41 44.5	H = 21 29 42.4 h = 48 km MAG=4.6 (USCGS)
				D = 77.4° (h = 38 km)
19.	eiP	A	22 28 33	South of Java 9.17°S 113.10°E
	e(pP)	A	28 42	H = 22 14 35.3 h = 80 km MAG=6.2
	ePP	A	32(52)	D = 104.3° Az = 319.9° (USCGS);
	eS	C	40 10	(h = 33 km)
	eSS	C	47 26	PV2:1.5s 33.3nm
				i 28 52.3 i 28 57.8 e 29 22.5
				(pP) is a clear phase and bigger than P.
19.	e	A	22 31 38	Near earthquake?
	e	A	32 28	
	e	A	35 41	
	i	A	33 09.0	
19.	e(P)	A	22 44 47	
	e	A	44 53.5	
19.	eP	A	23 42 22	Molucca Sea 0.03°S 124.17°E
	e	A	42 46.5	H = 23 28 28.0 h = 101 km MAG=5.7 (USCGS)
	ePP	A	46 46	D = 104.2°
	e	A	47 04	AN USSR gives: h = normal
	e	A	47 09.5	
20.	i	A	03 13 51.0	
20.	+iP	A	15 27 24.5	Eastern Kashmir 33.69°N 75.28°E
	i	A	27 29.5	H = 15 18 39.9 h = 24 km MAG=5.7
	eiPcP	A	28 47.5	D = 48.6° Az = 310.5° (USCGS)
	e(PP)	A	29 22	PV:1.6s 116nm
	LmH	C	52	MPV=5.7
	LmV	C	52	ei 28 34.5 ei 29 10
20.	eP	A	16 45 24	Tschagos Archipelago Region 5.20°S 68.68°E
				H = 16 33 49.2 h = normal MAG=4.8 (USCGS)
				D = 74.0°

Day	Phase		h m s	Remarks
20.	e	A	23 01 48	
22.	e	A	02 18(27)	
	e	A	18 42	
22.	eP	A	15 02 20	<u>Kurile Islands</u> 48.29°N 154.72°E
	e	A	02 37	H = 14 50 33.1 h = 45 km MAG=4.7 (USCGS)
				D = 76.3°
22.	+iPKP	A	18 46 11.7	<u>New Hebrides Islands</u> 19.52°S 169.04°E
	ipPKP	A	46 34.0	H = 18 26 46.7 h = 87 km MAG=5.6
	ePP	A	49 32	D = 144.2° Az = 335.3° (USCGS); h = 80 km
	ei(SKP)	A	49 43.3	PV1:1.3s 189nm PV2:1.3s 139nm
	eSS	C	19 02.5	PPV:1.7s 47.8nm (SKP)V:2s 120nm
	eSSS	C	08.4	MPPV=5.4
	LmH	C	43.5	e 49 55 e 50 21 e 51 44
23.	ePKP	A	06 18 20	<u>Tonga Islands</u> 21.61°S 174.44°W
	epPKP	A	18 32	H = 05 58 28.5 h = 21 km MAG=4.8
	e	A	18 46.5	D = 150.6° Az = 352.1° (USCGS); h = 43 km
23.	eP	A	19 02 18	<u>Nevada</u>
	e	A	02 27	H = 18 50 00 h = 702 m Yield: 200 kt (UPP)
				D = ca. 81.2°
				PV:1.2s 46.2nm
				MPV=5.5
				Underground nuclear explosion.
23.	eP	A	20 51 33.5	<u>Ryukyu Islands</u> 26.06°N 128.50°E
	epP	A	51 41.5	H = 20 38 56.3 h = 30 km MAG=5.4
	eS	C	21 02.0	D = 85.5° Az = 325.3° (USCGS); h = 30 km
	LmH1	B	27.5	PV:2.0s 155nm pPV:2.0s 155nm
	LmH2	B	35.0	LmH1:20s 4.0 µm LmH2:15.5s 4.8 µm
	LmV	B	35.0	LmV:16s 5.4 µm
				MPV=5.8 MLH1=5.8 MLH2=6.0 MLV=6.0
23.	iPn	A	22 40 40.2	<u>Yugoslavia</u> 43.7°N 15.9°E
	e(Pg)	A	41 13	H = 22 38 50 (BCIS)
	iSn	A	42 04.5	D = 7.6°

Day	Phase		h m s	Remarks
cont.				
23.	eiSg	A	22 42 54	e 41 07 e 41 14 e 42 44
	eLg2	A	43 07	
24.	e(Pn)	A	15 23 08.5	Yugoslavia (after Zagreb felt at Čapljina)
	e	A	23 34	D = ca. 8.3°
	e	A	24 04	
	e(Sn)	A	24 40	
	e(Sg)	A	25 37	
24.	ipP	A	18 29 38.0	Southern Alaska 60.28°N 153.69°W
	i	A	29 50.8	H = 18 18 10.5 h = 165 km MAG=4.0 (USCGS)
	i	A	30 23.5	D = 68.7°
25.	i(P)	A	00 00 32.0	
25.	e	A	08 29 40.5	
	e	A	30 09.5	
25.	eP	A	11 34 44	Northern Celebes 0.04°S 123.93°E
	ePP	A	39 06	H = 11 20 47.4 h = 70 km MAG=5.8 (USCGS)
	eSS	C	54.0	D = 104.1°
	e(Sa)	C	12 03.9	e 38 35.5 e 38 41.5 e 39 19 e 40 09
	LmH	C	(42)	e 40 19
				AN USSR gives: h = normal
25.	i	A	11 50 34.6	
25.	iP	A	11 52 40	Northern Celebes 0.11°S 123.93°E
	eipP	A	53 07	H = 11 38 46.0 h = 105 km MAG=5.7 (USCGS)
	e	A	56 28	D = 104.2° h = 104 km
	ePP	A	56 58.5	e 57 35 e 58 14.5
25.	e	A	12 02 14	
25.	eP	A	12 06 42	
	e	A	06 54	
	i	A	07 10.0	

Day	Phase		h m s	Remarks
26.	+iP	A	04 05 46.5	Eastern Kazakh SSR 49.78°N 78.12°E
	ePn	A	07 18.5	H = 03 57 57.7 h = 0 km MAG=6.0
	ePP	A	07 22	D = 41.3° Az = 297.7° (USCGS)
	LmH	C	20	PV:0.9s 305nm
				MPV=6.2
				Underground explosion.
26.	iPKHKP	A	12 16 51.5	South of Fiji Islands 24.34°S 179.77°E
	i	A	16 56.0	H = 11 57 54.8 h = 535 km MAG=4.4 (USCGS)
				D = 152.1°
26.	e	A	15 31 32	
	e	A	31 40	
26.	eP	A	15 32 05.5	Hindu Kush 35.8°N 68.9°E
				H = 15 24 03 MAG=4 1/2 (AN USSR)
				D = 43.2°
27.	iP	A	02 19 12.3	Colombia 2.92°N 74.79°W
	e(pP)	A	19 34.5	H = 02 06 42.5 h = 69 km MAG=5.2 (USCGS)
	e	A	19 50	D = 85.5°
27.	ePKIKP	A	02 48 16	Off Coast of Central Chile
				38.30°S 74.67°W
				H = 02 29 34.8 h = normal MAG=4.9 (USCGS)
				D = 116.3°
27.	i	A	04 13 11.0	
	i	A	13 50.0	
27.	eP	A	21 03 29	Rumania 44.9°N 26.7°E
	ePP	A	03 38	H = 21 00 41 h = normal MAG=5
	ei	A	06 23	D = 11.6° (AN USSR)
	LmH	B	09.2	e 06 48 e 07 28 e 07 40
	LmV	B	10	
28.	eP	A	09 49 57.5	South of Honshu/Japan 32.65°N 141.66°E
	e(pP)	A	50 08	H = 09 37 18.0 h = 23 km MAG=5.5
	ePP	A	53 15	D = 86.1° Az = 330.8° (USCGS);

Day	Phase		h m s	Remarks
cont.				
28.	iS	B	10 00 30	(h = 43 km)
	ePS	C	01.5	PV:2.0s 66.6nm SH:14s 1.7 µm
	LmH	B	34.2	LmH:14s 3.7 µm LmV:15s 4.6 µm
	LmV	B	38.5	MPV=5.1 MSH=6.1 MLH=5.9 MLV=6.0
				i 50 10.0 i 50 20.0 e 51 07
28.	i	A	11 26 54.0	
28.	eP	A	14 25 23.5	<u>Southern Greece</u> 37.45°N 21.29°E
	e	A	25 27	H = 14 21 53.5 h = 77 km MAG=4.6
	ePP	A	25 37	D = 14.9° Az = 335.4° (USCGS)
	e	A	26 07	PV:0.9s 19nm
	eiS	C	28 06	e 25 29 e 25 46.5 e 25 52.5 e 26 31.5
28.	eP	A	15 27 25	<u>Near East Coast of Kamchatka</u>
				53.12°N 159.92°E
				H = 15 15 56.0 h = normal MAG=4.6
				D = 73.2° Az = 339.5° (USCGS)
				PV:1.2s 15.4nm
				MPV=5.0
28.	e(P)	A	16 17 31	Explosion?
	e	A	17 35	
	e	A	17 42	

Day	Phase		h m s	Remarks
1.	i(P)	A	03 45 38.5	
1.	ei(P)	A	09 11 25	
1.	eP	A	10 20 28	Southern Iran 28.29°N 57.10°E
	epP	A	20 34	H = 10 12 49.4 h = 39 km MAG=4.9
				D = 40.8° Az = 316.0° (USCGS); h = 28 km
				AN USSR gives: 26.9°N 57.1°E
				H = 10 12 38 h = 23 km MAG=4½
				D = 41.8°
1.	e(P)	A	18 50 42	
1.	+eP	A	22 28 26.5	Andreanof Is./Aleutian Is.
				51.35°N 179.33°W
				H = 22 16 30.4 h = normal MAG=5.3
				D = 78.0° Az = 352.9° (USCGS)
				PV:1.2s 18.0nm
				MPV=5.1
2.	-iP	A	03 00 21.0	Ecuador 0.29°S 78.65°W
	e	A	00 28.5	H = 02 47 31.7 h = 121 km MAG=5.8
	i(pP)	A	00 43.5	D = 90.4° Az = 39.5° (USCGS); (h = 85 km)
	e(sP)	A	00 54	PV:1.8s 56.0nm
				MPV=5.4
				AN USSR gives: H = 02 47 22 h = normal
2.	e	A	06 17 07.5	
2.	e	A	09 56 43	Explosion?
2.	iPg	A	14 09 34.0	Explosion
	iSg	A	09 50.0	D = 1.2°
2.	eSg	A	15 27 02	Explosion
2.	eP	A	20 59 14	Off East Coast of Kamchatka
	e	A	59 22	52.39°N 160.53°E
	epP	A	59 24	H = 20 47 37.6 h = 18 km MAG=4.6 (USCGS)

Day	Phase		h m s	Remarks
cont.				
2.	LmH	B	21 38	D = 74.0° h = 38 km
				PV:1.1s 14.6nm
				MPV=5.0
2.	eiP	A	23 15 07.5	Near East Coast of Kamchatka
	i	A	15 09.2	53.75°N 160.54°E
	epP	A	15 26	H = 23 03 39.7 h = 21 km MAG=5.0 (USCGS)
	i	A	15 34.8	D = 72.7° h = 74 km
	i	A	15 49.5	PV1:1.0s 16.7nm PV2:1.2s 64.1nm
	ei	A	16 05	pPV:1.0s 19.0nm
	LmH	C	50.5	MPV1=5.1 MPV2=5.6
				AN USSR gives: h = 77 km
3.	e	A	01 36 39	
3.	eP	A	06 00 28.5	Chagos Archipelago Region 5.28°S 68.51°E
	epP	A	00 36.5	H = 05 48 54.8 h = 35 km MAG=4.8
				D = 74.0° Az = 326.3° (USCGS); h = 30 km
3.	eiPKP	A	13 05 40	South of Australia 50.34°S 139.81°E
				H = 12 45 54.9 h = 12 km MAG=5.1 (USCGS)
				D = 147.5°
3.	ei	A	15 00 12.5	
4.	e	A	00 02 55	Apennines/Italy 43.8°N 12.8°E
	e	A	03 12	H = 00 00 22 (BCIS)
	i(Sn)	A	03 23.5	D = 7.0°
	eSg	A	04 12	e 02 57 i 03 27.5 e 03 51.5 e 04 15
4.	+eP	A	05 21 49	Taiwan Region 21.40°N 121.77°E
	e(pP)	A	22 22	H = 05 09 24.2 h = 134 km MAG=5.5
	i	A	22 30.5	D = 85.7° Az = 323.2° (USCGS);
	ei	A	22 33	(h = 136 km)
	ei	A	22 47	PV:1.8s 107nm
	ei	A	23 06.5	LmH:18s 1.9 µm
	eS	C	32 07	MPV=5.4
	eSP	C	33(02)	AN USSR gives: h = 106 km, UPP: h = 170 km.

Day	Phase		h m s	Remarks
cont.				
4.	e(SPP)	C	05 33 30	If we interpret i 22 30.5 as ipP, the
	e(PPS)	C	33 42	focal depth will be h = 170 km. This
	LmH	B	06 03.7	would coincide with the focal depth gi-
	LmV	B	07	ven from Uppsala (UPP) with the aid of
				pP-readings of Swedish stations.
4.	+iPKIKP	A	06 35 38.8	Tonga Islands 18.49°S 175.45°W
	+iPKHKP	A	35 41.0	H = 06 16 21.9 h = 225 km MAG=5.7
	eipPKP	A	36 41	D = 147.4° Az = 351.6° (USCGS);
	ePP	A	39 06	h = ca. 235 km
	eSS	C	57(40)	PV1:1.8s 82nm PV2:1.8s 475nm
	esSS	B	59 07	PV3:2.0s 187nm
	eSSS	B	07 03 20	AN USSR gives: 18.4°S 174.6°W
	esSSS	B	04 48	H = 06 15 59 h = normal MAG=5½
	eLQ	C	15	D = 147.4°
				e 36 56 e 37 36
4.	e(Sg)	A	11 51 04.5	Explosion?
4.	eP	A	15 11 33	Fox Islands/Aleutian Is. 52.05°N 170.48°W
	e	A	11 41	H = 14 59 39.2 h = 42 km MAG=4.3 (USCGS)
	e	A	11 54	D = 77.6°
4.	-iP	B	18 01 32.0	Aegean Sea 39.24°N 24.64°E
	-eP1	A	01 32	H = 17 58 06.4 h = normal MAG=5.9
	-iP2	A	01 36.0	D = 14.6° Az = 325.4° (USCGS)
	iS	C	04 11	PV:8s 37 µm PH:8s 35.3 µm
	eSS	B	04 42	PV1:2.0s 327nm PV2:1.5s 580nm
	i(SSS)	B	04 57	LmH:10.5s 390 µm LmV:14s 440 µm
	iLg1	B	05 42	MLH=6.8 MLV=6.9
	iLg2	B	06 08	i 01 39.5 i 01 45.8 i 01 48 i 04 25
	iRg	B	06 56	e 04 28 i 05 32 i 05 50
	LmH	B	07.9	Multiple P. Well developed higher mode
	LmV	B	09.6	surface waves.
4.	eP	A	18 41 34.5	Aegean Sea 39.02°N 24.73°E
	e	A	41 40	H = 18 38 02.8 h = 35 km MAG=4.4 (USCGS)
	e	A	41 44	D = 14.9°

Day	Phase		h m s	Remarks
4.	e(P)	A	18 42 11	
	e	A	42 19	
	e	A	42 35.5	
4.	e(P)	A	18 46 19	
	e	A	46 31.5	
4.	eiP	A	18 49 32	
4.	LmH	C	21 16	Probably Bonin Islands Region
4.	LmH	C	23 37	Probably Caroline Islands Region
5.	e(P)	A	01 03(11)	Aegean Sea 39.0°N 24 3/4°E
	e	A	03 14	H = 00 59 38 (BCIS)
	e	A	03 17	D = 14.9°
				The first onset is doubtful.
5.	e(P)	A	02 42 38	
5.	e	A	03 03 09	
	e	A	03 20	
5.	eiP	A	10 07 08	Kurile Islands 46.8°N 152.7°E
	e	A	07 29	H = 09 55 15.4 h = normal MAG=4.4 (USCGS)
				D = 77.2°
5.	e(P)	A	11 43 38	
	e	A	43 45	
	e	A	43 49	
5.	eP	A	17 25 32	Rumania 45.8°N 26.8°E
	e(PP)	A	25 40	H = 17 22 54.1 h = 130 km MAG=4.4
	e(PPP)	A	25 47.5	D = 11.2° Az = 301.2° (USCGS)
	ei	A	26 56	
6.	ePg	A	02 34 02	Alps 47 1/2°N 14 3/4°E
	eSg	A	34 44	H = 02 32 43 (BCIS)
	e	A	34 50	D = 3.7°
	e	A	34 56	

Day	Phase		h m s	Remarks
6.	eP	A	04 52 07	South of Honshu/Japan 30.55°N 137.61°E
	epP	A	54 01	H = 04 40 17.8 h = 490 km MAG=5.1
	e	A	54 27.5	D = 86.1° Az = 328.9° (USCGS); h = 523 km
	e(pPP)	A	57 25	PV:2.0s 93.5nm
				MPV=5.1
				AN USSR gives: 31.2°N 137.6°E
				H = 04 40 26 h = 550 km
				D = 85.5°
6.	e	A	05 00 31.5	
6.	ePKIKP	A	08 31 20	South of Fiji Islands 22.72°S 177.48°W
	-iPKHKP	A	31 26.0	H = 08 11 58.8 h = 227 km MAG=4.7 (USCGS)
	ei	A	31 30	D = 151.0°
	+iPKP2	A	31 34.5	PV2:1.0s 52.5nm PV4:1.0s 21.4nm
	ei	A	31 39	AN USSR gives: 19.1°S 174.5°E
	i	A	31 50.2	H = 08 11 52 h = normal
				D = 145.7°
6.	eP	A	11 41 12.5	Off W. Coast of Northern Sumatra
	epP	A	41 29.5	3.74°N 95.82°E
				H = 11 28 49.4 h = 57 km MAG=5.1
				D = 83.4° Az = 320.4° (USCGS); h = 65 km
6.	e	A	11 48 23	
6.	iP	A	11 48 32.0	Fox Islands/Aleutian Is. 52.41°N 169.59°W
	e	A	48 35	H = 11 36 37.0 h = 12 km MAG=3.9 (USCGS)
				D = 77.2°
6.	e	A	18 32 25	Rock burst/Kladnov, ČSSR
	e(Sg)	A	32 32	
	e	A	32 35	
6.	e	A	19 49 13	
7.	e(Sg)	A	02 41 13.5	Explosion?
7.	eP	A	04 40 28	

Day	Phase		h m s	Remarks
7.	iPn	A	08 02 35	Yugoslavia 43.27°N 17.68°E
	e	A	02 39	H = 08 00 32.2 h = normal MAG=4.2
	eSn	A	04 06	D = 8.5° Az = 332.8° (USCGS)
	e	A	04 36	AN USSR gives: Adriatic Sea 42.2°N 15.9°E
	ei	A	04 51	H = 08 00 27 h = 116 km
	i	A	05 08.0	D = 8.9°
7.	iPg	A	14 03 21.8	Explosion
	eiSg	A	03 38	D = 1.2°
	LmH	A	03 55	i 03 22.6 i 03 37.4
	LmV	A	03 55	
7.	e	A	14 11 21.5	Explosion?
	e	A	11 24	
	e	A	11 25.5	
8.	e(P)	A	03 23 01.5	
8.	i(P)	A	04 19 06.5	
	e	A	19 08	
8.	e	A	08 06 19	Explosion
	eiSg	A	06 32	
8.	e	A	12 41 35	
8.	ei	A	20 41 36	Explosion?
	e	A	41 40	
	e	A	41(50)	
	e	A	42 04	
8.	e(P)	A	22 04 45.5	
	e	A	04 52.5	
	e	A	04 56.5	
8.	ePKIKP	A	22 33 13	New Hebrides Islands 15.61°S 167.63°E
				H = 22 13 56.8 h = 132 km MAG=4.5
				D = 140.1° Az = 336.2° (USCGS)

Day	Phase		h m s	Remarks
8.	ePKP	A	23 26 39	Fiji Islands Region 17.62°S 177.01°W
	e	A	26 49	H = 23 07 15.7 h = 182 km MAG=4.2
				D = 146.3° Az = 350.1° (USCGS)
9.	LmH	C	07(08)	Santa Cruz Islands
9.	ePKP	B	07 18(00)	Santa Cruz Islands 10.64°S 166.30°E
	eiPP	B	20 30	H = 06 58 35.7 h = 30 km MAG=6.0
	iSKP	B	21 20	D = 135.0° Az = 337.4° (USCGS)
	eiPKS	B	21 32	PV:10s 1.1 µm PPV:8s 2.6 µm
	eSS	C	38.5	SKPV:10s 1.8 µm
	eSSP	C	39.0	LmH:16s 3.1 µm LmV:20s 4.4 µm
	eSSS	C	43.6	MPPV=6.4 MLH=6.2 MLV=6.2
	LmH	B	08 11.0	PKIKP must be about 8 s earlier than the
	LmV	B	24.7	first given PKP-onset.
9.	LmH	B	09 06	Santa Cruz Islands 10.79°S 166.36°E
	LmV	B	07.5	H = 07 50 57.6 h = normal MAG=5.1 (USCGS)
				D = 135.2°
				LmV:17s 1.2 µm
				MLV=5.7
9.	e(P)	A	17 53 41	
9.	ePKP	A	18 18 09.5	Tonga Islands 15.54°S 175.48°W
				H = 17 58 33.2 h = 95 km MAG=4.3 (USCGS)
				D = 143.5°
9.	ePP	B	18 24 33	Santa Cruz Islands 10.70°S 166.31°E
	ePP	C	24 36	H = 18 02 45.7 h = 59 km MAG=6.4 (USCGS)
	eSKP	C	25(30)	D = 135.1°
	i(SKP)	B	25 36	PPV:7.5s 1.4 µm SKPV:8s 1.0 µm
	eSS	C	42.5	LmH:19s 1.3 µm LmV:19s 2.1 µm
	e	C	43.3	MPPV=6.1 MLH=5.7 MLV=5.9
	eSSS	C	47.5	
	LmH	B	19 28.8	
	LmV	B	28.8	
9.	e	A	20 08 48.5	
	e	A	09 27.5	

Day	Phase		h m s	Remarks
9.	eP	A	21 06 40	
9.	LmV	B	21 18.1	North Atlantic Ocean 55.84°N 34.33°W
	LmH	B	18.3	H = 20 59 42.6 h = normal MAG=4.4 (USCGS)
				D = 27.5°
				LmH:13.5s 1.6 µm LmV:12s 2.1 µm
				MLH=4.8 MLV=5.0
9.	eP	A	21 28 34	North Atlantic Ocean 56.07°N 34.42°W
	e	A	28 40	H = 21 22 48.9 h = normal MAG=4.9 (USCGS)
	LmH	B	41.4	D = 27.6° Az = 81.8° (USCGS)
	LmV	B	41.4	LmH:14.5s 1.8 µm LmV:13.5s 2.3 µm
				MLH=4.8 MLV=5.0
9.	e	A	21 30 39	
	e	A	30 49	
9.	ei	A	21 31 47.5	
9.	iPKHKP	A	21 44 55.5	Fiji Islands Region 21.53°S 176.32°W
				H = 21 25 34.6 h = 283 km MAG=4.8
				D = 150.3° Az = 349.8° (USCGS)
				PV:1.0s 33.4nm
10.	eP	A	00 43 15	Bonin Islands Region 28.72°N 138.66°E
	ePP	A	46 52	H = 00 31 17.0 h = 490 km MAG=5.1
				D = 88.2° Az = 329.4° (USCGS)
				AN USSR gives: h = 575 km
10.	LmH	B	02(00)	North Atlantic Ocean 56.13°N 34.68°W
	LmV	B	02.3	H = 01 43 55.6 h = normal MAG=4.7 (USCGS)
				D = 27.5°
				LmV:13s 0.7 µm
				MLV=4.5
10.	e(PKP2)	A	06 52 53	South of Fiji Islands 23.00°S 179.72°W
	e	A	53 08	H = 06 33 53.3 h = 551 km MAG=4.2 (USCGS)
				D = 151.0°

Day	Phase		h m s	Remarks
10.	e(P)	A	11 01 09	
10.	e(P)	A	11 21 49	
10.	LmH	C	11 30.1	Probably North Atlantic Ocean
10.	e	A	13 48 17	
10.	eP	A	14 28 18	Near S. Coast of Honshu/Japan
				34.40°N 137.63°E
				H = 14 16 26.8 h = 331 km MAG=4.5
				D = 82.9° Az = 328.7° (USCGS)
10.	eP	A	20 52(24)	North Atlantic Ocean 55.94°N 34.59°W
	e	A	52 27.5	H = 20 46 33.6 h = normal MAG=4.7 (USCGS)
				D = 27.8°
11.	iP	A	06 38 56.6	Hindu Kush Region 36.39°N 70.75°E
	epP	A	39 43	H = 06 31 09.0 h = 220 km MAG=5.0 (USCGS)
	esP	A	40 08	D = 44.0° h = 222 km
	iPP	A	40 37.0	e 38 59.5 e 41 53
11.	ePP	A	08 55 18	Santa Cruz Islands 10.67°S 166.20°E
	e(pPP)	A	55 29	H = 08 33 27.4 h = 49 km MAG=6.1 (USCGS)
				D = 135.1° (h = ca. 40 km)
11.	e	A	12 04 56	Explosion?
	e(Sg)	A	05 11	
11.	eP	A	14 57 39	Vera Cruz/Mexico 19.08°N 95.83°W
	e	A	57 44.5	H = 14 44 59.2 h = 33 km MAG=5.5 (USCGS)
	i	A	57 50.0	D = 86.0° (h = 50 km)
	i(pP)	A	57 52.5	PV1:1.8s 143nm PV2:1.6s 49.2nm
	ePP	A	15 00 58	PV3:1.6s 49.2nm PV4:1.8s 117nm
	e	A	01 16	MPV=5.8
	e	A	01 22	AN USSR gives: 20.8°N 96.0°W
	e	A	01 25	H = 14 45 44 h = 360 km
	e	A	01 47	D = 84.7°

Day	Phase		h m s	Remarks
11.	eP	A	17 07 26	India-China Border Region
	e(pP)	A	07 30	28.36°N 94.40°E
	e	A	07 39	H = 16 56 48.7 h = 7 km MAG=5.3
	e	A	07 49.5	D = 64.2° Az = 315.9° (USCGS); (h = 15 km)
				PV:1.5s 40.0nm
				MPV=5.4
11.	eP	A	19 45 35	Red Sea 19.65°N 38.86°E
	e(pP)	A	45 39	H = 19 38 21.5 h = normal MAG=5.2
				D = 37.7° Az = 331.5° (USCGS); (h = 19 km)
12.	eP	A	03 03 51	Hokkaido/Japan Region 42.58°N 143.04°E
	i(pP)	A	03 56.0	H = 02 51 54.7 h = normal MAG=5.3
	e	A	04 26	D = 77.9° Az = 330.8° (USCGS); (h = 19 km)
	e	A	06 35	PV1:1.6s 35.6nm PV2:1.5s 46.6nm
	ePP	A	06 47	MPV=5.2
				AN USSR gives: h = 132 km
12.	e(Sg)	A	06 49 28	Probably explosion.
	e	A	49 35	
12.	e	A	06 58 50	
12.	e	A	06 59 30	Explosion?
12.	eP	A	21 51 46	Red Sea 19.71°N 38.89°E
	e	A	51 50	H = 21 44 32.8 h = normal
				D = 37.7° Az = 331.5° (USCGS)
13.	LmH	C	06 05.5	Bismarck Sea 3.17°S 147.60°E
	LmV	C	05.5	H = 04 53 30.0 h = normal MAG=4.9 (USCGS)
				D = 119.9°
				LmH:20s 0.8 µm LmV:(20s 1.2 µm)
				MLH=5.4 (MLV=5.5)
13.	eP	A	07 35 18.5	Red Sea 19.66°N 38.82°E
	ei(pP)	A	35 29	H = 07 28 05.7 h = 41 km MAG=5.3
	e	A	35 43	D = 37.7° Az = 331.6° (USCGS); (h = 50 km)
				PV:1.8s 75.0nm
				MPV=5.3

Day	Phase		h m s	Remarks
13.	iPKHKP	A	07 56 20.0	Fiji Islands Region 20.58°S 178.42°W
	iPKP2	A	56 26.0	H = 07 37 37.5 h = 586 km MAG=4.6
	e(pPKP)	A	58 12	D = 149.0° Az = 347.6° (USCGS);
				(h = ca. 455 km)
				PV:1.3s 53.5nm
13.	eP	A	08 18 10	Red Sea 19.63°N 38.88°E
	e	A	18 14.5	H = 08 10 56.3 h = normal MAG=5.0
	e	A	18 22	D = 37.7° Az = 331.5° (USCGS)
				PV:1.8s 37.4nm
				MPV=5.0
13.	iPn	A	11 23 19.8	Explosion/ČSSR 50.58°N 14.05°E
	iPg	A	23 21.8	H = 11 22 52 (BCIS)
	i	A	23 27.5	D = 1.5°
	eSn	A	23 38	i 23 27.5 is a very clear phase.
	iSg	A	23 41.3	
13.	eP	A	11 53 43	PV:1.7s 22.2nm
13.	eP	A	14 55 54	Fox Islands/Aleutian Is. 53.72°N 165.36°W
	i	A	56 02	H = 14 44 07.2 h = normal MAG=5.2
	i	A	56 19.5	D = 76.0° Az = 2.0° (USCGS)
				PV1:1.4s 23.8nm PV2:0.6s 19.6nm
				MPV=5.1
13.	iPKIKP	A	16 25 37.0	Off Coast of Southern Chile
	ePP	A	26 58	40.09°S 74.45°W
	LmH	B	17 15.5	H = 16 06 54.3 h = normal MAG=6.0 (USCGS)
	LmV	B	15.5	D = 117.4°
				LmV:18s 1.3 µm
				MLV=5.6
13.	iPg	A	16 42 16.5	Rock burst/Augsdorf, GDR
	iSg	A	42 30.2	H = 16 41 57
				D = 1.1°
13.	iPg	A	16 50 41.2	Rock burst/Augsdorf, GDR
	iSg	A	50 55.7	H = 16 50 21
	e	A	50 59	D = 1.1°

Day	Phase		h m s	Remarks
13.	iPg	A	16 51 17.2	Rock burst/Augsdorf, GDR
	iSg	A	51 32.2	H = 16 50 57
	i	A	51 35.7	D = 1.1°
	LmH	A	51 56	LmH:1.4s 163nm LmV:1.4s 154nm
	LmV	A	51 56	
13.	iPg	A	16 58 15.7	Rock burst/Augsdorf, GDR
	iSg	A	58 30.7	H = 16 57 55
	LmH	A	58 54	D = 1.1°
	LmV	A	58 54	LmH:1.4s 94nm LmV:1.5s 100nm
				e 58 33 e 58 42
13.	iPg	A	17 08 07.2	Rock burst/Augsdorf, GDR
	eSg	A	08 21.5	H = 17 07 46
	e	A	08 23.5	D = 1.1°
	LmH	A	08 49	
13.	eP	A	17 41 03	Talaud Islands 3.56°N 126.48°E
				H = 17 27 11.3 h = 63 km MAG=5.4
				D = 102.7° Az = 323.7° (USCGS)
13.	eP	A	17 47(00)	Southern Greece 37.61°N 22.63°E
	i	A	47 02.7	H = 17 43 23.5 h = 89 km MAG=4.5 (USCGS)
	i	A	47 06.2	D = 15.3°
13.	eP	A	19 14 57.5	Talaud Islands 3.57°N 126.46°E
				H = 19 01 01.4 h = 35 km MAG=5.4
				D = 102.7° Az = 323.7° (USCGS)
13.	-eP	A	19 29 32	Red Sea 19.70°N 38.85°E
	+iP	A	29 33.0	H = 19 22 15.4 h = 7 km MAG=5.8
	+ipP	A	29 36.5	D = 37.7° Az = 331.5° (USCGS); h = 17 km
	i	A	29 39.5	PV2:1.9s 412nm pPV:2.0s 313nm
	ePcP	A	31 50	PcPV:1.8s 81.7nm
	eS	C	35 28	MPV2=6.0
	iS	B	35 30	e 29 55 ei 30 10 e 31 18
	LmH	C	(51)	The P-onsets of Red Sea-earthquakes are
	LmV	C	(54)	strong at our station, but have relatively
				long periods (T = 1.8 - 2.0 s) in the

Day	Phase		h m s	Remarks
cont. 13.				shortperiod seismographs. PcP is clearly developed. Surface waves are abnormally small.
13.	iP	A	20 56 18.0	North Atlantic Ocean 56.39°N 35.20°W
				H = 20 50 30.8 h = normal MAG=4.6 (USCGS)
				D = 28.0°
13.	eP	A	21 50 41	Franz Josef Land 82.16°N 39.66°E
	e	A	50 49	H = 21 44 03.6 h = 11 km MAG=4.4 (USCGS)
	e	A	50 57	D = 32.7°
14.	+eP	A	07 08 38.5	India-China Border Region 28.37°N 94.31°E
	-i	A	08 39.7	H = 06 58 04.6 h = 24 km MAG=5.9 (USCGS)
	eipP	A	08 45	D = 64.2° h = 25 km
	ePcP	A	09 15.5	PV2:1.6s 144nm SH(B):14s 1.6 µm
	ePP	C	11 00	LmH:13.5s 7.5 µm LmV:14s 9.4 µm
	ePPP	AB	12 34	MPV2=6.0 MSH(B)=5.9 MLH=6.2 MLV=6.3
	eS	BC	17 10	e 08 57 e 09 10 e 09 46 e 09 56
	eScS	C	18(27)	e 10 47 e 20 50 e 34.9
	eSS	C	21 22	
	eSSS	C	24 28	
	LmH	B	40.0	
	LmV	B	40.0	
14.	eP	A	07 56 50.5	Franz Josef Land 82.46°N 36.24°E
	e	A	56 58.5	H = 07 50 19.3 h = normal MAG=4.7
	e	A	57 06.5	D = 32.8° Az = 209.4° (USCGS)
	e	A	57 21	PV1:1.2s 20.5nm PV3:1.7s 43.5nm
	e(PcP)	A	59 44	LmH:13s 2.6 µm LmV:13s 3.4 µm
	LmH	B	08 12.0	MPV=4.9 MLH=5.3 MLV=5.3
	LmV	B	12.0	e 58 23 e 58 38 e 59 03
14.	e	A	08 04 31	Near earthquake
	e	A	05 46	
	e	A	06 08	
	i	A	06 45.5	
	i	A	07 36.7	

Day	Phase		h m s	Remarks
cont.				
14.	e	A	08 08 29	
	LmH	A	09.3	
14.	LmH	B	08 19.5	LmH:11s 2.1 μm LmV:11s 2.1 μm
	LmV	B	19.5	
14.	e	A	11 37 27	Explosion?
14.	e	A	11 37 44	
14.	iP	A	21 59 21.0	Red Sea 19.37°N 38.67°E
				H = 21 52 05.3 h = normal MAG=4.6
				D = 37.9° Az. = 331.8° (USCGS)
14.	ePKIKP	A	23 43 24	South of Fiji Islands 23.02°S 178.73°E
	-ePKHKP	A	43 29	H = 23 24 47.8 h = 650 km MAG=4.9
	ePKP2	A	43(36.5)	D = 150.6° Az = 343.2° (USCGS)
				PV2:0.9s 18.8nm
15.	e	A	03 42 58	Near earthquake/Probably Yugoslavia
	e	A	43 30	Zagreb gives: eP 03 40 12.0 iS 40 35.7
	e(Sg)	A	43 44	Trieste gives: D = 320 km
	e	A	43 55	
15.	iPKP	A	06 59 03.5	New Hebrides Islands 18.86°S 169.27°E
	ei	A	59 10	H = 06 39 59.8 h = 253 km MAG=4.9
	epPKP	A	07 00 06	D = 143.7° Az = 335.9° (USCGS);
	ei	A	00 27	h = 250 km
				PV:1.2s 28.2nm
				AN USSR gives: h = 160 km
15.	i(Sg)	A	16 19 12.0	Explosion?
	e	A	19 14	
	i	A	19 16.0	
15.	e	A	21 05 45	Near earthquake
	e	A	06 31.5	Trieste gives: ePn 21 03 19.5 iSg 03 56
	e	A	06 37	D = 260 km

Day	Phase		h m s	Remarks
16.	eP	A	03 19 13.5	Red Sea 19.50°N 38.88°E
	epP	A	19 18.5	H = 03 11 59.0 h = normal MAG=5.4
	e	A	19 20.5	D = 37.9° h = 24 km
	e	A	20 32	PV:1.6s 45.5nm
	ePP	A	20 40	MPV=5.2
	ePcP	A	21 33.5	e 19 50
16.	eP	A	11 48 41	Red Sea 19.71°N 39.03°E
	e(pP)	A	48 47	H = 11 41 28.9 h = normal (USCGS)
	e	A	49 02.5	D = 37.7° (h = 28 km)
16.	+iPKIKP	A	12 29 14.0	Loyalty Islands Region 22.06°S 170.45°E
	ePKHKP	A	29 16.5	H = 12 09 37.7 h = 66 km MAG=5.4
	eiPKP2	A	29 19.3	D = 147.0° Az = 335.0° (USCGS)
	e	A	29 34	PV1:1.1s 19.5nm PV2:1.1s 43.8nm
	LmH	A	13(34)	
16.	eP	A	14 52 26	Red Sea 19.73°N 38.88°E
	epP	A	52 31	H = 14 45 12.6 h = normal MAG=5.1 (USCGS)
				D = 37.7° h = 24 km
				PV:1.9s 70.5nm
				MPV=5.3
16.	eiP	A	16 07 30	Red Sea 19.79°N 38.90°E
	epP	A	07 36	H = 16 00 16.0 h = normal MAG=5.0
	e	A	07 42	D = 37.6° Az = 331.4° (USCGS); h = 28 km
	e	A	07 45	AN USSR gives: 18.3°N 38.5°E
				H = 16 00 06
				D = 38.8°
16.	ePKIKP	A	17 51 27	New Hebrides Islands Region
	e	A	51 43.5	13.63°S 170.65°E
	ePP	A	54 28	H = 17 33 07.5 h = 637 km MAG=4.8
				D = 139.3° Az = 339.5° (USCGS)
17.	e	A	02 34 48	
17.	e	A	07 00 20	

Day	Phase		h m s	Remarks
17.	e(Sg)	A	09 02 17	Explosion?
17.	e(Sg)	A	09 18 25	Explosion?
17.	e(Sg)	A	10 00 32	Explosion?
	e	A	00 45	
	e	A	00 52	
17.	i(Pg)	A	10 37 49	Explosion
	e(Sg)	A	38 07	(D = 1.4°)
	e	A	38 12	
	e	A	38 20	
17.	e(P)	A	11 35 33	PV:0.9s 11.9nm
17.	ePKP	A	11 43 45	New Ireland Region 3.60°S 150.92°E
	ePP	A	45 12	H = 11 24 45.7 h = normal MAG=5.4
	e	A	45 15	D = 122.0° Az = 330.7° (USCGS)
	ePS	C	54 58	PPV2:2.3s 100nm
	eSS	C	12 02 00	LmH:18s 5.1 μm LmV:17s 6.1 μm
	eSSS	C	05 52	MPPV2=5.9 MLH=6.2 MLV=6.3
	e	C	09 30	PKIKP must be 8 s earlier than the given
	e(Sa)	C	15.5	PKP-onset.
	LmV	B	40.3	AN USSR gives: New Guinea 4.6°S 152.5°E
	LmH	B	40.6	H = 11 24 31 h = normal
				D = 123.6°
17.	e	A	12 15 48	
17.	e(Pn)	A	12 09 31	Near earthquake
	e	A	10 47.5	Trieste gives: ePn 12 08 23 iSg 09 06
	e	A	11 18	D = 300 km
	e	A	11 47.5	e 09 54 e 10 33 e 10 38.5 e 10 52
	e	A	11 52	e 10 54 e 11 05 e 12 05 e 12 18
17.	ePn	A	14 47 22	Explosion
	ePg	A	47 24	D = 1.5°
	eiSn	A	47 41	ei 47 29 e 47 45
	eiSg	A	47 43.5	

Day	Phase		h m s	Remarks
17.	e	A	15 45 59	Explosion?
	e	A	46 01	
	e	A	46 11.5	
18.	ePKHKP	A	09 46 21	Fiji Islands Region 20.72°S 179.36°W
	ePKP2	A	46 27	H = 09 27 42.7 h = 650 km MAG=4.9
	ei	A	46 32	D = 148.9° Az = 346.4° (USCGS)
	i	A	47 31.0	
	i	A	47 33.5	
18.	e	A	10 01 14	Probably explosion
	e	A	01 17.5	
	e	A	01 21	
	e	A	01 26	
18.	e	A	18 02 06	Honshu/Japan 36.29°N 139.76°E
	e	A	02 10	H = 17 49 50.8 h = 105 km MAG=5.0 (USCGS)
	e(pP)	A	02 19	D = 81.3°
	e	A	02 25.5	The first onset of P must be about 9 s
	e	A	02 36	earlier.
	e	A	02 38.5	AN USSR gives: Japan 36.1°N 140.1°E
	e	A	04 52.5	H = 17 49 41 h = normal
				D = 82.5°
18.	ei	A	19 35 31.5	
18.	ei	A	21 32 30.5	
19.	ePKIKP	A	01 29 17	Banda Sea 6.66°S 129.85°E
	i	A	29 20.0	H = 01 10 45.8 h = 60 km MAG=5.9
	iPP	A	30 09.5	D = 112.8° Az = 322.5° (USCGS)
	e	A	30 16	PV:1.3s 25.0nm PPV:1.6s 49.2nm
	e	A	30 17.5	MPPV=5.9
	e	A	30 25.5	AN USSR gives: h = 95 km
19.	i	A	03 48 51.0	

Day	Phase		h m s	Remarks
19.	eP1	A	04 13 34	Kurile Islands 45.41°N 151.27°E
	+iP1	B	13 35.5	H = 04 01 36.7 h = normal MAG=5.7
	-iP2	AC	13 37.5	D = 78.1° Az = 335.1° (USCGS)
	eiS	C	23 22	PV1:1.1s 51.2nm PV1(B):10.5s 8.0 µm
	eS	B	23 28	PV2(A):1.8s 418nm SH(B):13s 5.5 µm
	eScS	B	23 54	LmH:18s 68 µm LmV:18.5s 87.2 µm
	ePS	B	24 12	MPV1=5.6 MPV2(A)=6.3 MPV2(B)=6.8
	eiSS	C	28 34	MSH(B)=6.5 MLH=7.0 MLV=7.1
	eSS	B	28 38	e 13 44 e 13 53 e 15 36
	eSSS	C	32 04	
	LmH	B	51.2	
	LmV	B	51.4	
19.	e	A	17 37 17	Probably Rat Islands/Aleutian Is.
	e	A	38 04.5	51.91°N 179.97°E
				H = 17 25 10.5 h = 18 km MAG=4.9
				D = 77.3° Az = 352.4° (USCGS)
				P must be 12 s earlier than the first
				given onset.
19.	e(P)	A	18 20 04	
	e	A	20 24	
20.	eP	A	05 25 53	Kurile Islands 45.53°N 150.93°E
				H = 05 13 57.6 h = normal MAG=4.5
				D = 77.9° Az = 334.9° (USCGS)
				PV:1.1s 14.6nm
				MPV=5.1
20.	e(P)	A	07 08 28	
20.	eP	A	09 42 32.5	Kurile Islands 45.65°N 151.46°E
				H = 09 30 36.4 h = normal MAG=4.4
				D = 78.0° Az = 335.2° (USCGS)
20.	ei	A	09 45 48	Explosion?
	e	A	46 06.5	

Day	Phase		h m s	Remarks
20.	eP	A	13 43 28.5	Kurile Islands 45.61°N 151.38°E
	-iP	A	43 29.5	H = 13 31 34.0 h = 51 km MAG=5.7
	e	A	43 35.5	D = 78.0° Az = 335.1° (USCGS)
	e(PcP)	A	43 40	PV2:1.2s 174nm PV3:1.3s 66.6nm
	i	A	43 49.0	(PcP)V:1.2s 66.6nm
	LmH	B	14 17.6	LmH:18s 6.6 μm LmV:18s 5.1 μm
	LmV	B	21	MPV2=6.1 MLH=6.0 MLV=5.9
20.	eP	A	13 50 47.5	Kurile Islands 45.75°N 151.63°E
				H = 13 38 53.7 h = 50 km MAG=4.6
				D = 77.9° Az = 335.3° (USCGS)
				PV:1.2s 12.8nm
				MPV=4.9
20.	+iP	A	13 52 48.3	Kurile Islands 45.55°N 151.52°E
	ei(PcP)	A	52 59	H = 13 40 52.8 h = 53 km MAG=5.3 (USCGS)
	e	A	53 04	D = 77.9°
	e	A	53 10.5	PV:1.4s 97.0nm
	i	A	53 20.2	MPV=5.7
	e	A	53 38	
	LmH	B	14 25	
20.	+eiP	A	14 04 02	Kurile Islands 45.64°N 151.53°E
	e	A	04 06.5	H = 13 52 05.5 h = 32 km MAG=5.4
	ei(pP)	A	04 08.5	D = 78.0° Az = 335.2° (USCGS);
	i	A	04 13.5	(h = 24 km)
	i	A	04 20.5	PV1:1.9s 276nm PV2:1.2s 20.5nm
	e	A	05 03	PV3:1.6s 114nm
	LmH	B	38.1	LmH:16s 4.9 μm
				MPV=6.1 MLH=5.9
				AN USSR gives: h = 61 km
20.	ei(Sg)	A	14 10 54.5	Explosion?
20.	LmH	B	14 17.6	LmH:17.5s 6.2 μm LmV:18s 5.1 μm
	LmV	B	21	

Day	Phase		h m s	Remarks
20.	eP	A	14 56 13.5	Kurile Islands 45.44°N 151.48°E
				H = 14 44 17.8 h = 50 km MAG=4.6
				D = 78.2° Az = 335.2° (USCGS)
				PV:1.3s 16.6nm
				MPV=5.0
20.	+iP	A	15 58 23.0	Kurile Islands 45.58°N 151.23°E
	e	A	58 33	H = 15 46 29.4 h = 60 km MAG=5.1
	e	A	58 54.5	D = 78.0° Az = 335.1° (USCGS)
				PV:1.6s 38.0nm
				MPV=5.3
20.	eP	A	16 23 57	Kurile Islands 45.54°N 151.13°E
				H = 16 12 00.7 h = normal MAG=4.0 (USCGS)
				D = 77.8°
20.	+eP	A	17 23 32	Kurile Islands 45.47°N 151.45°E
	e(pP)	A	23 51	H = 17 11 34.8 h = normal MAG=5.0
	i(sP)	A	23 59	D = 78.1° Az = 335.2° (USCGS); h = 74 km
				PV:1.4s 42.5nm
				MPV=5.4
20.	+eiPKIKP	A	19 27 06.7	Loyalty Islands Region 22.14°S 170.55°E
	-iPKHKP	A	27 08.8	H = 19 07 25.2 h = 28 km MAG=5.5
	i	A	27 11.0	D = 147.1° Az = 335.1° (USCGS)
	e	A	27 31	PV1:1.2s 46.1nm PV2:1.3s 89.0nm
	i	A	27 48.5	ei 27 13 i 27 18.5 e 27 26 i 27 33.5
20.	eP	A	22 07 03.5	Honshu/Japan 36.27°N 139.74°E
				H = 21 54 48.0 h = 57 km MAG=4.7 (USCGS)
				D = 82.0°
21.	i(P)	A	00 33 45.5	
21.	ePKIKP	A	11 44 31.5	Tonga Islands 23.83°S 175.18°W
	ePKHKP	A	44 39.5	H = 11 24 44.6 h = normal MAG=5.4 (USCGS)
	e(pPKIKP)	A	44 43	D = 152.7° (h = ca. 38 km)
	e(pPKHKP)	A	44 49.5	e 44 58.5 e 45 08 e 45 13
	e(PKP2)	A	44 52.5	

Day	Phase		h m s	Remarks
21.	e	A	13 06 19	
21.	e	A	13 10 03	
21.	e	A	17 40 25	
21.	eP	A	18 23 44.5	Northern Columbia 6.84°N 72.96°W
	epP	A	24 23.5	H = 18 11 42.2 h = 151 km MAG=5.4 (USCGS)
	e	A	24 28	D = 81.4° h = 160 km
	e	A	24 35	pP is the greatest onset on the short-
	e	A	24 42	period vertical components.
22.	eP	A	03 01 30	Red Sea 20.30°N 38.69°E
				H = 02 54 22.6 h = normal
				D = 37.1° Az = 331.3° (USCGS)
				PV:1.9s 35nm
				MPV=5.0
22.	eP	A	06 10 16.7	Near Islands/Aleutian Is. 51.61°N 173.86°E
	e	A	10 30	H = 05 58 25.5 h = normal MAG=4.9 (USCGS)
				D = 77.0°
22.	e	A	07 59 46.5	
22.	e	A	08 28 53	
22.	e(Sg)	A	11 04 06	Explosion?
22.	LmH	C	14(13)	East New Guinea Region
22.	iPg	A	15 04 35.8	Explosion/GDR
	iSg	A	04 51	D = 1.2°
22.	iPn	A	19 15 57.5	Alps/Italy 46.3°N 12.6°E
	ePg	A	16 16	H = 19 14 51 (BCIS)
	eSn	A	16 47.5	D = 4.4°
	e(Sg)	A	17(12)	
	i	A	17 14.5	
	iLg2	A	17 16.5	

Day	Phase		h m s	Remarks
22.	ei	A	19 20 42	Aftershock of the preceding Italy-earthquake. After Trieste felt at Sacile.
	ei(Sg)	A	20 53.5	
	i	A	21 00	
22.	LmH	C	22 14	South Sandwich Islands
22.	eP	A	23 07 03.5	Red Sea 19.73°N 38.79°E H = 22 59 49.8 h = normal D = 37.6° Az = 331.5° (USCGS) PV:1.2s 15.4nm MPV=4.8
23.	ePKP	A	00 55(18)	Fiji Islands Region 16.99°S 177.11°W H = 00 35 43.1 h = 55 km MAG=4.8 D = 145.7° Az = 350.1° (USCGS) PV2:1.4s 21.2nm The second onset is much greater than the first.
	e	A	55 24	
	e	A	55 30	
23.	LmV	B	13 25.5	LmV:24s 1.45 μm
23.	eP	A	13 53 54	Kurile Islands 45.76°N 151.85°E H = 13 41 57.5 h = normal MAG=4.7 (USCGS) D = 77.8° (h = 50 km) PV:1.2s 20.5nm MPV=5.1
	e(pP)	A	54 07	
	e(sP)	A	54 12	
	LmH	C	14 27.5	
23.	LmH	C	16 06.8	Southern Sumatra 4.71°S 101.84°E H = 15 00 27.8 h = normal MAG=5.2 (USCGS) D = 83.8° LmH:23s 0.4 μm MLH=4.9
23.	e	A	20 17 33	
24.	i(Sg)	A	00 49 16	Explosion?
24.	eP	A	01 35 22	Kurile Islands 45.29°N 150.96°E H = 01 23 24.7 h = normal MAG=4.4 (USCGS) D = 78°

Day	Phase		h m s	Remarks
24.	eP	A	02 04 57	<u>Red Sea</u> 20.21°N 38.32°E
	e	A	05 04	H = 01 57 49.0 h = normal MAG=5.0
	e	A	05 10	D = 37.0° Az = 331.6° (USCGS)
				PV:1.9s 41.1nm
				MPV=5.0
24.	eP	A	04 23 40	<u>Off East Coast of Honshu/Japan</u>
				40.24°N 144.60°E
				H = 04 11 29.6 h = 27 km MAG=5.0 (USCGS)
				D = 80.5°
24.	eP	A	06 45 18	<u>Red Sea</u> 20.02°N 38.66°E
	e(PP)	A	46 50	H = 06 38 08.8 h = 42 km MAG=5.1 (USCGS)
	e	A	47 24	D = 37.3°
24.	eP	A	09 13 10	<u>Java Sea</u> 5.98°S 112.35°E
	+ipP	B	15 19	H = 09 00 19.5 h = 600 km MAG=6.0 (USCGS)
	epP	A	15 19.5	D = 101.5° h = 593 km
	esP	B	16 18	pPV(B):5s 1.4 μm pPV(A):2.5s 200nm
	iPP	A	17 27.5	PPV:1.5s 66.6nm pPPV:1.9s 153nm
	epPP	A	19 16	SKSH(B):10s 3.8 μm SH:13s 3.2 μm
	eSKS	C	22 45	SSH:(13s) 4 μm SSPH:12s 1.9 μm
	iSKS	B	22 50	SaH:42s 6 μm
	iS	B	23 55	LmH:19s 2.9 μm LmV:20s 4.1 μm
	eSP	C	25 28	MPPV=5.7 MSH=6.2
	iSPP	C	26 28	e 15 21.5 e 15 28 e 16 24.5 e 16 42.5
	isS	C	27 50	ei 17 15 i 17 24.0 e 17 57.5 e 18 34.5
	iPKKP	A	29 15.0	e 19 27.5 e 21 00
	iSS	B	31 10	pPP is the greatest onset in the short-
	iSSP	B	31 20	period vertical components, Sa the
	epPKKP	A	31 31.5	strongest phase in the ultra long-period
	e(sSS)	C	34 42	N-component.
	iSSSS	C	38 40	
	iSa(4.52)	C	41 55	
	LmH	B	10 08.3	
	LmV	B	08.3	
24.	e	A	12 04 27	Probably Java Sea 6.0°S 112.3°E
	e(pPP)	A	05 12.5	H = 11 46 13.9 h = 600 km MAG=5.3 (USCGS)
				D = 101.5°

Day	Phase		h m s	Remarks
24.	ePn	A	17 39(30)	Switzerland 46.4°N 7.4°E
	e	A	39 35.5	H = 17 38 15 (BCIS)
	iPb	A	39 42.5	D = 5.1°
	iPg	A	39 50.5	PbV:0.8s 37.7nm PgV:0.7s 152nm
	eiSg	A	40(56.5)	LmH:9s 3.1 µm
	LmH	B	41.2	MLH=4.0
				Pn is unusualy small with an ambiguous beginning.
24.	ePKP	A	23 18 28.5	Fiji Islands Region 20.16°S 179.05°W
	epPKP	A	20 54.5	H = 22 59 47.3 h = 654 km MAG=4.7 (USCGS)
				D = 148.4° h = 658 km
				PV1:1.6s 30.3nm PV2:1.9s 58.7nm
25.	+iP	A	06 05 47.5	Kazakh SSR 50.0°N 78.0°E
	e	A	05 51	H = 05 58 00 (BCIS)
	e	A	05 52	D = 41.2°
	ePn	A	07 19.5	PV:0.8s 54.3nm
	ePP	A	07 23.5	MPV=5.5
	i	A	07 29.5	Probably underground explosion.
25.	eP	A	14 41(50)	South of Panama 7.38°N 79.74°W
	e	A	41 55.5	H = 14 29 13.6 h = 35 km MAG=4.7 (USCGS)
	e	A	42 51.5	D = 85.2°
25.	LmH	C	18 53.5	LmH:16s 0.35 µm
25.	-eiP	A	22 59 55.5	Kurile Islands Region 45.45°N 151.41°E
	eiPcP	A	23 00 06	H = 22 47 58.4 h = 41 km MAG=5.5
	epP	A	00 09	D = 78.1° Az = 335.2° (USCGS); h = 50 km
	isP	A	00 15.0	PV:1.4s 121nm
	eiS	C	09 48	LmH:16s 13 µm LmV:18s 6.0 µm
	eSS	C	15 05	MPV=5.8 MLH=6.4 MLV=6.0
	LmH	B	34.1	i 00 01.0 i 01 44 e 10 18
	LmV	B	(38)	
26.	e	A	03 17 12	West Pakistan 27.18°N 67.50°E
	e	A	17 19.5	H = 03 08 26.9 h = 21 km MAG=4.5
	e	A	17 22	D = 48.1° Az = 314.9° (USCGS)
	e	A	17 34.5	P must be 5 s earlier than the first given onset.

Day	Phase		h m s	Remarks
26.	eP	A	04 34 52	Central Alaska 64.11°N 147.23°W H = 04 24 13.5 h = 33 km MAG=4.4 (USCGS) D = 64.5°
26.	e	A	16 02 31.5	
26.	e	A	16 29 41	
26.	e	A	22 08 15	
	e	A	08 17	
26.	ePKIKP	A	22 58 03	East New Guinea Region 9.27°S 148.58°E H = 22 39 01.5 h = 14 km MAG=5.3 (USCGS) D = 125.8°
27.	eP	A	08 20 23.5	Gulf of Bengal 15.1°N 80.5°E
	e	A	20 27.5	H = 08 09 44 h = normal MAG=4¾
	e	A	20 36	D = 65.1° (AN USSR)
	ePoP	A	20 56.5	PV:1.0s 19.0nm
	e(PP)	A	22 52	MPV=5.2
	LmH	C	(48)	e 20 29 e 21 09.5
27.	eP	A	08 38 43	Western Brazil 8.89°S 71.31°W
	+i	A	38 44.0	H = 08 26 34.5 h = 603 km MAG=5.3
	epP	A	40 53.5	D = 92.4° Az = 39.2° (USCGS); h = 603 km
	e	A	41 16	PV:2.0s 120nm MPV=5.6 AN USSR gives: h = normal
27.	eiP	A	09 09 31	Northeastern China 38.42°N 116.49°E
	e(pP)	A	09 37.5	H = 08 58 25.5 h = 61 km MAG=5.4
	e(sP)	A	09 40.5	D = 69.6° Az = 319.0° (USCGS); h = 25 km
	eiPP	B	12 12	PPV:7s 0.9 µm
	eiS	C	18 36	LmH:18.5s 27.3 µm LmV:15s 24.7 µm
	eSS	C	23 08	MPPV=6.2 MLH=6.5 MLV=6.6
	eSSS	C	26 20	i 09 45.0 i 09 48.0 e 10 03.5 e 23 34
	LmH	B	36.8	e 26 42
	LmV	B	43.8	Successive P-onsets with increasing amplitudes.

Day	Phase		h m s	Remarks
27.	ePKIKP	A	10 21 15	New Hebrides Islands 16.50°S 168.08°E
	eSS	C	42 48	H = 10 01 42.0 h = 11 km MAG=5.5
	eSSS	C	48.0	D = 141.0° Az = 336.2° (USCGS)
	eLQ	C	59.5	LmH:22s 6.9 μm LmV:24s 8.6 μm
	LmV	B	11 20.1	MLH=6.4 MLV=6.4
	LmH	B	20.5	i 21 24 i 21 29
27.	eP	A	18 33 25	Mindanao/Philippine Islands
				5.78°N 126.39°E
				H = 18 19 31.9 h = normal (USCGS)
				D = 101.5°
27.	+eP	A	20 00 52	Red Sea 20.02°N 38.60°E
	epP	A	00 57	H = 19 53 40.6 h = 23 km MAG=5.1
	ePP	A	02 22	D = 37.3° Az = 331.5° (USCGS); h = 24 km
	ePcP	A	03 12	PV1:1.8s 92nm pPV:1.8s 71.5nm
				PPV:2.0s 60nm PcPV:1.9s 41.2nm
				MPV=5.4 MPPV=5.1
28.	e(P)	A	00 08 10.5	Aegean Sea 38.48°N 25.33°E
	e	A	08 12.5	H = 00 04 27.3 h = 34 km MAG=4.3
	e	A	08 37.5	D = 15.6° Az = 325.8° (USCGS)
	LmH	B	13 04	LmH:11.5s 1.3 μm
				MLH=4.3
				P must be 5 s earlier than e(P) for the USCGS epicentre data, but coincides exactly with e(P) for the epicentre calculation of AN USSR: 39.0°N 25.8°E
				H = 00 04 34 h = normal
				D = 15.3°
28.	eP	A	02 48 45	Red Sea 19.87°N 38.61°E
	i	A	48 50.0	H = 02 41 33.5 h = normal (USCGS)
	e	A	48 58.5	D = 37.4°
28.	e(P)	A	06 26 12.5	
28.	ePn	A	15 50 36.5	Belgium 50.55°N 4.09°E
	e(Pb)	A	50 49.5	H = 15 49 23.4 h = 18 km MAG=3.9

Day	Phase		h m s	Remarks
cont.				
28.	ePg	A	15 50 58	D = 4.8° Az = 85.9° (USCGS)
	iSn	A	51 28.0	LmH:7.5s 4.9 /um LmV:5.5s 1.8 /um
	ei(Sb)	A	51 44	MLH=4.2
	iSg	A	51 57.0	i 50 37.5 e 50 46 e 50 53.5 e 51 02
	LmH	B	52.2	i 51 11.5 i 51 19.2 ei 51 23.5
	LmV	B	52.4	i 51 49.0
28.	eP	A	19 45 18	Luzon/Philippine Islands 17.05°N 122.41°E
	e	A	45 25	H = 19 32 25.4 h = 51 km MAG=5.3 (USCGS)
				D = 89.6°
				PV:1.3s 11.1nm
				MPV=5.0
28.	e(P)	A	20 08 41	PV:2.0s 40.0nm
28.	e(LQ)	C	20 20.0	LmV:15s 0.9 /um
	LmH	B	31.4	
	LmV	B	31.4	
29.	ePKP	A	02 09 08	Kermadec Islands (UPP)
29.	eP	A	09 43 13	Chagos Archipelago Region 7.36°S 68.05°E
	i	A	43 15.0	H = 09 31 31.1 h = normal MAG=5.0 (USCGS)
				D = 75.5°
29.	LmH	B	09 46.5	LmH:18s 0.7 /um LmV:18s 1.1 /um
	LmV	B	46.5	
29.	LmH	C	14 33	New Hebrides Islands
29.	e(PKIKP)	A	17 28 01	Fiji Islands Region 20.05°S 179.00°W
	e(PKHKP)	A	28 06.5	H = 17 09 21.9 h = 610 km MAG=4.6 (USCGS)
				D = 148.3°
				PV2:0.8s 9nm
				AN USSR gives: Western of Tonga Islands
				19.0°S 176.2°W
				H = 17 08 20 h = normal

Day	Phase		h m s	Remarks
30.	i	A	00 05 52	
30.	eP	B	02 22 16	South of Bali Island 10.99°S 115.48°E
	e	B	22 53	H = 02 08 02.4 h = normal MAG=6.0 (USCGS)
	eiPKIKP	A	26 15	D = 107.2°
	ePP	B	26 42	PV1:7s 0.35 µm PV2:8s 0.5 µm
	e(SKS)	C	32.9	PPV:9s 1.0 µm
	eS	B	34 16	LmH:18s 2.4 µm LmV:22s 2.4 µm
	ePS	C	36 06	MPPV=6.5 MLH=5.8 MLV=5.7
	ePPS	C	37 08	e 26 46 e 26 50 e 26 53 e 27 05
	eSS	C	41 40	e 49 24
	LmV	B	03 14.8	
	LmH	B	17.0	
30.	eP	A	03 35 02	North of Severnaya Zemlya 85.69°N 85.99°E
	epP	A	35 07	H = 03 27 41.5 h = normal MAG=4.4
	e	A	36 20	D = 38.6° Az = 259.6° (USCGS); h = 23 km
				PV:0.8s 9.4nm pPV:1.2s 12.8nm
				MPV=4.8
30.	eP	A	08 48 29	North of Severnaya Zemlya 86.43°N 71.65°E
	epP	A	48 34.5	H = 08 41 13.9 h = normal MAG=4.8
	e	A	48 41	D = 37.9° Az = 244.0° (USCGS); h = 25 km
	e	A	48 49	PV:1.6s 37.9nm pPV:1.2s 15.4nm
				MPV=5.1
30.	ePg	A	12 58 59	Explosion/GDR 51.25°N 12.66°E
	iSg	A	59 13.5	D = 1.1°
30.	ePn	A	13 51 28.5	Yugoslavia 43.4°N 20.8°E
	ePg	A	52 23	H = 13 49 11 (BCIS)
	e	A	53 45	D = 9.6°
	eSg	A	54 31	e 51 37 e 52 52 e 54 19
30.	e	A	16 20 20	
	e	A	20 32	
30.	e(P)	A	21 54 07	

Day	Phase		h m s	Remarks
30.	ePKP	A	23 24 22.5	Fiji Islands Region 16.87°S 176.87°W
	e	A	24 29	H = 23 04 45.8 h = normal MAG=5.1
	ePPS	C	40 36	D = 145.6° Az = 350.4° (USCGS)
	eSS	C	46(40)	PV1:1.8s 51.0nm
	eLQ	C	00 04	LmH:20s 1.3 /um LmV:22s 1.4 /um
	LmV	B	28.8	MLH=5.7 MLV=5.7
	LmH	B	30.8	e 24 36 e 24 55 ei 25 16
31.	eP	A	02 24 13	Fox Islands/Aleutian Is. 52.09°N 169.70°W
	e	A	24 18	H = 02 12 17.8 h = 28 km MAG=4.8
	e	A	24 22.5	D = 77.6° Az = 359.1° (USCGS)
	e	A	24 30	PV:1.5s 46.7nm
	LmH	C	10.3	MPV=5.4
	LmV	C	10.3	
31.	eSg	A	03 07 15	Probably near earthquake.
	e	A	07 19	Trieste gives: iSg 03 05 54
	e	A	07 24	
31.	eP	A	03 25 33	Red Sea 20.22°N 38.64°E
	epP	A	25 38	H = 03 18 24.2 h = normal MAG=4.8
	e	A	25 40	D = 37.1° Az = 331.3° (USCGS); h = 23 km
	e(PP)	A	27 07	PV:1.9s 47.0nm
	ePcP	A	27 53	MPV=5.1
31.	LmH	C	07(45)	Probably Luzon/Philippine Islands
31.	e(P)	A	08 49 26	
31.	LmH	C	09 08.5	Aleutian Islands
31.	eSS	C	11 09.0	Southern Pacific Ocean 35.74°S 102.71°W
	LQ	C	24	H = 10 29 43.5 h = normal MAG=4.9 (USCGS)
	LR	C	31	D = 131.5°
	LmH	C	45	LmV:20s 0.4 /um
	LmV	B	45	MLV=5.1

Day	Phase		h m s	Remarks
31.	eSS	C	11 13.5	Southern Pacific Ocean 35.65°S 102.96°W
	LmH	C	(49.5)	H = 10 34 00.5 h = normal MAG=4.8 (USCGS)
	LmV	B	50.4	D = 131.5°
				LmV:(16s 0.3 µm)
31.	iPg	A	14 11 41.0	Explosion
	iSg	A	11 56.8	D = 1.2°
31.	iPg	A	15 06 24.5	Explosion
	ei	A	06 39.5	D = 1.3°
	iSg	A	06 41.5	
31.	e	A	20 24 26	New Hebrides Islands 15.38°S 167.51°E
	ePKHKP	A	24 28	H = 20 05 18.9 h = 132 km MAG=5.3
	ePKIKP	A	24 32	D = 139.8° Az = 336.2° (USCGS)
	e	A	24 35	PV3:1.4s 36.4nm PV4:1.8s 61.2nm
	eSS	C	46.4	e 27 14 e 27 57
	LmH	C	21(21)	AN USSR gives: h = normal
31.	eiSg	A	23 25 29	Explosion?

Day	Phase		h m s	Remarks
1.	eP	A	06 06 14	Kurile Islands 45.82°N 151.76°E
	-iP	A	06 15.5	H = 05 54 19.1 h = 40 km MAG=5.7
	iS	C	16 06	D = 77.9° Az = 335.3° (USCGS)
	eScS	C	16 31	PV2:1.9s 425nm
	iPS	B	16 52	LmH:16.5s 15 µm LmV:17s 11.1 µm
	eSS	C	21 40	MPV2=6.3 MLH=6.4 MLV=6.3
	eSSS	C	24 40	i 06 20.5 e 06 31 e 07 15.5 e 09 06.5
	LmH	B	40.4	e 09 17 e 16 22 e 18(00)
	LmV	B	45.5	
1.	+iP	A	06 09 02.0	Kurile Islands 46.26°N 151.96°E
	ei	A	09 07	H = 05 57 09.1 h = 40 km MAG=5.5 (USCGS)
	ipP	A	09 17.0	D = 77.4° h = 57 km
	eS	C	18 48	PV:2.2s 421nm pPV:2.0s 291nm
	eSS	C	24 24	MPV=6.2
1.	e(P)	A	07 49 42	Kurile Islands 45.89°N 151.83°E
	e(pP)	A	49 58	H = 07 37 41.6 h = 40 km MAG=4.7 (USCGS)
				D = 74.6°
				The first onset of P must be about 5 s earlier.
1.	iP	A	08 00 23.0	Kurile Islands Region 45.89°N 152.01°E
	i	A	00 29.0	H = 07 48 27.8 h = 40 km MAG=5.0
	e(pP)	A	00 36.5	D = 77.9° Az = 335.5° (USCGS);
				(h = 50 km)
				PV:1.5s 26.7nm
				MPV=5.0
1.	iP	A	08 01 01.5	Kurile Islands Region
	e	A	01 07	H = 07 49 07 (UPP)
	epP	A	01 15	D = ca. 77.9° h = 50 km
	LmH	B	39.3	PV:2.0s 73.3nm
				LmH:18s 0.5 µm
				MPV=5.5 MLH=4.9
1.	e(Pg)	A	08 09 46	Explosion
	i(Sg)	A	10 04	(D = ca. 1.4°)
	LmH	A	10 16	

Day	Phase		h m s	Remarks
1.	ePP	A	11 00 12	Northern Easter Is. Cordillera
	eSKS	C	06(12)	4.56°S 105.81°W
	ePS	C	09 35	H = 10 41 00.2 h = normal MAG=5.0 (USCGS)
	eiSS	C	15 48	D = 110.7°
	LmH	B	41.5	LmH:20s 1.9 /um LmV:20s 2.3 /um
	LmV	B	43.8	MLH=5.7 MLV=5.8
				AN USSR gives: 17.4°S 111.9°W
				H = 10 40 40 h = normal MAG=5 1/2
				D = 124.4°
1.	eP	A	12 35 30.5	Kurile Islands 45.70°N 151.79°E
	+iP	A	35 31.2	H = 12 23 35.5 h = 40 km MAG=5.9
	e(pP)	A	35 42.5	D = 78.0° Az = 335.4° (USCGS);
	eS	C	45 24	(h = 45 km)
	eScS	C	45 50	PV2:1.7s 505nm
	ePPS	C	46 24	LmH:17.5s 13.8 /um LmV:16s 9.4 /um
	eSS	C	51 00	MPV=6.4 MLH=6.5 MLV=6.2
	LmH	B	09.5	ei 35 37 e 47 12
	LmV	B	14.7	
1.	eP	A	12 46 25.5	Iceland 63.69°N 18.94°W
	e(sP)	A	46 34	H = 12 41 40.6 h = 2 km MAG=4.8
				D = 20.8° Az = 114.5° (USCGS);
				(h = 19 km)
				PV:1.6s 45.5nm
				MPV=4.5
1.	+eiP	A	14 12 31.2	Kurile Islands 45.84°N 151.74°E
				H = 14 00 33.8 h = 23 km MAG=5.4
				D = 77.9° Az = 335.3° (USCGS)
				PV:1.9s 100nm
				MPV=5.6
1.	eP	A	15 47 33	
1.	eP	A	17 27 42	Kurile Islands 45.87°N 151.95°E
	epP	A	27 54.5	H = 17 15 45.7 h = normal MAG=4.7
				D = 77.9° Az = 335.4° (USCGS); h = 46 km

Day	Phase		h m s	Remarks
1.	+iP	A	17 33 05.0	Kurile Islands 45.62°N 151.93°E
	e	A	33 11	H = 17 21 09.3 h = 40 km MAG=4.7
	epP	A	33 17	D = 78.1° Az = 335.5°(USCGS); h = 45 km
	e	A	33 30	PV:1.5s 36.8nm
	LmH	B	18 05	MPV=5.3
	LmV	B	12	AN USSR gives: 46.4°N 150.8°E
				H = 17 21 25 h = 129 km
				D = 77.1°
1.	eP	A	21 53 55.5	
1.	iP	A	22 53 15.0	
	e	A	54 24	
1.	iP	A	23 32 20.5	Alaska Peninsula 58.41°N 154.95°W
	e	A	32 26.5	H = 23 21 12.2 h = 96 km MAG=4.3
	epP	A	32 32.5	D = 70.8° Az = 9.0° (USCGS); h = 46 km
				PV:0.8s 11.8nm
				MPV=5.1 (by h = 46 km) MPV=4.8 (by h = 96 km)
				AN USSR gives: Kodiak Island
				57.3°N 154.6°W
				H = 23 20 58 h = normal
				D = 71.8°
2.	ePKHKP	A	02 13 39	Tonga Islands 20.43°S 173.76°W
	ePKP2	A	13 45	H = 01 53 50.5 h = normal MAG=4.6
	e	A	13 48	D = 149.5° Az = 353.2° (USCGS)
2.	e(P)	A	09 03 03.5	
2.	iP	A	15 29 19.5	Northern Yukon Territory/Canada
				66.58°N 136.10°W
				H = 15 19 09.8 h = normal MAG=4.5 (USCGS)
				D = 60.6°
2.	ePS	C	18 11.1	New Britain Region 6.28°S 148.82°E
	e	C	13.3	H = 17 40 38.8 h = 37 km MAG=5.0 (USCGS)
	eSS	C	18.1	D = 123.3°

Day	Phase		h m s	Remarks
cont.				
2.	LmH	B	18 53.8	LmH:22s 3.6 µm LmV:21s 4.9 µm
	LmV	B	54.3	MLH=6.0 MLV=6.1
3.	e	A	01 25 23	Near earthquake?
	e(Sg)	A	26 19	
3.	+eP	A	07 45 39	<u>Red Sea</u> 19.91°N 38.46°E
	ei	A	45 43	H = 07 38 28.4 h = normal MAG=5.1
	e	A	45 55	D = 37.3° Az = 331.7° (USCGS)
	ePP	A	47 10	PV:1.7s 56.5nm
	ePcP	A	48 00	MPV=5.2
3.	ePKIKP	A	08 23 16	<u>New Britain Region</u> 6.06°S 151.50°E
	e	A	23 37	H = 08 04 15.4 h = 16 km MAG=4.9 (USCGS)
	e	C	34.0	D = 124.4°
	ePS	C	34 53	LmH:14s 0.6 µm LmV:14s 1.1 µm
	eSS	C	41.8	MLH=5.4 MLV=5.6
	LmH	B	09 23	
	LmV	B	24.6	
3.	ePKIKP	A	13 18 23	<u>Tonga Islands</u> 20.21°S 173.74°W
	ePKHKP	A	18 27.5	H = 12 58 40.9 h = 48 km MAG=5.3
	ePKP2	A	18 32.5	D = 149.3° Az = 353.3° (USCGS);
	epPKP	A	18 40	h = ca. 45 km
	e	A	18 55	PV2:2.0s 133nm
	e	A	19 20	LmH:18s 1.9 µm LmV:18s 1.9 µm
	LmH	B	14 27	MLH=6.0 MLV=5.9
	LmV	B	32.5	
3.	ePn	A	16 37 45	<u>Northern Italy</u> 44.91°N 10.57°E
	e	A	37 49.5	H = 16 36 19.8 h = normal MAG=4.7
	iPg	A	38 10.0	D = 5.8° Az = 6.6° (USCGS)
	iSn	A	38 49.0	LmH:9.5s 1.7 µm LmV:9.5s 2.0 µm
	iSg	B	39 26	MLH=3.8
	iLg2(3.30)	B	39 35	
	LmH	B	40.5	
	LmV	B	40.5	

Day	Phase		h m s	Remarks
4.	eSS	C	01 13.0	West New Guinea 2.34°S 138.71°E
	LmH	B	48.5	H = 00 37 26.1 h = normal MAG=5.6 (USCGS)
	LmV	B	50.6	D = 114.5°
				LmH:18s 1.3 μm LmV:19s 2.5 μm
				MLH=5.6 MLV=5.8
4.	eP	A	02 45 51	Greece (UPP)
	i	A	45 57.0	
4.	i	A	03 51 10.5	
4.	iP	A	04 06 23.5	Kurile Islands 45.48°N 152.16°E
	e	A	06 28	H = 03 54 26.2 h = 42 km MAG=5.0
	eipP	A	06 36	D = 78.3° Az = 335.6° (USCGS); h = 46 km
	e	A	07 00	PV:1.5s 33.2nm pPV:1.6s 91.0nm
	LmH	B	44.5	LmH:14s 0.5 μm LmV:17s 0.6 μm
	LmV	B	46.2	MPV=5.2 MLH=5.2 MLV=5.2
				e 07 05.5 e 07 13
4.	eP	A	09 17 51	Near S. Coast of Honshu/Japan
	i	A	17 55.5	33.37°N 137.47°E
	epP	A	19 16	H = 09 06 01.1 h = 353 km MAG=5.2
	e	A	19 29.5	D = 83.7° Az = 328.7° (USCGS); h = 360 km
	ePP	A	21 08	PV:1.5s 16.7nm
				MPV=4.6
4.	iPg	A	16 08 46.0	Explosion
	eSg	A	09 07	D = ca. 1.6°
4.	eP	A	17 03 05	Crete 35.43°N 23.60°E
	e	A	03 11.5	H = 16 59 04.1 h = 71 km MAG=4.8
	e	A	03 22.5	D = 17.5° Az = 333.9° (USCGS)
	e	A	03 29.5	e 04 00
	e	A	03 33.5	AN USSR gives: Aegean Sea 36.6°N 24.0°E
	e	A	03 43	H = 16 59 22 h = 166 km
	LmH	B	10.5	D = 16.6°
4.	e	A	17 58 42	Aegean Sea 39.06°N 24.51°E
	LmH	C	18 03.5	H = 17 55 02.2 h = normal (USCGS)
				D = 14.8°

Day	Phase		h m s	Remarks
4.	e	A	18 07 02	Near earthquake?
	e(Sg)	A	07 06	
	i	A	07 22.5	
4.	eP	A	18 08 41	Rumania 45.73°N 26.22°E
	e	A	08 51	H = 18 06 04.3 h = 131 km MAG=4.7 (USCGS)
	e	A	09 13	D = 10.9°
4.	eP	A	19 44 48	
4.	e(P)	A	22 05 46	
4.	e	A	22 22 08	
5.	-eP	A	02 47 50.5	Mariana Islands 20.00°N 147.12°E
	i(pP)	A	48 05	H = 02 34 11.1 h = 50 km MAG=5.9
	e(sP)	A	48 08.5	D = 99.5° Az = 333.1° (USCGS);
	ePP	B	51 54	(h = 52 km)
	e(pPP)	A	52 08.5	PV:1.4s 30.7nm PPV:9s 1.1 µm
	eSKS	C	58.5	LmH:15s 1.3 µm LmV:15s 1.9 µm
	eS	C	59 24	MPV=5.7 MPPV=6.3 MLH=5.5 MLV=5.7
	ePS	C	03 00 48	e 47 56 e 48 02 e 51 44 e 51 48
	eSS	C	06.0	e 52 05 e 52 15
	LmH	B	31.4	
	LmV	B	35.8	
5.	eP	A	03 01 34	Mariana Islands 19.98°N 147.20°E
	e	A	01 41	H = 02 47 55.4 h = 50 km MAG=5.7
	e	A	05 32	D = 99.5° Az = 333.1° (USCGS)
	ePP	B	05 37	PPV:7.5s 0.9 µm
	e	B	06 00	MPPV=6.3
	ePS	C	14 24	
5.	ePP	A	03 07 41	Mariana Islands
				H = 02 49 59.8 (UPP)
5.	iP	A	08 08 49.0	Honshu/Japan 36.03°N 137.23°E
	e	A	09 10	H = 07 57 04.0 h = 272 km MAG=4.2 (USCGS)
				D = 81.3°

Day	Phase		h m s	Remarks
5.	e(P)	A	08 34 13	
	e	A	34 32	
5.	eP	A	08 34 50	Algeria 35½°N 4.0°E
	e	A	35 08	H = 08 31 05 (BCIS)
				D = 16.2°
5.	e(P)	A	19 02 11	
5.	ePKP	A	21 49 33	Fiji Islands Region 17.65°S 178.41°W
				H = 21 30 53.2 h = 546 km MAG=4.4
				D = 146.1° Az = 348.5° (USCGS)
5.	ePKHKP	A	22 49 18.5	West of Macquarie Is. 53.16°S 140.56°E
	ePKP2	A	49 23.5	H = 22 29 35.0 h = normal MAG=5.0
	e	A	49 35	D = 148.9° Az = 286.3° (USCGS)
5.	eP	A	23 10 26	Southern Sumatra 5.36°S 102.45°E
				H = 22 57 06.1 h = normal MAG=5.3 (USCGS)
				D = 94.6°
5.	ePKP2	A	23 53 37	Kermadec Islands 31.08°S 178.23°W
				H = 23 33 06.0 h = 60 km MAG=5.2 (USCGS)
				D = 159.1°
				PV:1.2s 20.5nm
6.	eP	A	02 46 49	Ryukyu Islands 29.58°N 129.78°E
	e	A	47 03	H = 02 34 24.1 h = 31 km MAG=5.2
	e	A	47 06	D = 83.3° Az = 325.6° (USCGS)
	e	A	47 26	LmH:16.5s 2.4 µm LmV:16s 2.6 µm
	LmH	B	03 28.2	MLH=5.7 MLV=5.7
	LmV	B	28.2	
6.	eP	A	06 29 57.5	Near S. Coast of Honshu/Japan
	e	A	30 00	34.35°N 139.03°E
	e	A	30 03	H = 06 17 29.3 h = 13 km MAG=5.3
	LmH	B	07 09.9	D = 83.5° Az = 329.4° (USCGS)
	LmV	B	11.6	PV2:1.6s 26.3nm PV3:1.7s 38.9nm
				LmH:14s 1.8 µm LmV:14s 1.1 µm

Day	Phase		h m s	Remarks
cont.				
6.				MPV2=5.1 MPV3=5.3 MLH=5.6 MLV=5.4
				Successive P-onsets with increasing
				amplitude.
6.	eP	A	09 02 07.5	Near S. Coast of Honshu/Japan
	e	A	02 12.5	34.39°N 139.07°E
	LmH	B	41.8	H = 08 49 41.3 h = normal MAG=5.0
	LmV	B	45.7	D = 83.5° Az = 329.4° (USCGS)
				PV:1.6s 19.0nm
				LmH:16s 1.2 µm LmV:14s 0.7 µm
				MPV=5.0 MLH=5.4 MLV=5.2
6.	iP	A	09 19 11.2	Near S. Coast of Honshu/Japan
	LmH	B	59	34.39°N 139.09°E
				H = 09 06 44.0 h = 25 km MAG=4.9 (USCGS)
				D = 83.5°
6.	iPg	A	11 01 29.5	Explosion
	iSg	A	01 48.0	D = ca. 1.4°
6.	eP	A	11 12 14	PV:1.4s 18.2nm
6.	e	A	12 17 28	Near earthquake?
	e	A	17 46.5	e 17 40 e 17 48.5
6.	e(Sg)	A	12 27 50	
6.	eP	A	12 35 39	Mariana Islands Region 20.13°N 147.18°E
	e	A	36 01	H = 12 21 57.0 h = 22 km MAG=5.7
	e	A	39 31	D = 99.4° Az = 333.1° (USCGS)
	ePP	B	39 42	PV:1.8s 30.6nm PPV:9s 0.65 µm
	e	A	40 00	LmH:16s 1.7 µm LmV:16s 1.6 µm
	e	C	54(32)	MPV=5.6 MPPV=6.1 MLH=5.7 MLV=5.7
	LmH	B	13 19.2	e 39 55 e 40 04 e 40 11 e 40 18
	LmV	B	22.5	PP shows a distinct normal dispersion
				with periods from 2.7 s to 1.4 s in our
				short-period records.

Day	Phase		h m s	Remarks
6.	eP	A	13 04 15	Iran 30.12°N 50.92°E
	e	A	04 33	H = 12 57 14.0 h = 10 km MAG=5.4
	e	A	04 51	D = 35.7° Az = 316.3° (USCGS)
	e	A	05 50	PV:1.9s 46.7nm
				MPV=5.1
6.	e	A	13 50 50	Yugoslavia 42.4°N 19.0°E
	eSg	A	51 15	H = 13 46 08 (BCIS)
	e	A	58 28	D = 9.4°
6.	iPg	A	14 07 15.2	Explosion
	iSg	A	07 35.0	D = ca. 1.5°
6.	eP	A	23 41 20.5	Near S. Coast of Honshu/Japan
	epP	A	41 25	34.29°N 139.09°E
	e	A	41 31	H = 23 28 51.0 h = 15 km MAG=5.1
				D = 83.6° Az = 329.4° (USCGS); h = 17 km
				PV:1.4s 15.2nm
				MPV=4.9
6.	eP	A	23 44 29.5	Near E. Coast of Honshu/Japan
	e	A	44 37	36.35°N 140.54°E
	e(pP)	A	44 43.5	H = 23 32 10.6 h = 44 km MAG=5.2
	e(sP)	A	44 47	D = 82.4° Az = 330.0° (USCGS);
	LmH	B	00 21.4	(h = 52 km)
	LmV	B	22.9	PV:1.4s 30.3nm
				LmH:14.5s 1.6 μm LmV:15s 1.0 μm
				MPV=5.2 MLH=5.6 MLV=5.3
7.	iPg	A	09 55 22.5	Explosion/GFR 49.57°N 12.36°E
	i	A	55 32.5	D = 1.1°
	iSg	A	55 38.5	
7.	ePg	A	13 01 45.5	Explosion/GDR 51.29°N 12.73°E
	iSg	A	02 01.5	D = 1.0° Yield: 1.05 t
7.	eP	B	17 12 06	Turkey 37.40°N 36.13°E
	eP	A	12 07	H = 17 07 16.2 h = 49 km MAG=4.8
	e	A	13 16	D = 21.9° Az = 315.0° (USCGS)

Day	Phase		h m s	Remarks
cont.				
7.	eS	C	17 16 10	PV(A):2.4s 114nm SH(B):10s 1.3 µm
	iS	B	16 19	LmH:11s 1.1 µm LmV:12s 1.1 µm
	e	B	17 26	MPV(A)=4.9 MSH(B)=5.2 MLH=4.5 MLV=4.6
	LmH	B	22.2	Higher-mode surface waves are registered.
	LmV	B	22.6	
7.	-eiP	B	18 38 23	Turkey 37.36°N 36.19°E
	+iP	A	38 24.6	H = 18 33 31.3 h = 39 km MAG=5.0
	eS	C	42 30	D = 22.0° Az = 315.0° (USCGS)
	iS	B	42 34	PV(B):5.5s 0.65 µm PV(A):2.6s 174nm
	eLi(3.70)	B	44 32	SH(B):11.5s 2.7 µm
	eLg1(3.56)	B	44 55	LmH:12s 1.8 µm LmV:14s 1.7 µm
	eLg2(3.30)	B	45 50	MPV(B)=5.3 MPV(A)=5.0 MSH(B)=5.4
	LmH	B	48.5	MLH=4.7 MLV=4.7
	LmV	B	48.9	e 38 30 e 38 36.5 e 39 03 e 40 11
				i 43 40
				Well developed higher-mode surface waves.
7.	+iP	A	19 50 23.5	Sea of Okhotsk 47.02°N 145.96°E
				H = 19 39 12.8 h = 296 km MAG=5.0
				D = 75.2° Az = 58.5° (USCGS)
				PV:1.3s 25.0nm
				MPV=4.8
				AN USSR gives: 46.7°N 146.2°E
				H = 19 39 16 h = 355 km
				D = 75.4°
8.	iPKIKP	A	05 53 51.5	Fiji Islands Region 19.87°S 178.59°W
	eiPKHKP	A	53 55.5	H = 05 35 17.1 h = 616 km MAG=5.3
	+eiPKP2	A	54 02	D = 148.2° Az = 347.6° (USCGS);
	e	A	55 55	h = 623 km
	epPKHKP	A	56 16	PV2:1.6s 121nm PV3:1.5s 67.5nm
	e	A	57 38	PV5:1.8s 150nm
				AN USSR gives: 19.9°S 177.3°W
				H = 05 35 02 h = 472 km
				D = 148.6°

Day	Phase		h m s	Remarks
8.	eP	A	09 07 26	Kurile Islands 47.29°N 153.37°E
				H = 08 55 39.8 h = 60 km MAG=4.6 (USCGS)
				D = 77.0°
8.	e(Sg)	A	11 02 26	Explosion
8.	i(Pg)	A	13 00 11.5	Probably Upper Silesia
	iSn	A	00 35.5	D = ca. 3.0°
	iSg	A	00 48	
9.	e	A	00 24 15	West New Guinea Region 3.97°S 135.78°E
	e	A	24 37	H = 00 05 07.0 h = 15 km MAG=5.1 (USCGS)
	e(PP)	B	24 40	D = 114.2°
	ePP	A	24 48	LmH:18s 2.1 µm LmV:20s 2.2 µm
	ePS	C	34 18	MLH=5.8 MLV=5.7
	eSS	C	40.5	
	eSSS	C	44.4	
	LmH	B	01 09.0	
	LmV	B	17.8	
9.	e	A	01 23 29	
9.	ePKIKP	A	01 46 53.5	Fiji Islands 19.66°S 177.99°W
	eiPKHKP	A	46 57	H = 01 27 57.6 h = 415 km MAG=4.5
	iPKP2	A	47 01.2	D = 148.2° Az = 348.4° (USCGS)
				PV2:1.6s 37.9nm
9.	iPg	A	07 08 10.5	Italy 45.9°N 10.9°E
	eSn	A	08 40	H = 07 06 39 (BCIS)
	i(Sg)	A	09 07	D = 4.8°
	e	A	09 11	
	i	A	09 27.5	
9.	ePKIKP	A	09 16 01.5	Solomon Islands 7.24°S 155.85°E
	e	A	16 12.5	H = 08 56 59.7 h = 40 km MAG=5.1
	LmH	B	10 15.5	D = 127.5° Az = 332.0° (USCGS)
	LmV	B	16.0	PV:1.3s 27.8nm
				LmH:20s 0.5 µm LmV:19s 0.5 µm
				MLH=5.2 MLV=5.3

Day	Phase		h m s	Remarks
cont. 9.				AN USSR gives: Solomon Is. 5.5°S 155.6°E
				H = 08 57 55 h = 700 km
				D = 127.5°
9.	LmH	C	16 15.5	
	LmV	C	15.5	
9.	ePKIKP	A	21 37 38.5	<u>Solomon Islands</u> 7.27°S 155.71°E
	LmH	B	22 25	H = 21 18 36.5 h = 44 km MAG=5.6 (USCGS)
				D = 127.5°
9.	eP	A	22 04 31	<u>Kurile Islands</u> 45.55°N 151.6°E
	e	A	04 47	H = 21 52 34.0 h = normal MAG=4.7 (USCGS)
				D = 78.0°
10.	ePKP	A	00 17 01	<u>Tonga Islands Region</u> 17.67°S 172.97°W
	epPKP	A	17 11.5	H = 23 57 24.9 h = 70 km MAG=4.9
	e(sPKP)	B	17 18	D = 146.9° Az = 354.6° (USCGS);
	e	A	17 29	h = ca. 40 km
	e	A	17 39	PV1:1.9s 141nm PV2:2.0s 107nm
	LmH	B	01 26	LmH:18s 0.5 μm LmV:18s 0.5 μm
	LmV	B	26	MLH=5.3 MLV=5.3
10.	ePKIKP	A	05 18 56.5	<u>Solomon Islands</u> 7.41°S 155.68°E
	epPKIKP	A	19 05	H = 04 59 53.9 h = 37 km MAG=5.5
	ePP	A	20 58	D = 127.6° Az = 331.9° (USCGS);
	epPP	A	21 06	h = 30 km
	ePS	C	30 56	PV:1.4s 28.6nm
	ePPS	C	32 36	LmH:16s 1.6 μm LmV:17s 0.7 μm
	eSS	C	38.0	MLH=5.8 MLV=5.4
	LmH	B	06 17.7	
	LmV	B	17.7	
10.	e(P)	A	07 36 25	
10.	ePKIKP	A	15 21 45	<u>Solomon Islands</u> 7.30°S 155.80°E
	-iPKIKP	A	21 46.5	H = 15 02 42.2 h = 29 km MAG=5.6
	e(pPKIKP)	A	21 55	D = 127.6° Az = 332.0° (USCGS);
				(h = 30 km)

Day	Phase		h m s	Remarks
cont.				
10.	ePP	B	15 23 47	PV2:1.4s 54.5nm PPV:13s 0.8 /um
	ePPP	B	26 30	LmH:18s 3.5 /um LmV:19s 4.9 /um
	ePS	C	33 50	MPPV=5.9 MLH=6.1 MLV=6.2
	eSKKP	A	35 07	i 21 49.5 e 25 00
	ePPS	C	35 20	
	eSS	C	40 55	
	eSSS	C	45.5	
	LmH	B	16 21.5	
	LmV	B	21.7	
10.	eSS	C	17 34.0	South Pacific Ocean 63.56°S 167.29°W
	eSSS	C	40.4	H = 16 47 49.7 h = normal MAG=5.4 (USCGS)
	e	C	46.2	D = 167.0°
	LmH	B	18 27.7	LmH:21.5s 3.3 /um LmV:19s 4.5 /um
	LmV	B	28.2	MLH=6.1 MLV=6.3
10.	iP	A	19 12 04.5	Colorado 39.92°N 104.78°W
				H = 19 00 25.6 h = 5 km MAG=4.8 (USCGS)
				D = 74.1°
10.	+iP	A	20 08 42.0	Alaska Peninsula 58.61°N 154.25°W
	+i	A	08 43.6	H = 19 57 34.4 h = 86 km MAG=5.5 (USCGS)
	e	A	08 47	D = 70.5° h = 98 km
	-ipP	A	09 06.0	PV:0.9s 47.0nm pPV:1.2s 41.0nm
	esP	A	09 18	SH:4.5s 0.5 /um
	iS	B	17 50	MPV=5.3 MSH=5.7
				e 09 15.5 e 09 22 e 09 32
10.	e(PKP2)	A	21 23(15)	South of Fiji Islands 24.11°S 176.30°W
	e	A	23 20	H = 21 03 32.2 h = 205 km MAG=4.7 (USCGS)
	e	A	23 30	D = 152.8°
10.	ePKIKP	A	21 26 43.5	Solomon Islands 7.39°S 155.68°E
				H = 21 07 47.5 h = 103 km MAG=5.4
				D = 127.6° Az = 331.9° (USCGS)
				AN USSR gives: 6.0°S 155.9°E
				H = 21 08 54 h = 713 km
				D = 127.6°

Day	Phase		h m s	Remarks
10.	ePKIKP	A	22 08 22.5	Solomon Islands 7.28°S 155.90°E
	e	A	08 32	H = 21 49 19.5 h = 39 km MAG=5.3 (USCGS)
	LmH	B	23 07.8	D = 127.6°
	LmV	B	08.7	PV:1.2s 18.0nm
				LmH:20s 0.6 µm LmV:20s 0.8 µm
				MLH=5.3 MLV=5.4
11.	ePKIKP	A	05 11 48	Solomon Islands 7.42°S 155.73°E
				H = 04 52 48.3 h = 86 km MAG=5.5 (USCGS)
				D = 127.5°
11.	ePP	A	05 27 32	Celebes 3.31°S 119.23°E
	eSS	C	42.0	H = 05 09 12.1 h = 21 km MAG=5.2 (USCGS)
	e	B	44.6	D = 103.7°
	e	C	45.4	LmH:18s 1.4 µm LmV:20s 1.7 µm
	LmH	B	06 15	MLH=5.5 MLV=5.6
	LmV	B	15.0	
11.	eN	B	05 59 48	
	eE	B	59 56	
11.	e(Pg)	A	12 27 42	Probably explosion
	e	A	29 08	
	e	A	29 25	
11.	eP	A	12 53 32	Leeward Islands 18.84°N 62.66°W
	e	A	53 43	H = 12 42 47.7 h = 49 km MAG=5.2 (USCGS)
	eiS	C	13 02 16	D = 65.7°
	eScS	C	03 28	LmH:17s 1.4 µm LmV:17s 2.4 µm
	LmH	B	20.0	MLH=5.2 MLV=5.5
	LmV	B	20.0	e 54 24.5
11.	iPg	A	14 10 46.0	Explosion/GDR 51.37°N 12.89°E
	iSg	A	11 02.0	D = 1.0° Yield:4.54 t
12.	LmV	B	03 11.5	Bismarck Sea 3.10°S 148.09°E
	LmH	B	12.0	H = 02 00 15.9 h = normal MAG=4.9 (USCGS)
				D = 120.2°
				LmH:20s 0.6 µm LmV:22s 1.1 µm
				MLH=5.3 MLV=5.5

Day	Phase		h m s	Remarks
12.	eiPKP	A	04 50 56	New Hebrides Islands 19.21°S 168.90°E
				H = 04 31 45.3 h = 200 km MAG=4.7
				D = 143.8° Az = 335.4° (USCGS)
				PV:1.0s 28.6nm
				AN USSR gives: H = 04 31 24 h = normal
12.	eP1	A	05 03 59	Northern Sumatra 5.29°N 96.50°E
	+iP2	A	04 05.5	H = 04 51 40.2 h = 55 km MAG=6.1
	-iP3	B	04 09	D = 82.7° Az = 320.3° (USCGS)
	iS1	B	14 14	PV2:1.5s 153nm PV3:9.5s 5.1 μm
	iS3	B	14 22	SH3:15s 6.7 μm PKPPKPV:2.0s 66.6nm
	ePKPPKP	A	30 31	LmH:17s 12.7 μm LmV:19s 14.0 μm
	eSKPPKP	A	33 55	MPV2=5.9 MPV3=6.6 MSH3=6.6 MLH=6.4
	LmV	B	47.7	MLV=6.4
	LmH	B	48.6	Multiple P-onsets with successive increasing amplitude. P1 is a small-amplitude precursor. P3 is recorded in our long-period records only, followed by abnormally long-period P-waves (T = 16 - 22 s). PKPPKP and SKPPKP are very clear.
12.	eP	A	05 23 35	Northern Sumatra 5.53°N 96.66°E
	e	A	23 40	H = 05 11 14.1 h = normal MAG=5.7 (USCGS)
				D = 82.6°
				PV:1.6s 19.0nm
12.	eP	A	05 30 24	Northern Sumatra 5.63°N 96.66°E
	e	A	30 28	H = 05 18 11.9 h = 102 km MAG=5.0 (USCGS)
				D = 82.5°
12.	iPg	A	12 56 18.0	Explosion/GDR 51.25°N 12.66°E
	iSg	A	56 32.3	D = 1.0° Yield: 8.45 t
	LmV	A	56 46	LmH:1.3s 29.4nm LmV:1.6s 30.3nm
	LmH	A	56 48	
12.	ePKIKP	A	14 13 59	Solomon Islands 7.34°S 155.63°E
	e	A	14 05.5	H = 13 54 57.2 h = 52 km MAG=5.2
	ePP	B	16 00	D = 127.5° Az = 331.9° (USCGS)
	e(SKP)	B	17 44	PV1:1.4s 28.6nm PV2:1.2s 20.5nm

Day	Phase		h m s	Remarks
cont.				
12.	eSPP	B	14 27(30)	LmH:20s 0.8 µm
	eSS	C	33.1	MLH=5.4
	LmH	B	15 01.3	e 14 22 e 18 32
12.	eP	A	14 45 27	Near Coast of Oaxaca/Mexico
				15.21°N 94.02°W
				H = 14 32 39.1 h = normal MAG=4.7 (USCGS)
				D = 88.0°
12.	ePKP	A	15 10 50	Solomon Islands 7.45°S 155.72°E
	e(PKIKP)	A	10 55	H = 14 51 49.4 h = 21 km MAG=5.3 (USCGS)
	e	A	10 58.5	D = 127.7°
	ePP	A	12 58	
12.	eP	A	16 08 18	Gulf of Alaska 59.42°N 144.51°W
	e	A	08 22	H = 15 57 09.6 h = 10 km MAG=4.1 (USCGS)
	e	A	08 39	D = 68.7°
12.	LmH	C	22 27	Off Coast of Northern Chile
	LmV	C	27	35.52°S 73.29°W
				H = 21 22 09.3 h = 12 km MAG=5.3 (USCGS)
				D = 113.5°
13.	eiPKP	A	04 33 51.5	New Hebrides Islands 18.74°S 168.85°E
				H = 04 14 33.6 h = 123 km MAG=5.2
				D = 143.4° Az = 335.6° (USCGS)
				PV:1.2s 41.0nm
				AN USSR gives: 19.7°S 169.4°E
				H = 04 14 14 h = normal
				D = 144.4°
13.	eP	A	08 38 01	Northern Sumatra 5.45°N 96.65°E
	e	A	39 35	H = 08 25 43.8 h = 68 km MAG=5.2 (USCGS)
	ei	A	39 40	D = 82.7°
	e	A	40 11	
	e	A	40 44	
13.	e(P)	A	16 48 58	

Day	Phase		h m s	Remarks
13.	iPKP	A	17 32 18.0	Fiji Islands Region 17.96°S 178.56°W
				H = 17 13 43.5 h = 610 km MAG=5.0
				D = 146.4° Az = 348.3° (USCGS)
13.	eP	A	18 51 35	Kamchatka 52.08°N 157.57°E
	i	A	51 57.0	H = 18 40 07.7 h = 50 km MAG=5.3 (USCGS)
				D = 73.6°
				AN USSR gives: 52.0°N 157.7°E
				H = 18 40 15 h = 145 km
				D = 72.8°
13.	+iP	A	20 06 13.0	Ryukyu Islands 27.33°N 128.69°E
	e	A	07 20	H = 19 53 42.4 h = 38 km MAG=6.0
	e	A	07 35	D = 84.6° Az = 325.3° (USCGS)
	ePP	A	09 29	PV:2.1s 455nm PPV:1.5s 33.3nm
	eS	B	16 40	MPV=6.3 MPPV=5.6
13.	-iP	A	20 12 39.0	Guerrero/Mexico 18.48°N 100.21°W
	e	A	12 45.5	H = 19 59 51.9 h = 86 km MAG=5.6
	e(pP)	A	12 57	D = 88.9° Az = 36.2° (USCGS);
	e	A	13 02	(h = 68 km)
	eiS	B	23 28	PV:1.6s 68.2nm SH:8s 0.8 µm
	LmH	B	41.1	LmH:19s 1.4 µm LmV:16s 0.8 µm
	LmV	B	49.0	MPV=5.6 MSH=6.0 MLH=5.4 MLV=5.2
13.	LmH	B	20 48.8	LmH:18s 0.9 µm LmV:16s 0.6 µm
	LmV	B	55.0	
14.	eP	A	05 31 29.5	Guerrero/Mexico 17.53°N 100.22°W
	e(pP)	A	31 50.5	H = 05 18 35.8 h = 62 km MAG=4.9 (USCGS)
				D = 89.7° (h = 79 km)
14.	iPg	A	08 33 18.2	Explosion
	eSg	A	33 34.5	D = ca. 1.2°
14.	e(Sg)	A	10 32 43	Explosion?
	e	A	32 47	
	LmH	A	33 04	

Day	Phase		h m s	Remarks
14.	LmH	C	13 53	LmH:20s 0.3/um LmV:20s 0.4/um
	LmV	B	53	
14.	iPg	A	15 05 20.5	Explosion
	i	A	05 35	D = ca. 1.2°
	iSg	A	05 36.0	
14.	e(P)	A	20 40 59	
	i	A	41 34	
14.	iP	A	23 48 04.5	LmH:16s 0.3/um
	e	A	48 08	
	i	A	51 49.5	
	LmH	C	53.5	
	LmV	B	55	
15.	eiPn	A	02 09 48	<u>Yugoslavia</u> 44.3°N 16.3°E
	eiSn	A	11 05.5	H = 02 08 04 (BCIS)
	iLg1(3.48)	A	11 50.5	D = 7.1°
	e(Sg)	A	11 55.5	e 10 02 e 10 36.5 e 11 38.5 i 11 47.0
	iLg2(3.29)	A	12 03.5	
15.	eP	A	17 06(13)	<u>Red Sea</u> 20.75°N 38.73°E
	e	A	06 18	H = 16 59 07.3 h = normal MAG=4.9 (USCGS)
	LmH	C	25	D = 36.6°
	LmV	B	25.0	LmH:20s 0.5/um LmV:20s 0.4/um
				MLH=4.3 MLV=4.3
15.	eP	A	21 07 42	<u>Andreanof Is./Aleutian Is.</u>
	e	A	07 52	H = 20 55 46.9 h = 49 km MAG=4.9 (USCGS)
	e	A	07 55	D = 78.0°
15.	e(P)	A	22 34 06.5	
15.	+iP	A	23 47 45.5	<u>Hokkaido/Japan Region</u> 41.92°N 142.27°E
	iPcP	A	47 56.0	H = 23 35 51.1 h = 66 km MAG=4.9
	ei	A	47 59	D = 78.2° Az = 330.4° (USCGS)
	ei	A	48 06	PV:1.0s 16.6nm
	LmH	B	00 25	MPV=5.1
	LmV	B	25	

Day	Phase		h m s	Remarks
16.	e(Sg)	A	07 17 10	Explosion?
16.	+ePKIKP	A	07 37 50	South of Fiji Islands 19.41°S 175.87°E
	iPKHKP	A	37 51.8	H = 07 18 11.8 h = 38 km MAG=5.3
	eiPKP2	A	37 54	D = 146.4° Az = 341.8° (USCGS)
	e	A	38 04	AN USSR gives: Sea of Okhotsk
	e	A	38 35	52.0°N 151.5°E
				H = 07 26 26 h = normal
				D = 72.2°
16.	ePKIKP	A	07 49 31.5	Tonga Islands Region 21.6°S 176.1°W
	e	A	49 35	H = 07 29 48 h = normal MAG=5
	ePKP2	A	49 44	D = 150.4° (AN USSR)
	LmH	C	09 08	e 49 40 e 49 57.5
16.	eP	A	10 04 40	North of Aszension Island 0.12°N 17.32°W
	e	A	05 32.5	H = 09 55 00.9 h = normal MAG=4.8
				D = 56.0° Az = 21.8° (USCGS)
16.	+iP1	A	10 22 03.5	Kurile Islands 46.43°N 153.32°E
	iP2	A	22 07.5	H = 10 10 06.7 h = 24 km MAG=5.3
	ei	A	22 14.5	D = 77.8° Az = 336.2° (USCGS)
	e	A	23 07	PV2:1.7s 87.0nm
	LmH	B	58.3	LmH:16s 1.1 /um LmV:15s 1.3 /um
	LmV	B	11 03.5	MPV2=5.6 MLH=5.3 MLV=5.4
				Multiple P-onsets.
17.	e(P)	A	05 05 17.5	
17.	i(Sg)	A	09 31 19	Explosion/ČSSR 50.18°N 14.40°E
				D = 1.8° Yield: 8.6 t
17.	iPKIKP	A	11 37 37	Santa Cruz Islands 12.51°S 166.31°E
	e	A	37 41.5	H = 11 18 19.3 h = 45 km MAG=4.9 (USCGS)
	ePP	A	40 18	D = 136.8°
	eSKP	A	41 13	LmV:20s 0.7 /um
	e	C	53.4	MLV=5.4
	LmV	B	12 40.2	e 37 50 e 38 07.5 e 40 27 e 41 23

Day	Phase		h m s	Remarks
17.	LmH	B	11 57.3	Taiwan Region 24.91°N 122.15°E
	LmV	B	12 00.9	H = 11 07 12.9 h = 31 km MAG=5.0 (USCGS)
				D = 83.1°
				LmH:14s 0.9 μm LmV:12s 0.6 μm
				MLH=5.3 MLV=5.2
17.	e	A	15 25 55	Probably near earthquake
	e	A	26 15	e 26 47
	e(Sg)	A	26 43	
17.	ePKIKP	A	18 04 33	Fiji Islands Region 20.63°N 178.18°W
	ePKHKP	A	04 38.5	H = 17 45 52.2 h = 570 km MAG=4.5
	ePKP2	A	04 45	D = 149.1° Az = 347.8° (USCGS)
				PV2:1.3s 22.2nm
17.	eP1	A	21 42 50.5	Cyprus 34.61°N 32.96°E
	iP2	A	42 53.0	H = 21 37 54.5 h = normal MAG=4.8
	e	A	43 08.5	D = 22.3° Az = 322.4° (USCGS)
	e	A	43 13.5	PV1:1.0s 19.1nm PV2:1.1s 61.0nm
				MPV1=4.5 MPV2=5.0
17.	eP	A	22 10 13	Cyprus 34.51°N 33.10°E
	i	A	10 14.2	H = 22 05 17.6 h = 40 km
				D = 22.5° Az = 322.4° (USCGS)
				PV2:1.2s 30.8nm
				MPV2=4.7
18.	e(PKP)	A	20 32 53	PV2:1.0s 23.6nm
	ei	A	32 54.5	
	e	A	33 10	
19.	e(Sg)	A	04 25 22	Explosion?
19.	e	A	10 59 18	
19.	+iPKHKP	A	17 33 21.2	Fiji Islands Region 20.63°S 177.99°W
	ePKP2	A	33 27.5	H = 17 14 24.1 h = 449 km MAG=4.9
	e	A	33 33	D = 149.1° Az = 348.1° (USCGS)
				PV:1.4s 24.2nm

Day	Phase		h m s	Remarks
19.	eP	A	22 08(09)	Dominican Republic Region 18.84°N 69.63°W
	epP	A	08 35.5	H = 21 57 05.1 h = 103 km MAG=5.0 (USCGS)
	esP	A	08 47.5	D = 70.2° h = 108 km
	e	A	08 59	P is smaller than pP and sP.
				AN USSR gives: 18.1°N 70.0°W
				H = 21 56 54 h = normal
				D = 71.0°
20.	+iP	A	04 15 47.2	Eastern Kazakh SSR 49.74°N 78.12°E
	i	A	15 54.0	H = 04 07 57.6 h = 0 km MAG=5.7
	ePn	A	17 19	D = 41.3° Az = 297.7° (USCGS)
	ePP	A	17 24.5	PV:0.8s 80.2nm
	e	A	17 29	MPV=5.7
				Probably underground explosion.
20.	iPg	A	14 10 09.5	Explosion/GDR 51.37°N 12.89°E
	iSg	A	10 25.5	D = 1.0°
21.	ePKIKP	A	08 32(56)	Banda Sea 5.45°S 126.86°E
	ePP	A	33 33	H = 08 14 25.0 h = normal MAG=5.5
	eS	C	41 04	D = 110.1° Az = 322.2° (USCGS)
	ePPS	C	44 06	LmH:20s 2.8 µm LmV:17s 2.7 µm
	eSS	C	49(02)	MLH=5.8 MLV=5.9
	eSSS	C	56 18	e 33 16 e 33 38 e 34 19 e 35 02
	eLQ	C	59.2	e 36 06 e 52 20
	LmH	B	09 25.5	
	LmV	B	34	
21.	e(Sg)	A	11 06 22	Explosion?
21.	iPKHKP	A	14 57 01.0	Fiji Islands Region 20.59°S 177.55°W
	ePKP2	A	57 06.5	H = 14 38 02.8 h = 454 km MAG=4.9
				D = 149.1° Az = 348.6° (USCGS)
				PV:1.6s 22.7nm
21.	e	A	21 49 42.5	Probably Northeast of Jan Mayen
	e	A	49 48	H = 21 44 13 (UPP)

Day	Phase		h m s	Remarks
22.	e(P)	A	00 10 43	Probably Northeast of Jan Mayen
	e	A	10 49	H = 00 05 46 (UPP)
22.	ePP	B	08 56 36	Banda Sea 5.6°S 126.82°E
	ePS	C	09 05.9	H = 08 37 25.5 h = normal MAG=5.2 (USCGS)
	ePPS	C	07 00	D = 110.2°
	eLQ	C	22.3	LmH:19s 0.65 µm LmV:18s 0.75 µm
	LmH	B	55.6	MLH=5.2 MLV=5.3
	LmV	B	55.6	e 56 45 e 56 54 e 01 15
22.	e(P)	A	12 22 41	Greenland Sea 73.45°N 8.41°E
	e	A	22 53	H = 12 17 34.5 h = normal MAG=4.7
	e	A	23 12	D = 22.9° Az = 174.8° (USCGS)
				PV:1.2s 13.6nm
				MPV=4.4
				The first onset of P must be about 5 s earlier.
22.	+eP	A	13 19 59	Northern Sumatra 5.13°N 96.43°E
	e	A	20 19	H = 13 07 38.1 h = 42 km MAG=5.4
	e	A	20 43	D = 82.8° Az = 320.3° (USCGS)
	ePP	A	23 12.5	PV:2.0s 81.5nm
	LmH	B	14 03	LmH:19s 0.9 µm LmV:18s 0.9 µm
	LmV	B	03	MPV=5.5 MLH=5.2 MLV=5.2
22.	eP	A	14 56(02)	Panama-Costa Rica Border Region
	e	A	56 06.5	8.30°N 82.80°W
	e	A	57 09	H = 14 43 21.4 h = 40 km MAG=5.0 (USCGS)
	LmH	B	15 35.5	D = 86.4°
	LmV	B	35.5	LmH:16s 0.4 µm LmV:16s 0.5 µm
				MLH=4.9 MLV=5.0
23.	eP	A	06 50(11)	North Atlantic Ocean 57.02°N 33.78°W
	e	A	50 16	H = 06 44 26.2 h = normal MAG=4.3 (USCGS)
	e	A	50 40	D = 27.2°
23.	e(Pg)	A	08 39 31	Near earthquake
	e	A	40 14.5	(D = ca. 5.5°)
	e(Sg)	A	40 43	e 39 33 e 39 37

Day	Phase		h m s	Remarks
23.	eP1	A	09 34 03.5	Algeria 36.31°N 2.45°E
	eiP2	A	34 06.2	H = 09 30 22.0 h = normal MAG=4.8
	e	A	34 11	D = 15.8° Az = 21.9° (USCGS)
	i	A	34 20.5	PV2:1.3s 41.6nm PV3:1.1s 24.4nm
	i	A	34 46.5	LmH:12s 1.5 µm LmV:14s 0.7 µm
	LmH	B	40.3	MLH=4.4 MLV=4.1
	LmV	B	40.3	eP1 is a small-amplitude precursor.
23.	eP	A	14 04 01	Mindanao/Philippine Is. 8.60°S 126.49°E
	e	A	04 24	H = 12 50 24.9 h = 43 km MAG=5.3 (USCGS)
	LmH	B	52	D = 98.7°
	LmV	B	52	e 04 49
23.	eP	A	15 12 49	North Indian Rise 1.59°N 80.23°E
	epP	A	12 56.5	H = 15 01 06.8 h = normal MAG=5.1
	e	A	13 04	D = 75.3° Az = 322.2° (USCGS); h = 29 km
	e	A	13 26	pPV:1.2s 15.4nm
	LmH	C	46.5	LmH:22s 0.2 µm
				The amplitude of pP is much greater than that of P.
23.	ePP	A	18 11 12	South of Mariana Islands 13.50°N 146.14°E
	LmH	B	19 03.5	H = 17 52 51.0 h = 56 km MAG=5.6 (USCGS)
	LmV	B	07.5	D = 104.8°
				PPV:1.4s 15.2nm
				LmH:16s 0.5 µm LmV:14s 0.4 µm
				MPPV=5.4 MLH=5.1 MLV=5.1
23.	eP	A	22 38 08.5	Costa Rica 8.10°N 83.26°W
	e	A	38 14	H = 22 25 27.4 h = 46 km MAG=4.5
				D = 86.9° Az = 39.4° (USCGS)
				PV:0.9s 11.8nm
				MPV=5.1
24.	-eiP	A	08 59 22.5	Tadzhik SSR 37.40°N 72.66°E
	ePP	A	09 01 02	H = 08 51 10.9 h = 31 km MAG=5.6
	e(Pn)	A	01 30	D = 44.6° Az = 307.5° (USCGS)
	ePPP	A	01 43	PV:1.3s 69.5nm
	eS	C	06.2	LmH:14s 1.6 µm LmV:14s 2.1 µm

Day	Phase		h m s	Remarks
cont.				
24.	eSS	C	09 09.3	MPV=5.5 MLH=5.1 MLV=5.3
	eLQ	C	09 40	e 59 26.5 . e 00 22
	LmH	B	20.1	Higher-mode surface waves are registered.
	LmV	B	20.1	
24.	+eP	A	11 57 56	Mid-Indian Rise 24.00°S 69.60°E
	e	A	58 03	H = 11 44 58.1 h = normal MAG=5.0
	e	A	58 07.5	D = 90.2° Az = 327.3° (USCGS)
				PV1:2.0s 53.3nm PV2:1.6s 34.1nm
				MPV=5.4
24.	+iP	A	15 23 27.0	E. Russia-N.E China Border Region
				42.43°N 131.03°E
				H = 15 12 48.2 h = 521 km MAG=4.2
				D = 73.3° Az = 324.6° (USCGS)
				PV:1.0s 23.8nm
				MPV=4.6
25.	e	A	04 03 09	
25.	+iP	A	10 39 29.8	Northern Sinkiang Prov./China
	+ipP	A	39 39.5	43.33°N 87.04°E
	esP	A	39 43.5	H = 10 30 37.8 h = 34 km MAG=5.2
	e	A	40 28	D = 49.9° Az = 306.3° (USCGS); h = 40 km
	LmH	B	11 01.5	PV:1.4s 33.4nm pP:1.4s 27.2nm
	LmV	B	01.5	sP:1.6s 45.5nm
				LmH:16s 0.7 µm LmV:18s 0.9 µm
				MPV=5.2 MLH=4.8 MLV=4.8
25.	ePKIKP	A	10 54 43	Mendoza Prov./Argentina 32.88°S 69.01°W
	e	A	56 43	H = 10 36 14.3 h = 39 km MAG=5.7 (USCGS)
	e(SKP)	A	58 25	D = 109.3°
	LmH	C	11 38.0	e 57 28 e 59 29.5
25.	ePKP2	A	15 44 29	Kermadec Islands Region 29.09°S 178.16°W
				H = 15 24 25.9 h = 210 km MAG=4.5 (USCGS)
				D = 157.1°

Day	Phase		h m s	Remarks
26.	-ePKP	A	06 54 46	Tonga Islands 15.60°S 173.84°W
				H = 06 35 24.7 h = 140 km MAG=4.4
				D = 144.8° Az = 354.0° (USCGS)
				PV:1.0s 14.3 µm
26.	e(Sg)	A	13 05 53	Explosion?
26.	eP1	A	13 24 07	South Indian Ocean 1.25°S 89.45°E
	+iP2	A	24 16.5	H = 13 11 42.3 h = normal MAG=5.1
	e	A	24 36	D = 83.3° Az = 321.2° (USCGS)
	e	A	24 41	PV2:1.3s 30.6nm PV3:1.3s 19.5nm
	ei	A	24 53	PV4:1.4s 24.2nm
	LmH	C	14 04.5	MPV2=5.3
	LmV	C	06	The amplitude of P2 is much greater than that of P1. Two separated shocks in the same focus?
26.	iPg	A	14 16 12.5	Explosion/GDR 51.37°N 12.89°E
	iSg	A	16 28.5	D = 1.0° Yield: 1.8 t
26.	ePKP	A	22 06 04	Fiji Islands Region 16.5°S 175.6°E
				H = 21 46 41.2 h = 116 km MAG=4.8 (USCGS)
				D = 143.6°
				AN USSR gives: H = 21 46 31 h = normal
27.	LmH	B	01 48.5	LmH:22s 0.6 µm LmV:20s 1.0 µm
	LmV	B	49.5	
27.	e(Sg)	A	07 50 38.5	
27.	e(PP)	A	08 29(37)	Near N. Coast of West New Guinea
	e	A	29 53	1.84°S 138.69°E
	eSKKS	C	37.0	H = 08 09 47.9 h = normal MAG=5.3 (USCGS)
	e	C	45.2	D = 114.2°
	LmH	B	09 22	LmH:18s 1.4 µm LmV:18s 2.2 µm
	LmV	B	22.2	MLH=5.6 MLV=5.8
27.	iPg	A	13 13 46.0	Explosion/GDR 51.29°N 12.73°E
	iSg	A	14 02.0	D = 1.0° Yield: 2.45 t

Day	Phase		h m s	Remarks
27.	iPg	A	14 03 21.5	Explosion/GDR
	eiSg	A	03 37.5	D = 1.0°
27.	i	A	14 17 56	
27.	i	A	17 46 46	
27.	e	A	21 37 14	Near earthquake
	e(Sg)	A	38 31	e 37 46 e 38 25 e 38 37.5
27.	+iP	A	23 23 56.8	Southern Sinkiang Prov./China
	epP	A	24 00.5	41.65°N 82.29°E
	e	B	39 35	H = 23 15 19.7 h = normal MAG=5.0
	eLg1(3.47)	B	41 00	D = 48.0° Az = 306.1° (USCGS); h = 16 km
	eLg2(3.34)	B	42 00	PV:1.1s 26.7nm pPV:1.4s 33.3nm
	e	B	43 50	LmH:10s 1.0 μm LmV:11s 1.0 μm
	LmH	B	44.8	MPV=5.2 MLH=5.1 MLV=5.1
	LmV	B	44.8	Clear developed higher-mode surface waves.
28.	ePKIKP	A	08 03 58	Santa Cruz Islands 11.48°S 165.81°E H = 07 44 37.3 h = 30 km MAG=4.7 (USCGS) D = 135.7°
28.	e	A	08 20 12	
28.	e(Sg)	A	09 02 26	Explosion (PRUHONICE)
	e	A	02 44	
28.	e(Sg)	A	10 32 12	
	e	A	32 16	
28.	e(P)	A	12 41 55.5	
28.	i(Sg)	A	12 17 48.8	Near earthquake?
28.	iPg	A	19 15 33.0	Explosion
	iSg	A	15 47.5	D = ca. 1.1°

Day	Phase		h m s	Remarks
28.	eP	A	19 46 12	Southern Iran 28.48°N 57.49°E
	e(PPP)	A	48 25	H = 19 38 28.9 h = 24 km MAG=4.8 (USCGS)
	LmH	B	20 06.5	D = 40.9°
	LmV	B	06.5	e 46 35 e 47 25
29.	eiP	A	00 16 21	Queen Charlotte Is. Region
	e	A	16 24.5	51.16°N 130.37°W
	eS	C	25 55	H = 00 04 41.8 h = 6 km MAG=5.1
	eSS	C	30(36)	D = 73.5° Az = 24.1° (USCGS)
	LmH	B	47.8	PV:1.0s 28.6nm
	LmV	B	53.3	LmH:20s 2.0 µm LmV:14.5s 2.1 µm
				MPV=5.4 MLH=5.4 MLV=5.6
29.	+iP	A	04 07 15.1	Andreanof Is./Aleutian Is.
	e	A	07 19.5	51.44°N 178.25°W
	ePcP	A	07 24.5	H = 03 55 20.8 h = 50 km MAG=6.0
	epP	A	07 30.5	D = 77.9° Az = 353.6° (USCGS); h = 59 km
	eS	C	17(10)	PV:1.5s 180nm PcPV:1.2s 30.8nm
	e(SoS)	C	17 40	LmH:18s 1.3 µm LmV:19s 1.8 µm
	ePPS	C	18 08	MPV=6.0 MLH=5.3 MLV=5.5
	eSS	C	22 20	i 08 32.5 e 09 26 e 10 07 e 10 24
	LmH	B	46	e 10 30
	LmV	B	46	
29.	e(P)	A	05 05 17	Probably Southern Sinkiang Prov./China
	e	A	05 37	(USCGS)
29.	eiP	A	12 37 26	Andreanof Is./Aleutian Is.
	epP	A	37 41.5	51.48°N 178.24°W
				H = 12 25 32.7 h = 51 km MAG=5.3
				D = 77.9° Az = 353.6° (USCGS); h = 59 km
29.	ePKP	A	12 50 40	Tonga Islands Region 15.6°S 173.8°W
				H = 12 31 09.4 h = 59 km MAG=4.6 (USCGS)
				D = 144.8°
29.	eP	A	22 14 33.5	Near E. Coast of Honshu/Japan
	e	A	14 40	35.81°N 140.79°E
	epP	A	14 47	H = 22 02 09.5 h = 49 km MAG=4.9 (USCGS)
	LmH	B	(49)	D = 83.0° h = 50 km

Day	Phase		h m s	Remarks
30.	ePP	A	07 27 44	<u>New Britain Region</u> 6.35°S 150.54°E
	e	C	48 56	H = 07 07 00.5 h = 32 km (USCGS)
	e	C	51(24)	D = 124.2°
	LmH	B	08(29)	LmV:17s 0.4 µm
	LmV	B	29.5	MLV=5.2
30.	e	A	14 05 59	
	e	A	06 03	
30.	LmH	C	17 22.4	Probably Guatemala (USCGS)
30.	eSS	C	17 34	<u>Near N. Coast of W. New Guinea</u>
	LmH	B	18 11.5	1.77°S 138.74°E
	LmV	B	11.5	H = 16 59 00.6 h = normal MAG=5.5 (USCGS)
				D = 114.1°
				LmH:20s 1.3 µm LmV:20s 1.9 µm
				MLH=5.5 MLV=5.7

Day	Phase		h m s	Remarks
1.	iP	A	06 54 08.8	<u>Unimak Island Region</u> 53.90°N 163.34°W
				H = 06 42 23.7 h = normal MAG=4.0
				D = 75.7° Az = 3.3° (USCGS)
1.	eP1	A	07 12 06	<u>Greece</u> 39.72°N 21.34°E
	+ePL	B	12 07	H = 07 09 00.5 h = 15 km MAG=5.6
	eP2	A	12 08	D = 12.9° Az = 331.2° (USCGS)
	-iP3	A	12 12.8	PLV:13.5s 4.2 μm PLH:15s 3.1 μm
	i(PP)	A	12 15.5	PV2:1.2s 28.1nm PV3:1.5s 168nm
	eiSS	B	14 54	SSH:10s 9.8 μm
	LmH	B	16.6	LmH:13.5s 136.5 μm LmV:8.5s 62.5 μm
	LmV	B	17.8	MLH=6.2 MLV=6.2
				Successive P-onsets with increasing amplitude in our short-period records. Clear PL-waves in the long period registrations.
1.	eP	A	08 31 29	<u>Greece</u> 39.46°N 21.29°E
				H = 08 28 22.3 h = normal MAG=4.5 (USCGS)
				D = 13.1°
1.	eP	A	09 50 49	<u>Greece</u> 39.76°N 21.49°E
	i	A	51 03	H = 09 47 42.9 h = 27 km MAG=4.7
				D = 12.9° Az = 330.7° (USCGS)
1.	eP	A	09 53 13	<u>Greece</u> 39.64°N 21.40°E
	e(PP)	A	53 22	H = 09 50 06.6 h = 19 km MAG=4.9
	e(PPP)	A	53 28.5	D = 13.0° Az = 331.2° (USCGS)
	e	A	53 36.5	(PP)V:1.0s 11.8nm (PPP)V:1.2s 25.5nm
	e	A	54 29	LmH:10.5s 3.4 μm LmV:12s 1.9 μm
	e	A	54 33	MLH=4.6
	e	A	54 38	e(PP) and e(PPP) are much bigger than P.
	e	A	54 50	After Bath the short-period record (Z) at
	e	A	54 58	Umea in several shocks in this series from
	LmV	B	59	Greece exhibits also a clear phase, about
	LmH	B	59.3	10 - 12 s later than P and much bigger
				than the proper P. Our e(PP) probably correspond to this phase.

Day	Phase		h m s	Remarks
1.	eP	A	14 41 13	Greece 39.45°N 21.25°E
	LmH	C	47.3	H = 14 38 04.3 h = normal MAG=4.5
	LmV	B	47.3	D = 13.1° Az = 331.7° (USCGS)
				LmH:11.5s 0.5 µm LmV:12s 0.4 µm
				MLH=3.8
1.	i	A	16 43 27.5	Greece 39.78°N 21.70°E
	e	A	43 52	H = 16 40 06.6 h = 33 km MAG=4.3
	LmH	C	47.6	D = 13.0° Az = 330.2° (USCGS)
1.	LmH	C	18 17.8	Greece (UPP)
1.	eP	A	22 49 59	Greece (UPP)
2.	eP	A	01 30 25	Greece 39.73°N 21.20°E
	e(PP)	A	30 36	H = 01 27 20.2 h = normal MAG=4.4 (USCGS)
	e	A	30 48	D = 12.9°
	LmH	B	34.8	LmH:7s 0.3 µm
				MLH=3.8
2.	eP	A	08 15(02)	Greece 39.56°N 21.26°E
	e	A	15 14.5	H = 08 11 53.7 h = 19 km MAG=4.5
	e	A	15 40.5	D = 13.0° Az = 331.7° (USCGS)
	LmH	B	21.1	LmH:8s 0.4 µm LmV:12s 0.3 µm
	LmV	B	21.2	MLH=3.9
				AN USSR gives: Aegean Sea 37.8°N 20.0°E
				H = 08 11 40 h = normal
				D = 14.2°
2.	e(Sg)	A	08 24 57	
	e	A	25 03	
	e	A	25 25	
2.	e	A	17 28(42)	East New Guinea Region 5.60°S 147.25°E
	e	A	30(16)	H = 17 10 04.7 h = 148 km MAG=5.4 (USCGS)
	LmH	C	18 12.4	D = 121.8°
				LmH:26s 0.4 µm
				AN USSR gives: New Guinea Region
				5.6°S 147.4°E
				H = 17 10 22 h = 315 km

Day	Phase		h m s	Remarks
2.	i(P)	A	19 32 32.5	Greece 39.99°N 21.44°E
	ei(PP)	A	32 41.5	H = 19 29 24.3 h = 7 km MAG=4.4 (USCGS)
	LmH	C	36.9	D = 12.6°
				The first onset of P must be 5 s earlier than i(P).
3.	e(Sg)	A	11 38 40	Explosion?
	e	A	38 43	
	e	A	38 46	
	e	A	38 53	
3.	-i(P)	A	14 02 14.7	(P)V:0.9s 14nm
3.	eP	A	18 44 52	Greece 39.66°N 21.54°E
	e(PP)	A	45 02	H = 18 41 47.0 h = 35 km MAG=4.8
	e	A	45 06	D = 13.0° Az = 330.8° (USCGS)
	e	A	45 22.5	LmH:11s 1.8 μm LmV:12s 1.6 μm
	LmH	B	51	MLH=4.4
	LmV	B	51.1	
3.	e(Sg)	A	21 40 12	Explosion?
4.	e(Sg)	A	00 16 37	Near earthquake?
	e	A	16 43	
	e	A	17 04.5	
4.	eS	C	08 44(30)	South Sandwich Islands Region
	eSS	C	52 10	55.68°S 27.89°W
	eLQ	C	09 02.5	H = 08 17 32.1 h = normal MAG=5.8
	LmH	B	16.9	D = 112.3° Az = 294.8° (USCGS)
	LmV	B	16.9	LmH:22s 1.6 μm LmV:22s 2.4 μm
				MLH=5.6 MLV=5.7
4.	iPg	A	09 26 38.4	Explosion/GDR
	iSg	A	26 55.9	D = ca. 1.2°
4.	ePKP	A	10 38 43.5	Fiji Islands Region 19.70°S 176.16°W
	e	A	38 47.5	H = 10 18 58.0 h = normal MAG=4.9
	ei	A	38 55	D = 148.5° Az = 350.5° (USCGS)

Day	Phase		h m s	Remarks
cont.				
4.	e	A	10 38 59.5	PV1:1.7s 34.8nm PV2:1.2s 15.4nm
	LmH	C	11 41	PV3:1.2s 20.5nm
				LmH:24s 0.5 µm
				MLH=5.1
4.	eP	A	13 34 11.5	Greece 39.83°N 21.46°E
	e	A	34 31.5	H = 13 31 08.5 h = 40 km MAG=4.7
	e	A	34 40.5	D = 12.8° Az = 330.7° (USCGS)
	LmH	C	38.5	LmH:16s 0.9 µm
				MLH=3.9
4.	eP	A	23 37 46	Honshu/Japan 36.30°N 138.30°E
	LmH	B	00 10.9	H = 23 25 33.2 h = 8 km MAG=4.9
				D = 79.7° Az = 47.4° (USCGS)
				LmH:18s 1.2 µm
				MLH=5.3
5.	eP	A	01 39 14.5	PV:1.6s 22.7nm
5.	eP	A	06 29 46	Greece 39.65°N 21.54°E
	e	A	29 53.5	H = 06 26 37.9 h = 55 km MAG=4.6
	e	A	30 15	D = 13.0° Az = 330.9° (USCGS)
	LmH	B	35.9	LmH:11s 0.7 µm
				MLH=4.0
5.	e(P)	A	07 35 09	
5.	e(P)	A	08 53 48	(P)V:0.8s 18.9nm
5.	e(P)	A	14 53 09	Greece 39.60°N 21.24°E
	e	A	53 14	H = 14 50 02.9 h = normal MAG=4.4 (USCGS)
	LmH	B	59.2	D = 12.9°
	LmV	B	59.2	
5.	e(P)	A	14 55 55	
	e	A	56 05	

Day	Phase		h m s	Remarks
5.	eP	A	15 18 53	Kenai Peninsula/Alaska 59.32°N 151.38°W
	e	A	19 05	H = 15 07 49.5 h = 57 km MAG=5.4 (USCGS)
				D = 69.5°
5.	ePP	A	15 21 48	Solomon Islands 10.49°S 161.34°E
	e	A	21 57	H = 15 00 07.7 h = 41 km MAG=5.4 (USCGS)
	eSKP	A	22 43	D = 132.8°
	eSS	C	39.6	LmH:22s 1.8 μm LmV:24s 2.1 μm
	eSSS	C	44.6	MLH=5.7 MLV=5.7
	LmV	B	16 18.0	AN USSR gives: 10.4°S 160.9°E
	LmH	B	20.0	H = 15 00 16 h = 120 km
5.	eP	A	17 16 44	Central Alaska 63.72°N 148.50°W
	epP	A	17 09	H = 17 06 14.9 h = 102 km MAG=4.9
	esP	A	17 19	D = 64.9° Az = 13.8° (USCGS); h = 100 km
				AN USSR gives: Alaska 62.4°N 147.1°W
				H = 17 05 58 h = normal
				D = 66.1°
5.	ePP	A	17 55 51	Java 7.95°S 107.24°E
	e	A	56 11	H = 17 38 05.3 h = normal MAG=5.3 (USCGS)
	LmH	C	18 38	D = 99.7°
	LmV	B	44	PPV:1.9s 64.8nm
				LmH:24s 0.8 μm LmV:22s 0.8 μm
				MPPV=5.7 MLH=5.2 MLV=5.2
5.	eP	A	22 07 40	Norwegian Sea 72.70°N 3.03°E
				H = 22 02 40.2 h = normal MAG=4.8
				D = 22.5° Az = 165.6° (USCGS)
				PV:1.8s 51.0nm
				MPV=4.7
5.	eP	A	23 23 22.5	Mid-Indian Rise 23.08°S 69.29°E
	e	A	23 32	H = 23 10 26.9 h = 23 km MAG=4.9
				D = 89.3° Az = 327.4° (USCGS)
				PV:1.8s 30.6nm
				MPV=5.2
6.	e(P)	A	04 13 40.5	

Day	Phase		h m s	Remarks
6.	e(P)	A	04 47 39	
6.	LmH	C	09 34	South Sandwich Is. Reg. 55.57°S 26.33°W
	LmV	C	34	H = 08 31 15.8 h = normal MAG=5.1 (USCGS)
				D = 110.4°
				LmH:18s 0.35 µm LmV:16s 0.3 µm
				MLH=5.0 MLV=5.0
6.	e	A	11 18 07	
6.	e(pP)	A	14 12 04	Dominican Republic Region 19.29°N 69.97°W
	LmH	B	(40)	H = 14 00 41.4 h = 39 km MAG=5.3
	LmV	B	(40)	D = 70.1° Az = 42.1° (USCGS)
				LmH:18s 0.7 µm LmV:18s 1.0 µm
				MLH=5.0 MLV=5.1
				The first onset of P must be about 13 s earlier than e(pP).
6.	LmH	C	20(40)	Japan (USCGS)
	LmV	C	(40)	
7.	eP	A	06 52 58	Fox Islands/Aleutian Is. 52.21°N 171.88°W
	e	A	53 05	H = 06 41 05.8 h = 52 km MAG=4.5 (USCGS)
	epP	A	53 13.5	D = 77.5° h = 59 km
	esP	A	53 20	
7.	eP	A	07 52 09	Fox Islands/Aleutian Is. 52.18°N 171.79°W
	epP	A	52 21.5	H = 07 40 15.9 h = 45 km MAG=4.3 (USCGS)
				D = 77.5° h = 48 km
7.	e	A	08 23 11.5	
	e	A	23 23	
7.	ePKIKP	A	10 35 50	New Britain Region 4.14°S 152.81°E
	e	A	36 03	H = 10 16 56.2 h = 47 km MAG=5.0
	e	A	36 50	D = 123.4° Az = 331.5° (USCGS)
	LmH	C	11(32)	
	LmV	C	(32)	

Day	Phase		h m s	Remarks
8.	LmH	B	04 18.5	Ryukyu Islands 28.45°N 130.18°E
	LmV	B	25.3	H = 03 31 08.8 h = 58 km MAG=4.7 (USCGS)
				D = 84.4°
				LmH:18s 0.8 µm LmV:15s 0.8 µm
				MLH=5.2 MLV=5.2
8.	e(Sg)	A	07 01 31	
	e	A	01 54	
8.	e(P)	A	14 07 29	
8.	eP	A	14 59 33.5	Kyushy/Japan 30.13°N 131.55°E
				H = 14 47 11.1 h = 64 km MAG=4.9
				D = 83.7° Az = 326.3° (USCGS)
				PV:0.8s 9.5nm
				MPV=5.0
8.	e(Sg)	A	16 48 17.5	
8.	eP	A	18 55 50	Hindu Kush Region 36.43°N 70.16°E
	i	A	56 29.5	H = 18 48 04.8 h = 215 km MAG=4.8
	epP	A	56 37	D = 43.6° Az = 308.1° (USCGS); h = 224 km
				PV:0.8s 9.5nm
				MPV=4.3
8.	+iPKP2	A	19 05 35	South of Kermadec Is. 33.15°S 178.39°W
	e	A	05 39	H = 18 44 56.8 h = 50 km MAG=5.3
	e	A	05 50	D = 161.0°
				PV1:1.1s 39.0nm PV2:1.2s 20.5nm
				PV3:1.2s 20.5nm
9.	+eP	A	04 08 54	Turkey 39.57°N 27.00°E
	e	A	08 58.5	H = 04 05 11.2 h = 29 km MAG=4.5
	e	A	09 04	D = 15.5° Az = 320.7° (USCGS)
	e	A	09 08.5	PV:1.1s 9.8nm
	LmH	B	14.1	LmH:13.5s 0.7 µm LmV:12s 0.4 µm
	LmV	B	15.5	MLH=4.0 MLV=4.0

Day	Phase		h m s	Remarks
9.	+eP1	A	06 26 55.5	Kurile Islands 44.21°N 149.01°E
	+iP2	A	26 58.2	H = 06 14 57.1 h = 40 km MAG=5.3
	+eipP1	A	27 08	D = 78.5° Az = 333.9° (USCGS); h = 45 km
	+ipP2	A	27 10.3	PV1:1.3s 41.8nm PV2:2.0s 111nm
	i	A	27 22.0	pPV1:1.1s 40.0nm pPV2:1.6s 94.7nm
	i	A	27 41.0	LmH:19s 2.4 µm LmV:16s 2.6 µm
	eS	C	36 48	MPV1=5.4 MPV2=5.6 MLH=5.6 MLV=5.7
	eSS	C	42.0	i 27 22.0 (V:1.8s 106nm) and i 27 41.0
	e	B	47 16	(V:1.7s 94.5nm) are two very clear sepe-
	LmH	B	07 00.0	rated onsets in the short-period vertical
	LmV	B	06.6	components. A seperated shock in the same
				focus?
9.	e(P)	A	07 28 12	Greece (UPP)
9.	e(P)	A	08 03 57	Greece 39.84°N 21.52°E
	e	A	04 24	H = 08 00 47.9 h = 54 km MAG=4.5 (USCGS)
	e	A	07 55	D = 12.7°
	LmH	B	08.3	LmH:13s 0.5 µm
				MLH=3.8
				The first onset of P must be about 9 s
				earlier than e(P).
9.	eP	A	11 11 28	Eastern Sea of Japan 44.76°N 140.65°E
				H = 11 00 09.8 h = 256 km MAG=4.8
				D = 74.6° Az = 327.3° (USCGS)
				PV:1.3s 8.3nm
				MPV=4.3
9.	eiP	A	12 48 01.5	Kodiak Island Region 56.56°N 152.57°W
	epP	A	48 08.5	H = 12 36 36.8 h = normal MAG=5.0
	e	A	48 32	D = 72.4° Az = 10.5° (USCGS); h = 26 km
	eS	B	57 26	PV:1.0s 28.2nm pPV:1.4s 47.6nm
	eSS	C	13 02 20	LmH:20s 0.6 µm LmV:19s 0.5 µm
	LmH	B	22.8	MPV=5.4 MLH=4.9 MLV=4.8
	LmV	B	23.5	e 48 12 e 49 01.5
9.	e(Sg)	A	13 05 38	

Day	Phase		h m s	Remarks
9.	e(pP)	A	15 18 32	Kodiak Island 56.62°N 152.27°W
	e	A	18 46	H = 15 06 58.6 h = 17 km MAG=4.8
				D = 72.3° Az = 10.7° (USCGS);
				(h = ca. 28 km)
				P must be about 8 s earlier than (pP).
9.	ei	A	15 44 03	
9.	i(P)	A	16 44 03.5	
	i	A	44 15.5	
9.	eP	A	16 44 56.5	Northern Pamir 39.2°N 71.4°E
				H = 16 37 02 h = normal MAG=4½ (AN USSR)
				D = 42.8°
9.	ePKP	A	20 32 59	Tonga Islands 15.54°S 173.41°W
	epPKP	A	33 25	H = 20 13 32.4 h =89 km MAG=4.8
				D = 144.7° Az = 354.5° (USCGS); h = 92 km
9.	eP	A	21 43 50	Philippine Islands Region 5.21°N 127.47°E
	e(pP)	A	44 18	H = 21 30 08.3 h = 119 km MAG=5.5
	e	A	44 24	D = 101.9° Az = 324.2° (USCGS);
	e	A	47 54.5	(h = 109 km)
	e	A	48 03	PV:1.7s 27.7nm PV3:2.0s 44.4nm
	ePP	A	48 08	LmH:16s 0.4 μm LmV:16s 0.3 μm
	eSKS	C	54 16	MPV=5.7
	eS	C	55 20	If we interpret e 44 24 as pP than the
	eSPP	C	57 40	focal depth would be h = 134 km
	eSS	C	22 02(24)	AN USSR gives: 5.2°N 127.6°E
	LmH	B	35.3	H = 21 30 18 h = 204 km
	LmV	B	35.3	D = 102.0°
10.	ePn	A	05 23 27.5	Italy 44.6°N 10.4°E
	e	A	23 33	H = 05 22 00 (BCIS)
	e	A	23 52	D = 6.1°
	eiSn	A	24 35	
	e(Lg2)	A	25 26	

Day	Phase		h m s	Remarks
10.	e	A	07 52 25	Austria 47.8°N 14.2°E
	eSn	A	52 34.5	H = 07 51 02 (BCIS)
	eSg	A	52 52	D = 3.3°
	ei	A	52 56	
10.	e(P)	A	12 39 34	(P)V:0.9s 7.1nm
10.	eP	A	13 52 17	Nevada
				H = 13 40 00 (UPP)
				Underground explosion.
10.	eP	A	17 52 32	Taiwan 23.74°N 121.52°E
	e	A	52 53	H = 17 40 06.7 h = 44 km MAG=5.1
	LmH	C	18 33	D = 83.7° Az = 323.0° (USCGS)
	LmV	C	33.4	PV:0.6s 9.5nm
				LmH:19s 0.3 µm LmV:20s 0.2 µm
				MPV=5.1 MLH=4.6 MLV=4.6
10.	eP	A	21 15 38	Greece (UPP)
	LmH	B	20 28	
11.	e(Sg)	A	14 36 35	Sg(V):1.3s 13.9nm
11.	eP1	A	14 59 08.5	Tadzhik-Sinkiang Border Reg.
	+iP2	A	59 11.8	39.38°N 73.78°E
	iPP	A	15 00 58	H = 14 50 58.8 h = 21 km MAG=5.6
	iPP	B	01 06	D = 44.1° Az = 306.1° (USCGS)
	eS	C	05 35	PV1:0.9s 18.9nm PV2:1.2s 125nm
	eSS	C	08 48	LmH:18s 40.0 µm LmV:14s 17.4 µm
	iScS	C	08 58	MPV1=5.0 MPV2=5.7 MLH=6.4 MLV=6.2
	LQ	C	09.1	ei 59 13.5 e 59 40.5 i 01 16.3 e 08 14
	iLg1	B	13 53	P is multiple. The amplitude of P2 is
	iLg2	B	15 16	much bigger than that of P1. Probably
	LmH	B	18.2	two shocks in the same focus. Well deve-
	LmV	B	19.0	loped higher mode surface waves.
11.	ePP	A	15 22 53	Chile-Bolivia Border Reg. 20.26°S 68.51°W
				H = 15 05 16.8 h = 67 km MAG=6.1 (USCGS)
				D = 99.4°

Day	Phase		h m s	Remarks
11.	eP	A	15 35 15	Chile-Bolivia Border Region?
	e(pP)	A	35 42	PV:0.8s 9.5nm (pP)V:1.5s 20.1nm
	ePP	A	39 20	
11.	e(P)	A	15 43 28	
12.	e(Sg)	A	03 00 23	
	e	A	00 30.5	
12.	eP	A	05 29 16	<u>Tadzhik-Sinkiang Border Reg.</u>
	LmH	B	48	39.49°N 73.83°E
	LmV	B	49.3	H = 05 21 04.8 h = 5 km MAG=4.9 (USCGS)
				D = 44.1° Az = 306.0°
				LmH:19s 0.9 μm LmV:13s 0.4 μm
				MLH=4.7 MLV=4.6
12.	i(P)	A	05 37 15	
12.	epPKP	A	06 34 03	<u>Tonga Islands</u> 17.76°S 174.04°W
				H = 06 13 59.7 h = 115 km MAG=4.8 (USCGS)
				D = 146.8°
12.	eP	A	10 27 00	
12.	LmH	C	11(35)	Probably off Coast of Southern Chile
	LmV	C	(35)	(USCGS)
12.	e	A	11 15 16.5	
12.	eP	A	17 10 24	<u>Fox Islands/Aleutian Is.</u> 52.89°N 167.03°W
				H = 16 58 33.2 h = 32 km MAG=4.9
				D = 76.8° Az = 0.9° (USCGS)
				MPV=5.1
12.	ePn	A	17 54 52	<u>Northern Italy</u> 44.67°N 10.45°E
	iPg	A	55 22.3	H = 17 53 23.1 h = 39 km MAG=4.2
	iSn	A	55 56.5	D = 6.0° Az = 7.1° (USCGS)
	iSg	A	56 34.8	LmH:14s 1.0 μm LmV:14s 0.8 μm
	iLg2	A	56 43	MLH=3.5

Day	Phase		h m s	Remarks
cont.				
12.	LmH	B	17 57.5	e 54 56 e 55 17.5
	LmV	B	57.5	
12.	ePKP2	A	19 24 52	Balleny Islands Region 62.67°S 167.72°E
	LmV	C	20(45)	H = 19 04 00.8 h = normal MAG=5.4 (USCGS)
	LmH	C	(46.5)	D = 162.4°
12.	eP	A	22 28 05.5	Southern Alaska 60.13°N 152.55°W
	eipP	A	28 30	H = 22 17 09.6 h = 93 km MAG=4.6
				D = 68.8° Az = 10.7° (USCGS); h = 100 km
				AN USSR gives: Alaska 63.7°N 154.4°W
				H = 22 17 25 h = normal
				D = 65.5°
13.	+eP	A	05 30 20.5	Kodiak Island Region 56.53°N 152.59°W
	e	A	30 26.5	H = 05 18 55.4 h = normal MAG=5.3
	eS	C	39 46	D = 72.4° Az = 10.5° (USCGS)
	LmH	B	06 06.3	PV:1.3s 41.7nm SH:16s 0.9 µm
	LmV	B	09.8	LmH:18s 1.4 µm LmV:18s 1.5 µm
				MPV=5.4 MSH=5.7 MLH=5.3 MLV=5.3
13.	e(P)	A	07 50 22	
13.	iPg	A	12 00 04.2	Explosion/ČSSR 50.58°N 14.00°E
	iSg	A	00 25.0	D = 1.5° Yield: 20 t
13.	e(P)	A	12 19 45	
13.	eSg	A	20 11 42	Probably Austria (aftershock to May 10.,
	e	A	11 52	H = 07 51 02)
14.	eP1	A	04 19 23	Southern Greece 37.73°N 21.20°E
	eP2	A	19 31.5	H = 04 16 01.7 h = 66 km MAG=4.8
	i	A	19 37.5	D = 14.6° Az = 335.2° (USCGS)
	i	A	19 45.0	PV1:1.0s 12nm PV2:1.0s 49.8nm
	eSS	B	22.4	LmH:13s 1.0 µm LmV:13s 1.3 µm
	ei	B	24 28	MLH=4.1 MLV=4.4
	LmH	B	26.2	Two clear separated P-onsets (P2 - P1 =

Day	Phase		h m s	Remarks
cont. 14.	LmV	B	04 26.2	8.5 s), the first much smaller than the second one.
14.	ePKP	A	05 22 04	Loyalty Islands Region 22.30°S 170.47°E H = 05 02 17.1 h = 10 km D = 147.2° Az = 334.9° (USCGS)
14.	i(P)	A	07 40 59.0	
14.	e(P)	A	08 36 13	
14.	ePP	A	08 56(08)	Chile-Bolivia Border Reg. 20.56°S 68.92°W H = 08 38 33.1 h = 109 km MAG=5.2 (USCGS) D = 99.8° AN USSR gives: h = normal
14.	eP	A	09 09 03	Tadzhik-Sinkiang Border Reg.
	e	A	09 49	39.23°N 73.90°E
	ePP	A	10 47	H = 09 00 54.8 h = normal MAG=5.0
	e	A	11 28.5	D = 44.3° Az = 306.2° (USCGS)
	eSa	B	18(45)	PV:1.0s 19.0nm
	eLg2	B	25 12	LmH:18s 3.1 µm LmV:16s 2.2 µm
	LmH	B	28	MPV=5.0 MLH=5.3 MLV=5.2
	LmV	B	29.3	Higher mode surface waves are registered.
14.	e(P)	A	17 38 00	
14.	e	A	21 16 30	
	e	A	16 51	
15.	eP	A	00 18 24	South of Honshu/Japan 32.86°N 141.33°E H = 00 05 53.3 h = 63 km MAG=4.5 (USCGS) D = 85.8°
15.	eP	A	00 26 08	South of Honshu/Japan 32.89°N 141.39°E H = 00 13 36.8 h = 58 km MAG=4.5 D = 85.8° Az = 330.6° (USCGS)

Day	Phase		h m s	Remarks
15.	eP	A	02 40 14	South of Honshu/Japan 32.84°N 141.44°E
	epP	A	40 27	H = 02 27 36.0 h = 40 km MAG=5.4
	esP	A	40 31.5	D = 86.1° Az = 330.7° (USCGS); h = 48 km
	ePP	A	43 33	PV:1.2s 25.6nm SH(B):16s 0.6 µm
	epPP	A	43 46	LmH:16s 1.5 µm LmV:16s 1.7 µm
	eS	B	50 38	MPV=5.3 MSH(B)=5.6 MLH=5.5 MLV=5.5
	eiS	C	50 44	e 40 19 e 40 35 e 40 44
	LmH	B	03 25	
	LmV	B	(26.5)	
15.	iPKHKP	A	02 47 04.2	Fiji Islands Region 20.06°S 177.65°W
	ePKP2	A	47 11.5	H = 02 28 16.8 h = 564 km MAG=4.9 (USCGS)
	e(pPKP)	A	48 40.5	D = 148.5° (h = 400 km)
15.	e(P)	A	04 21 51.5	
15.	e(P)	A	06 16 44	
15.	eP1	A	08 17 22.2	Crete 34.59°N 26.73°E
	-iP2	AB	17 25.2	H = 08 12 57.1 h = normal MAG=4.9
	eiP3	A	17 30	D = 19.5° Az = 330.1° (USCGS)
	e	A	17 44	PV2(B):5.5s 0.85 µm PH2(B):5.5s 0.7 µm
	e	A	17 52	PV2(A):1.8s 71.5nm PV3:1.8s 184.0nm
	eS	C	21 00	LmH:14s 5.3 µm LmV:15s 7.0 µm
	ei	B	21 10	MPV2(B)=5.2 MPH2(B)=5.4 MPV2(A)=4.6
	i(SS)	B	21 21	MPV3(A)=5.0 MLH=5.0 MLV=5.2
	LmV	B	26.4	e 18 08 e 18 22.5 e 22 50 e 23 38
	LmH	B	26.8	Successive P-onset with increasing amplitude in the short-period records.
15.	e(P)	A	08 36 43	Crete 34.53°N 26.76°E
	e	A	36 54	H = 08 32 08.9 h = normal MAG=4.2
				D = 19.5° Az = 330.2° (USCGS)
				The first motion of P must be about 7 s earlier.
15.	e	A	10 05 30.5	Apennines/Italy 44.6°N 10.4°E
	iPg	A	05 35	H = 10 03 34 (BCIS)
	eiSn	A	06 09.5	D = 6.1°

Day	Phase		h m s	Remarks
cont.				
15.	eiSg	A	10 06 55	LmH:14s 0.4 µm LmV:14s 0.4 µm
	LmH	B	07.8	MLH=3.0
	LmV	B	07.8	ei 05 49.5 i 06 12.5
16.	LmH	C	09 30	Easter Island Cordillera (USCGS)
	LmV	C	30	
16.	e	A	12 19 10	Explosion
	eiSg	A	19 18.5	
	LmH	A	19 32	
	LmV	A	19 32	
16.	eP	A	13 10 48	Near Coast of Guatemala 13.51°N 90.63°W
	eS	B	21 24	H = 12 58 09.5 h = 95 km MAG=4.8
	ePS	B	22 38	D = 87.3° Az = 38.5° (USCGS)
	LmH	B	52.5	LmH:17s 0.7 µm LmV:17s 1.1 µm
	LmV	B	54.2	MLH=5.1 MLV=5.4
16.	e	A	13 28 52.5	Near earthquake (probably Northern Italy
	e	A	29 02	or Northern Yugoslavia. Trieste gives:
	e	A	29 13.5	D = ca. 75 km).
16.	e	A	14 45 32	Probably near earthquake.
	e	A	45 48	
16.	eP	A	16 16 06	Iceland 63.75°N 19.14°W
	LmH	B	25.4	H = 16 11 22.2 h = 4 km MAG=4.2
	LmV	B	25.4	D = 20.9° Az = 114.3° (USCGS)
				LmH:15s 0.5 µm LmV:15s 0.4 µm
				MLH=4.0 MLV=4.1
16.	e(P)	A	16 32 16	(P)V:0.6s 7.1nm
16.	ePKP	A	16 33 57	Tonga Islands 15.23°S 173.52°W
				H = 16 14 22.9 h = normal MAG=5.2 (USCGS)
				D = 144.5°

Day	Phase		h m s	Remarks
16.	eP	A	19 37 37.5	South of Honshu/Japan 32.45°N 141.25°E
	epP	A	37 48.5	H = 19 24 58.6 h = 36 km MAG=5.3
	e	A	37 58.5	D = 86.1° Az = 330.6° (USCGS); h = 41 km
	LmV	B	20 23.9	PV:1.4s 24.5nm
	LmH	B	24.0	LmH:16s 0.4 μm LmV:16s 0.5 μm
				MPV=5.1 MLH=4.9 MLV=5.0
16.	e	A	19 46 50	Probably Kurile Islands (AN USSR)
	e	A	47 12	
16.	e	A	21 21 12	Yugoslavia 42.9°N 20.3°E
	eSg	A	21 56	H = 21 16 34 (BCIS) MAG=3.9 (Beograd)
				MAG=3.7 (Skopje)
				D = 9.8°
16.	ePKP	A	23 27 03	Fiji Islands Region 20.96°S 178.81°W
				H = 23 08 16.3 h = 551 km MAG=3.9
				D = 149.3° Az = 347.0° (USCGS)
17.	eP1	A	00 44 08	Southern Alaska 60.80°N 143.72°W
	iP2	A	44 10.6	H = 00 33 12.3 h = 15 km MAG=4.8
				D = 67.1° Az = 16.8° (USCGS)
				PV2:0.8s 42.5nm
				MPV2=5.6
				The amplitude of P2 is about five times greater than that of P1. Probably a second shock in the same focus.
17.	eP	A	04 34 24	Turkey-Iran Border Reg. 38.69°N 44.18°E
	e	A	34 51	H = 04 28 51.9 h = 39 km MAG=4.6
	LmH	B	46.1	D = 25.8° Az = 308.2° (USCGS)
	LmV	B	47.6	LmH:14s 0.7 μm LmV:13s 0.6 μm
				MLH=4.4 MLV=4.5
17.	e(pP)	A	09 47 06	East of Honshu/Japan 38.3°N 142.2°E
	e	A	47 11	H = 09 34 42 h = normal (AN USSR)
				D = 81.2°
				P must be about 10 s earlier than (pP).

Day	Phase		h m s	Remarks
17.	eP	A	10 02 33.5	Taiwan Region 24.36°N 122.09°E
	e	A	03 05	H = 09 50 09.4 h = 50 km MAG=4.9
	e	A	03 23	D = 83.5° Az = 323.1° (USCGS)
	LmH	B	38	LmH:15.5s 0.5 µm LmV:13s 0.4 µm
	LmV	B	42	MLH=5.0 MLV=5.0
17.	LmH	B	12(00)	Dominican Republic 19.68°N 69.67°W
	LmV	B	(00)	H = 11 19 38.2 h = 54 km MAG=4.4 (USCGS)
				D = 69.6°
				LmH:18s 0.3 µm LmV:18s 0.3 µm
				MLH=4.6 MLV=4.6
17.	e	A	13 05 45	Explosion/GDR 51.29°N 12.73°E
	eiSg	A	05 49	D = 0.9° Yield: 2.2 t
17.	e	A	13 16 52	
17.	LmH	B	14 19.0	Solomon Islands 9.70°S 159.76°E
	LmV	B	20.0	H = 12 56 55.4 h = 32 km MAG=5.1 (USCGS)
				D = 131.5°
17.	ePKP	A	16 33 05	Fiji Islands Region 16.62°S 175.52°E
				H = 16 13 37.6 h = 80 km MAG=4.8
				D = 143.7° Az = 342.7° (USCGS)
				PV:1.4s 9.2nm
17.	-iP	B	17 57 51.5	Red Sea 19.74°N 38.72°E
	+iP	A	57 52.5	H = 17 50 39.6 h = 38 km MAG=5.3
	e	B	58 18	D = 37.6° Az = 331.6° (USCGS)
	e	A	58 30	PV(B):3.5s 0.5 µm PV(A):2.5s 247nm
	ePP	B	59 17	SH:12s 0.45 µm
	ePcP	B	59 50	LmH:12s 0.35 µm LmV:12s 0.4 µm
	eS	B	18 03 44	MPV(B)=5.8 MPV(A)=5.6 MSH=5.3 MLH=4.4
	LmV	B	22.1	MLV=4.5
	LmH	B	22.1	The P-phases on the short-period records of the Red Sea earthquakes shows unusually long periods around 2.5 s. The surface waves of Red Sea earthquakes are abnormally small at our station for an earthquake with normal focus depth.

Day	Phase		h m s	Remarks
17.	LmH	C	22 55	Probably Easter Island Ridge (USCGS)
	LmV	C	55	
18.	eP	A	04 18 55.5	Hokkaido/Japan Region 41.93°N 144.62°E
	e	A	19 02	H = 04 06 54.7 h = 44 km MAG=4.7
	e	A	19 08	D = 79.1° Az = 331.7° (USCGS)
	eS	C	28 52	PV:1.5s 26.8nm
	LmH	B	50.5	LmH:22s 2.2 µm LmV:16s 1.2 µm
	LmV	B	57.3	MPV=5.2 MLH=5.5 MLV=5.3
18.	+iP	A	11 34 33.7	Hokkaido/Japan Region 41.93°N 144.66°E
	e	A	34 39	H = 11 22 31.6 h = 41 km MAG=4.9
	e	A	34 46	D = 79.1° Az = 331.7° (USCGS)
	e	A	34 56.5	PV:1.4s 76.7nm
	e	A	35 43	LmH:24s 2.9 µm LmV:19s 1.5 µm
	eS	C	44 28	MPV=5.6 MLH=5.5 MLV=5.4
	LmH	B	12 06.4	AN USSR gives: Japan 43.3°N 144.3°E
	LmV	B	12.9	H = 11 22 48 h = 120 km
				D = 77.7°
18.	e(P)	A	11 39(03)	Tadzhik SSR 40.61°N 70.39°E
	e	A	39 26	H = 11 31 20.4 h = 14 km MAG=4.9 (USCGS)
	e	A	40 40	D = 41.3°
	e	A	40 49	The P-onset is doubtful.
18.	iPg	A	14 08 33.6	Explosion/GDR 51.37°N 12.89°E
	iSg	A	08 50.1	D = 1.0°
18.	eP	A	14 12 58	Hokkaido/Japan Region 41.95°N 144.67°E
	epP	A	13 09	H = 14 00 56.1 h = 40 km MAG=5.1
	e	A	13 20	D = 79.1° Az = 331.7° (USCGS); h = 40 km
	LmH	B	45.0	PV:1.5s 25.0nm
	LmV	B	51.3	LmH:22s 1.3 µm LmV:18s 1.0 µm
				MPV=5.1 MLH=5.2 MLV=5.2
18.	ePg	A	15 43 56	Explosion/ČSSR
	iSg	A	44 09.6	D = ca. 1.1°

Day	Phase		h m s	Remarks
18.	+eP	A	23 51 34.5	Kyushu/Japan 31.10°N 130.73°E
	epP	A	51 44.5	H = 23 39 15.2 h = 43 km MAG=5.6
	LmH	B	00 29.4	D = 82.5° Az = 325.9° (USCGS); h = 37 km
	LmV	B	34.2	PV:1.3s 50.0nm
				LmH:16s 0.4 µm
				MPV=5.5 MLH=4.8
19.	ePKP2	A	05 29 57.5	South of Kermadec Islands 34.90°S 179.01°W
	e(pPKP2)	A	30 10.5	H = 05 09 10.9 h = 35 km MAG=5.2 (USCGS)
	i	A	30 19.5	D = 162.6° (h = 15 km)
				PV1:1.5s 20.1nm PV2:1.2s 12.8nm
				PV3:1.3s 27.8nm
				i 30 19.5 is either pPKP2, which would suggest a focal depth of h = 77 km, or it can be the PKP2-phase of a second shock in the same focus (see Bulletin of Seism. Inst. Uppsala).
19.	iPKP	A	07 59 22.7	West of Macquarie Island 54.28°S 143.63°E
				H = 07 39 27.2 h = normal
				D = 151.0° Az = 282.8° (USCGS)
19.	ePKP2	A	12 22 50	Kermadec Islands Region 30.28°S 177.94°W
				H = 12 02 19.9 h = 25 km MAG=4.7 (USCGS)
				D = 158.4°
19.	eiPKP	A	13 10 49	LmH:18s 0.3 µm LmV:18s 0.5 µm
	e	A	10 59	
	i	A	11 23.0	
	LmH	C	14 21	
	LmV	B	24.5	
19.	+iP	A	16 00 33.6	Ethiopia 14.53°N 40.26°E
	e	A	00 46.5	H = 15 52 34.2 h = 13 km MAG=5.1
	e	A	01 11	D = 42.8° Az = 333.3° (USCGS)
	e	A	01 31	PV:1.3s 38.9nm
	ePP	A	02 16	MPV=5.2
	iPcP	A	02 28.0	

Day	Phase		h m s	Remarks
20.	e	A	03 08 53	Mariana Islands Region 19.84°N 146.05°E
	e(PP)	A	09 09	H = 02 51 09.4 h = 42 km MAG=5.5 (USCGS)
	e	A	09 17.5	D = 99.1°
	LmV	C	04 51.5	LmH:23s 0.5 µm LmV:23s 0.5 µm
	LmH	C	52	MLH=4.9 MLV=5.0
20.	eP	A	08 55 23	Kirgiz SSR 39.20°N 72.75°E
	e	A	55 34.5	H = 08 47 19.8 h = normal MAG=5.1
	ePP	A	57 08	D = 43.6° Az = 306.0° (USCGS)
	LmH	C	09 12.9	LmH:18s 0.4 µm
				MLH=4.4
20.	iPg	A	11 45 33.6	Explosion/ČSSR 50.57°N 14.00°E
	iSg	A	45 53.9	D = ca. 1.5° Yield: 20.7 t
20.	ePKIKP	A	13 21 08	Drake Passage 59.18°S 65.68°W
	LmH	C	14 24.5	H = 13 02 09.3 h = normal MAG=5.5 (USCGS)
	LmV	C	24.5	D = 126.2°
				PV:1.1s 24.0nm
20.	+iP	A	15 12 17.6	Nevada
	i	A	12 27.1	H = 15 00 00
	e(PP)	A	15 20	D = ca. 81.2°
	iPP	A	15 22.5	PV:1.6s 87.2nm PPV:1.8s 35.7nm
	LmH	C	16 48.5	MPV=5.6 MPPV=5.3
	LmV	C	52.8	e 12 33 e 12 41
				Underground explosion.
20.	eP	A	23 22 39	Western Russia 66.41°N 33.37°E
	eS	A	26(00)	H = 23 18 11.7 h = 17 km MAG=4.6
	eLi	A	27 53.5	D = 19.3° Az = 225.6° (USCGS)
	eiLg1	A	28 29	ei 23 35 e 26 57 e 28 03
	LmH	A	28 55	
21.	e	A	07 31 34	Gulf of California 27.93°N 111.31°W
	eLQ	C	53.8	H = 07 18 12.8 h = normal MAG=4.7 (USCGS)
	LmV	B	08 08	D = 87.0°
	LmH	B	11	LmH:16s 1.9 µm LmV:20s 2.3 µm
				MLH=5.6 MLV=5.6

Day	Phase		h m s	Remarks
21.	e(P)	A	11 37 43.5	
21.	eP	A	18 42 42	Hindu Kush Region 35.55°N 69.70°E
	ipP	A	43 12.0	H = 18 34 46.4 h = 138 km MAG=4.7 (USCGS)
	e	A	43 44	D = 43.9° h = 135 km
21.	+iP	AB	18 57 55.9	Southern Sumatra 0.97°S 101.47°E
	-ipP	AB	58 38.4	H = 18 45 11.7 h = 173 km MAG=6.3
	iPP	A	19 01 31.7	D = 90.6° Az = 320.5° (USCGS); h = 172 km
	iSKS	B	08 07	PV(A):1.9s 833nm PV(B):4.0s 2.7 µm
	iS	B	08 34	PH(B):4.0s 0.9 µm pPV(A):2.0s 496nm
	i(sSKS)	B	09 28	pPV(B):7s 4.1 µm SKSH:12s 8.2 µm
	isS	B	09 50	SH:12s 15.6 µm sSH:12.5s 10.6 µm
	iSS	C	14 38	MPV(A)=6.4 MPV(B)=6.6 MPH(B)=6.4 MSH=7.1
				e 00 54 ei 01 22 e 01 55 e 09 35
22.	e(PKP)	A	12 27 41	Probably Fiji Islands (USCGS)
22.	-eiP	A	19 50 20.3	Turkey 36.68°N 29.52°E
	e	A	50 22	H = 19 46 01.9 h = 54 km MAG=4.4
	e	A	50 24.5	D = 19.0° Az = 323.0° (USCGS)
	e	A	50 44	PV2:1.0s 18.9nm PV3:1.2s 28.1nm
				Multiple P.
22.	eP	A	21 04 18	Southwestern Ryukyu Islands
				24.89°N 124.19°E
				H = 20 51 57.4 h = 127 km MAG=4.5 (USCGS)
				D = 84.2°
				AN USSR gives: h = normal
22.	e	A	23 19 42	
22.	LmH	C	23 34	Probably Mexico (USCGS)
23.	-eP	A	01 34 22	Kurile Islands Region 44.63°N 150.16°E
	e	A	34 34	H = 01 22 22.2 h = normal MAG=4.3
	e	A	34 44	D = 78.5° Az = 334.5° (USCGS)
	LmH	C	02 12	PV:1.5s 40.3nm
	LmV	C	12	LmH:18s 0.3 µm LmV:18s 0.35 µm
				MPV=5.3 MLH=4.8 MLV=4.8

Day	Phase		h m s	Remarks
23.	+eiP	A	02 04 39.8	Kurile Islands Region 44.62°N 150.45°E
	e	A	04 52	H = 01 52 39.1 h = 22 km MAG=4.9
	e	A	05 02	D = 78.6° Az = 334.7° (USCGS)
	LmH	C	42.5	PV1:1.6s 91.0nm PV2:1.5s 30nm
	LmV	B	46.4	PV3:1.5s 37nm
				LmH:18s 1.1 μm LmV:16s 1.6 μm
				MPV=5.7 MLH=5.2 MLV=5.5
23.	eP	A	08 47 41.5	Southern Sumatra 3.12°S 101.52°E
	e	A	48 57	H = 08 34 35.8 h = 59 km MAG=5.3
	e(PP)	A	52 23	D = 92.3° Az = 320.4° (USCGS)
23.	eiP	A	12 07 17.8	Norwegian Sea 72.82°N 5.57°E
	e	A	07 25	H = 12 02 18.3 h = normal MAG=4.7
	e	A	07 27.5	D = 22.4° Az = 169.9° (USCGS)
	LmH	C	17	PV1:1.8s 62.5nm PV2:1.8s 56.2nm
				MPV=4.8
23.	+iP	A	14 12 17.8	Nevada
	e	A	12 30	H = 14 00 00
	e	A	12 39	D = ca. 81.2°
	e	A	13 11	PV:1.4s 46.0nm
	ePP	A	15 22.5	MPV=5.4
				Underground explosion.
23.	e	A	15 45 56	Probably Kermadec Islands (USCGS)
23.	ePP	A	19 36 58	South Sandwich Islands Region
	e	A	37 11.5	56.19°S 27.27°W
	eSP	C	46 08	H = 19 17 47.5 h = 130 km MAG=5.9 (USCGS)
	ePPS	C	47(30)	D = 111.2°
	eSS	C	52 24	LmH:20s 0.4 μm LmV:18s 0.6 μm
	LmH	B	20 19	AN USSR gives: 56.4°S 29.9°W
	LmV	B	21.5	H = 19 17 33 h = normal MAG=5½
				D = 112.0°
23.	eP	A	21 08 27	Off East Coast of Kamchatka
	epP	A	08 38	52.26°N 160.32°E
				H = 20 56 52.6 h = normal MAG=4.5
				D = 74.1° Az = 339.9° (USCGS); h = 41 km

Day	Phase		h m s	Remarks
24.	epP	A	01 47 33	Kurile Islands Region 50.04°N 159.32°E
	i(sP)	A	47 38.0	H = 01 35 37.1 h = 43 km MAG=4.4 (USCGS)
				D = 76.0°
24.	e(pP)	A	17 29 18	Malawi 12.14°S 34.18°E
				H = 17 18 29.2 h = normal MAG=4.9 (USCGS)
				D = 65.6°
				The first onset of P must be about 5 s earlier.
24.	e(pP)	A	22 50 15	Off East Coast of Kamchatka
				52.26°N 160.24°E
				H = 22 38 28.8 h = 27 km MAG=4.2 (USCGS)
				D = 74.2°
25.	e(Sg)	A	12 33 07.5	
25.	ePg	A	12 55 04.5	Explosion
	i	A	55 15	D = ca. 1.2°
	eSg	A	55 20	
25.	+eP	A	19 03 25	Hokkaido/Japan Region 45.99°N 143.02°E
				H = 18 52 17.1 h = 325 km MAG=4.8
				D = 75.0° Az = 330.4° (USCGS)
				PV:1.3s 22.2nm
26.	e(Sg)	A	01 31 14	
	e	A	31 40	
26.	e(P)	A	04 57 55	
26.	eiSg	A	08 27 21.5	Near earthquake.
26.	eP	A	15 12 19	Nevada
	e	A	12 48	H = 15 00 00 (UPP)
				D = ca. 81°
				PV:1.2s 15.3nm
				MPV=5.0
				Underground explosion.

Day	Phase		h m s	Remarks
26.	eP	A	15 33 23	<u>Jan Mayen</u> H = 15 28.3 (UPP) D = ca. 23^0 PV:1.8s 31.2nm MPV=4.5
26.	eP	A	15 40 06	<u>Jan Mayen</u> H = 15 35.0 (UPP) D = ca. 23^0
26.	eP e	A A	15 42 37 42 47	<u>Jan Mayen</u> 71.17^{0}N 6.70^{0}W H = 15 37 43.3 h = normal MAG=4.3 D = 22.2^0 Az = 148.1^0 (USCGS) PV:2.0s 33.1nm MPV=4.4
26.	eP	A	16 16 40.5	PV:1.3s 11.1nm
26.	e(Sg) e e	A A A	16 45 55 46 07 46 12	
26.	e e	A A	16 57 09 57 17.5	
26.	eP	A	17 35 35	<u>Rumania</u> 45.52^{0}N 26.31^{0}E H = 17 32 58.8 h = 133 km MAG=4.2 (USCGS) D = 11.0^0 AN USSR gives: h = 160 km; BCIS gives: h = 145 km
27.	eP e eS eSS LmH LmV	A A C C B B	01 51 10 51 22.5 57.9 02 01.1 11.8 11.8	<u>Southern Sinkiang Prov./China</u> 39.91^{0}N 77.33^{0}E H = 01 42 47.1 h = normal MAG=5.4 D = 46.0^0 Az = 306.3^0 (USCGS) LmH:17s 2.0 µm LmV:17s 2.4 µm MLH=5.1 MLV=5.3 e 51 30 e 53 27

Day	Phase		h m s	Remarks
27.	+eP	A	01 58 27	Algeria 35.76°N 0.30°W
	e	A	58 36	H = 01 54 26.0 h = 28 km MAG=4.7
	e	A	58 39	D = 17.2° Az = 26.4° (USCGS)
	e(S)	C	02 01 36	PV:1.7s 87.6nm PH:1.8s 112.5nm
	e	C	03 00	LmH:13s 0.5 µm LmV:12.5s 0.75 µm
	LmH	C	05.4	MLH=4.0 MLV=4.3
	LmV	C	05.4	
27.	i(Sg)		03 18 44.0	
27.	i(P)		09 01 31.6	
27.	e(P)		11 06 26.5	(P)V:1.1s 9.6nm
27.	eP	A	12 50 58	Afghanistan-USSR Border Reg.
	ePP	A	52 39	36.19°N 71.49°E
	esPP	A	53 13	H = 12 42 54.1 h = 109 km MAG=4.9
				D = 44.6° Az = 308.4° (USCGS)
				PV:1.3s 11.1nm
				MPV=4.6
27.	+iP	B	17 34 49.5	Rat Islands/Aleutian Is. 51.89°N 176.08°E
	+iP1	A	34 49.5	H = 17 22 58.7 h = 34 km MAG=5.8
	-eiP2	A	34 51.0	D = 77.0° Az = 349.9° (USCGS)
	e(pP)	A	34 58	PV:6.5s 1.8 µm PV2:1.6s 220nm
	ePP	C	37 44	LmH:19s 6.3 µm LmV:21s 6.4 µm
	ePPP	C	39 36	MPV=6.3 MPV2=6.0 MLH=6.0 MLV=6.0
	iS	C	44 36	The short-period records exhibit two P
	eScS	C	45 10	onsets, the first much smaller than the
	eSP	C	45(20)	second.
	iPPS	B	45 32	
	eSS	C	49.5	
	LmH	B	18 11.9	
	LmV	B	11.9	
27.	-eiP1	A	19 14 30.9	Kashmir-Sinkiang Border Reg.
	+iP2	AB	14 33.2	36.12°N 77.79°E
	-iPP	B	16 27	H = 19 05 48.5 h = 35 km MAG=5.4
	+iPP	A	16 27.4	D = 48.6° Az = 309.1° (USCGS)

Day	Phase		h m s	Remarks
cont.				
27.	eS	B	19 21 30	PV1:1.2s 30.8nm PV2(B):4.8s 0.8 /um
	eSS	C	25(00)	PPV(B):5s 1.0 /um SaH(B):14.5s 5.1 /um
	eiSa	BC	25 16	LmV:13s 6.4 /um
	LmV	B	33.3	MPV1=5.3 MPV2(B)=6.1 MPPV=6.1 MLV=5.9
				Sa is the greatest phase in the long-period registrations, followed by clear developed higher mode surface waves.
28.	eP	A	01 43 44.5	Rat Islands 52.10°N 175.05°E
	e(PS)	C	54 20	H = 01 31 56.7 h = 45 km MAG=5.2
	eSS	C	58 35	D = 76.7° Az = 349.2° (USCGS)
	LmH	B	02 15.5	LmH:23s 0.5 /um LmV:25s 0.5 /um
	LmV	B	15.5	MLH=4.8 MLV=4.7
28.	+iP	A	04 15 46.0	Kazakh SSR 50.0°N 78.0°E
	ePn	A	17 18	H = 04 08 00 (BCIS)
	LmH	B	33.7	D = 41.1°
	LmV	B	33.7	PV:0.8s 66.0nm PH:1.0s 35.8nm
				LmH:8.5s 0.3 /um LmV:9s 0.4 /um
				MPV=5.6 MPH=5.6 MLH=4.5 MLV=4.7
				Probably underground explosion.
28.	ePKP	A	06 48 49	South of Fiji Islands 23.82°S 179.71°W
	e	A	48 53.5	H = 06 29 45.6 h = 441 km MAG=4.4
				D = 151.8° Az = 344.6° (USCGS)
28.	LmH	B	12 28.9	Tadzhik SSR 37.69°N 73.44°E
	LmV	B	31.7	H = 12 03 01.7 h = normal MAG=4.9 (USCGS)
				D = 45.0°
				LmH:18s 0.5 /um LmV:11s 0.2 /um
				MLH=4.5 MLV=4.4
28.	epP	A	12 28 47.5	Uganda 1.89°N 31.35°E
				H = 12 19 31.9 h = 40 km MAG=5.1 (USCGS)
				D = 51.4°
28.	ei(Pg)	A	15 34 13.8	Rock burst/Ölsnitz, GDR
	e(Sg)	A	34 21	

Day	Phase		h m s	Remarks
29.	ePP	A	05 04 10	South of Mariana Is. 19.93°N 143.31°E
	LmH	C	54	H = 04 45 43.9 h = normal MAG=5.6 (USCGS)
	LmV	C	54	D = 104.9°
				LmV:20s 0.3 μm
				MLV=4.9
29.	e(P)	A	07 10 15.5	
29.	ePKP	A	11 29 11	Fiji Islands 19.17°S 176.26°W
	e	A	29 13	H = 11 09 53.9 h = 236 km MAG=5.1
	e	A	29 50	D = 148.0° Az = 350.5° (USCGS)
	e	A	30 05	PV1:1.2s 18.6nm PV2:1.6s 45.5nm
29.	ePKP	A	19 39 43	Tonga Islands 20.17°S 173.86°W
				H = 19 19 54.5 h = normal MAG=4.2
				D = 149.3° Az = 353.2° (USCGS)
29.	+iP	A	21 13 36.5	Hokkaido/Japan Region 43.31°N 145.68°E
	ei	A	13 42.5	H = 21 01 44.3 h = 88 km MAG=5.3
	ipP	A	14 00.0	D = 78.2° Az = 332.1° (USCGS); h = 93 km
	ei	A	14 09	PV:1.5s 60.0nm pPV:1.6s 60.5nm
	eS	B	23 22	SH:9s 0.4 μm
	eSKS	B	23 40	LmH:13s 0.3 μm LmV:17s 0.3 μm
	LmH	B	52.2	MPV=5.2 MSH=5.6 MLH=4.9 MLV=4.7
	LmV	B	52.5	e 13 46.5 e 14 24
29.	ePKP	A	22 14 49	Samoa Islands Region 15.83°S 172.50°W
				H = 21 55 14.0 h = normal MAG=4.6
				D = 145.1° Az = 355.4° (USCGS)
30.	ePKP	A	07 25 43	Tonga Islands 19.40°S 175.76°W
				H = 07 06 20.0 h = 185 km MAG=4.2 (USCGS)
				D = 148.3°
30.	eP	A	10 06 43	Andreanof Islands/Aleutian Is.
				50.14°N 176.60°W
				H = 09 54 38.3 h = 30 km MAG=5.0
				D = 79.3° Az = 354.7° (USCGS)
				PV:1.3s 16.7nm
				MPV=5.0

Day	Phase		h m s	Remarks
30.	LmH	B	15 17.7	Gulf of California 24.16°N 108.68°W
	LmV	B	17.8	H = 14 27 37.4 h = normal MAG=4.8 (USCGS)
				D = 88.7°
				LmH:19s 0.4 μm LmV:20s 0.7 μm
				MLH=4.9 MLV=5.1
30.	eP	A	23 58 10.5	Eastern Mediterranean Sea 34.20°N 28.82°E
	ei	A	58 17	H = 23 53 30.9 h = normal MAG=4.5
	LmH	B	00 10	D = 20.7° Az = 327.8° (USCGS)
	LmV	B	10	LmH:20s 0.2 μm LmV:20s 0.3 μm
				MLH=3.6 MLV=3.8
31.	eP	A	11 49 39	Windward Islands 12.51°N 60.34°W
	e	A	50 33	H = 11 38 39.0 h = 60 km MAG=5.1 (USCGS)
	LmH	C	12 15	D = 69.0°
31.	LmH	C	17 19.0	Probably Philippine Islands (USCGS)
31.	+iPKP	A	18 07 49.5	Fiji Islands Region 15.77°S 179.67°W
				H = 17 49 07.5 h = 465 km MAG=4.1
				D = 144.0° Az = 347.8° (USCGS)
				PV:1.3s 19.4nm

Day	Phase		h m s	Remarks
1.	eP1	A	03 48 02	<u>Fox Islands/Aleutian Is.</u> 53.67°N 165.65°W
	+iP2	A	48 02.7	H = 03 36 19.0 h = 60 km MAG=5.7
	epP	A	48 18.5	D = 76.0° Az = 1.8° (USCGS); h = 60 km
	eiS	B	57 37	PV2:1.2s 56.0nm pPV:1.2s 20.4nm
	LQ	C	04 08.3	LmH:20s 0.7 μm LmV:22s 0.7 μm
	LR	C	12.5	MPV2=5.6 MLH=5.0 MLV=5.0
	ePKPPKP	A	15 14	Multiple P-onset in the short-period re-
	LmV	B	23.5	cords, the first much smaller than the
	LmH	B	25.0	second one.
1.	eP	A	10 27 35.5	<u>Near East Coast of Kamchatka</u>
				53.87°N 160.61°E
				H = 10 16 09.4 h = 28 km MAG=4.9
				D = 72.6° Az = 339.9° (USCGS)
				PV:1.5s 37.0nm
				MPV=5.3
1.	-eP1	A	10 43 40.5	<u>Turkey</u> 36.85°N 29.24°E
	+iP2	A	43 41	H = 10 39 22.8 h = 36 km MAG=5.0
	-eiP	B	43 42	D = 18.7° Az = 323.0° (USCGS)
	-iP3	A	43 44.0	PV2:1.4s 122.8nm PV3:1.4s 209nm
	eS	C	47 12	PV:5s 1.5 μm
	ei	B	47 19	LmH:13s 1.9 μm LmV:9s 2.4 μm
	eLi	B	48 45	MLH=4.6 MLV=5.0
	eLg1	B	49 06	e 44 13.5
	LmH	B	51.4	Multiple P-onsets with increasing ampli-
	LmV	B	52.1	tude in the short-period records. Clear
				higher-mode surface waves.
1.	iPg	A	11 04 09.3	Explosion
	i	A	04 28	D = 1.5°
	iSg	A	04 29.5	
1.	e(P)	A	11 06 19	
1.	eP	A	11 15 48	<u>Kurile Islands</u> 44.46°N 149.00°E
	e	A	16 07.5	H = 11 03 52.4 h = 58 km MAG=5.1
				D = 78.3° Az = 333.9° (USCGS)
				PV1:1.3s 11nm PV2:1.4s 22.7nm
				MPV=4.8

Day	Phase		h m s	Remarks
1.	e(P)	A	12 04 12	(P)V:1.8s 15.3nm
1.	e(P)	A	12 24 53.5	(P)V:0.8s 4.7nm
1.	e(Sg)	A	13 03 21	(Sg)V:1.6s 15.8nm
1.	e(Sg)	A	15 27 48.5	(Sg)V:0.9s 14.2nm
	e	A	27 49.5	
1.	e(Sg)	A	15 33 42.5	(Sg)V:0.9s 9.4nm
1.	e(Sg)	A	16 03 49	(Sg)V:1.0s 9.5nm
1.	e(Sg)	A	16 15 18	(Sg)V:0.8s 9.4nm
1.	-ePKIKP	A	21 06 47.5	<u>Solomon Islands</u> 6.82°S 154.97°E
	ePS	C	18 46	H = 20 47 45.6 h = 31 km MAG=5.6 (USCGS)
	ePPS	C	20 22	D = 126.8°
	eSS	C	25.9	PV:1.4s 21.5nm
	LmH	B	22(09)	
2.	eP	A	04 36 40.5	<u>Southern Sinkiang Prov./China</u>
	e	A	42 52	41.05°N 88.09°E
				H = 04 27 33.2 h = normal MAG=4.9
				D = 51.9° Az = 308.1° (USCGS)
				PV:1.2s 15.3nm
				MPV=5.0
2.	eP	A	05 16 20	<u>Eastern Caucasus</u> 43.55°N 47.47°E
	e	A	17 20	H = 05 10 54.9 h = 28 km MAG=5.0
	e	B	24 30	D = 25.2° Az = 299.0° (USCGS)
	eLg2	B	25 00	PV:1.2s 23.0nm
	LmH	B	28.4	LmH:14s 0.3 µm LmV:12s 0.4 µm
	LmV	B	29.1	MPV=4.7 MLH=4.0 MLV=4.2
				Higher-mode surface waves are registered.
2.	-eP	A	06 41 34	<u>Central Mid-Atlantic Ridge</u> 0.88°N 28.40°W
	e	A	41 36	H = 06 31 28.2 h = normal MAG=5.0
	epP	A	41 42	D = 60.0° Az = 28.2° (USCGS); h = 30 km

Day	Phase		h m s	Remarks
cont.				
2.	e	A	06 41 46	PV1:1.2s 20.4nm PV2:1.2s 17.9nm
	eS	B	49 50	pPV:1.3s 27.8nm SH:12s 0.6 µm
				MPV=5.1 MSH=5.5
2.	e(Sg)	A	10 34 15	Probably explosion.
	i	A	34 19	(Sg)V:0.7s 21.3nm
	e	A	34 34	
2.	eP	A	12 15 13	Kurile Islands 45.22°N 150.12°E
				H = 12 03 17.5 h = normal MAG=4.7 (USCGS)
				D = 77.9°
2.	e(Sg)	A	12 45 24.5	
2.	iSg	A	13 04 01.7	Explosion
2.	iPg	A	14 07 16.5	Explosion
	iSg	A	07 32.5	D = ca. 1.2°
2.	e(Sg)	A	16 03 14	(Sg)V:2.0s 22.2nm
2.	eP	A	20 23 09	Sicily 38.69°N 14.81°E
	ei	A	23 10.5	H = 20 20 21.9 h = 259 km MAG=4.0
	e	A	23 16.5	D = 12.2° Az = 350.3° (USCGS)
				PV2:1.0s 16.6nm PV3:1.0s 18.9nm
3.	eSKS	B	06 35 20	Near Coast of Peru 10.78°S 78.96°W
	ePS	C	37 44	H = 06 11 07.8 h = 33 km MAG=4.6 (USCGS)
	eSS	C	43.0	D = 98.7°
	LmH	B	07 15	LmH:18s 0.5 µm LmV:18s 0.3 µm
	LmV	B	15	MLH=5.1 MLV=4.9
3.	+iP	A	09 20 09.2	Kodiak Island Region 58.39°N 151.23°W
	+i(pP)	A	20 18.3	H = 09 08 56.4 h = 32 km MAG=5.5
	iPcP	A	20 30.5	D = 70.4° Az = 11.5° (USCGS); h = 35 km
	i	A	20 35.5	PV:1.1s 43.3nm (pP)V:1.4s 178nm
	eS	C	29 18	LmH:23s 0.6 µm LmV:23s 0.9 µm
	eiS	B	29 22	MPV=5.5 MLH=4.8 MLV=5.0

Day	Phase		h m s	Remarks
cont.				
3.	eSS	C	09 34.0	(pP) is much stronger than P. Another
	ePKPPKP	A	48 04	interpretation would be that (pP) is the
	LmH	B	50.0	P-onset of a second stronger shock in the
	LmV	B	50.0	same focus.
3.	eP	A	13 21 08	Peru-Brazil Border Region 8.49°S 74.44°W
				H = 13 08 06.8 h = 152 km MAG=5.2
				D = 94.0° Az = 39.5° (USCGS)
3.	LmH	B	16 35.7	Aegean Sea 39.0°N 24¾°E
	LmV	B	36.8	H = 16 27 02 (BCIS)
				D = 14.9°
				LmH:11s 0.6 μm LmV:10s 0.5 μm
				MLH=4.0 MLV=4.1
3.	e(P)	A	22 33 15	
4.	+iP	A	05 38 26.4	Off East Coast of Kamchatka
	i	A	38 28.0	51.39°N 159.28°E
	e	A	38 39.5	H = 05 26 44.6 h = 9 km MAG=4.8
	LmH	B	06 16.4	D = 74.7° Az = 339.3° (USCGS)
	LmV	B	16.4	PV:1.2s 28.0nm
				LmH:15.5s 0.9 μm LmV:17s 1.0 μm
				MPV=5.3 MLH=5.2 MLV=5.2
4.	eP	A	06 35 19	Off East Coast of Kamchatka
	e	A	35 37.5	51.46°N 159.28°E
				H = 06 23 38.4 h = 12 km MAG=4.6
				D = 74.6° Az = 339.3° (USCGS)
				PV:1.0s 11.8nm
				MPV=5.0
4.	eP	A	06 46 04	Off East Coast of Kamchatka
				51.47°N 159.24°E
				H = 06 34 26.3 h = normal MAG=4.5
				D = 74.6° Az = 339.3° (USCGS)

Day	Phase		h m s	Remarks
4.	LmH	B	19 28.6	Near Coast of Peru 15.51°S 75.72°W
				H = 18 28 39.6 h = 38 km MAG=4.7 (USCGS)
				D = 100.2°
				LmH:16s 0.2 /um
				MLH=4.8
5.	ePKIKP	A	01 41 04.5	Tonga Islands 21.26°S 174.53°W
	eiPKHKP	A	41 10.5	H = 01 21 20.2 h = normal MAG=5.2 (USCGS)
	eiPKP2	A	41 20.5	D = 150.2°
	ei	A	41 30	PV1:1.5s 16.8nm PV2:1.4s 85.9nm
	LmH	B	02 50.3	PV3:1.4s 98.1nm
				LmH:20s 1.1 /um
				MLH=5.6
5.	-i	A	14 31 29.5	Explosion?
	+i	A	31 46	
	e	A	31 50	
5.	eP	A	16 50 13	Off East Coast of Kamchatka
	epP	A	50 24.5	51.46°N 159.15°E
	e	A	50 52.5	H = 16 38 36.2 h = normal MAG=4.5
	LmH	B	17 28.2	D = 74.6° Az = 339.2° (USCGS); h = 43 km
	LmV	B	28.2	PV:1.1s 14.4nm
				LmH:16s 0.4 /um LmV:16s 0.5 /um
				MPV=5.0 MLH=4.8 MLV=4.9
5.	i(Sg)	A	16 55 15.0	Explosion?
	e	A	55 20.5	
5.	e(P)	A	18 09 15.5	
	e	A	09 26	
	e	A	09 38	
6.	ePKIKP	A	06 54 11	New Britain Region 6.23°S 152.03°E
	e	A	54 21.5	H = 06 35 16.1 h = 58 km MAG=5.4
	e	A	54 29.5	D = 124.8° Az = 330.4° (USCGS)
				PV:1.3s 20.9nm

Day	Phase		h m s	Remarks
6.	i	A	09 02 59.5	Probably explosion.
	e	A	03 02.5	
	e	A	03 06	
7.	e(P)	A	02 22 08	
	e	A	22 25	
7.	eP	A	03 02 30	Iceland 63.83°N 19.01°E
	LmH	B	11.5	H = 02 57 48.7 h = normal MAG=4.5
	LmV	B	11.5	D = 20.9° Az = 114.7° (USCGS)
				PV:1.4s 28.6nm
				LmH:15s 0.4 µm LmV:15s 0.3 µm
				MPV=4.4 MLH=3.9 MLV=4.0
7.	e(PKP2)	A	10 17 53.5	Tonga Islands 20.69°S 174.88°W
	e	A	18 05.5	H = 09 57 59.1 h = normal MAG=4.3
				D = 149.7° Az = 351.8° (USCGS)
7.	eP	A	15 58 57	Crete 34.21°N 26.43°E
	e	A	59 07	H = 15 54 26.4 h = 9 km MAG=4.7
	e	A	59 38	D = 19.7° Az = 334.6° (USCGS)
	LmH	B	16 08	
	LmV	B	08	
7.	ePn	A	16 20 16	Austria 47.95°N 14.33°E
	e(Pg)	A	20 27	H = 16 19 27.2 h = normal
	iSn	A	20 55.5	D = 3.2° Az = 327.6° (USCGS)
	i(Lg1)	A	21 08	PnV:0.7s 9.5nm
	i(Sg)	A	21 11.5	LmH:3s 0.8 µm LmV:4s 0.5 µm
	LmH	B	21 26	MLH=3.5
	LmV	B	21 32	
7.	eP	A	17 10 20.5	USSR-Mongolia Border Region
	e(pP)	A	10 24	49.42°N 97.18°E
	e	A	10 41	H = 17 01 12.9 h = normal MAG=5.0
	LmH	B	30.1	D = 52.0° Az = 306.3° (USCGS); h = 15 km
	LmV	B	34.6	PV:1.5s 26.8nm (pP)V:1.4s 21.5nm
				LmH:18s 0.5 µm LmV:12s 0.8 µm
				MPV=5.1 MLH=4.6 MLV=5.0

Day	Phase		h m s	Remarks
7.	eiP	A	18 28 27.5	Kurile Islands Region 47.45°N 155.44°E
	ei(pP)	A	28 40	H = 18 16 31.4 h = 29 km MAG=5.2
	ei	A	28 56.5	D = 77.4° Az = 337.4° (USCGS);
				(h = 47 km)
8.	e	A	07 17 47	Probably to Mindanao/Philippine Islands (USCGS)
8.	eP	A	12 15 46	Talaud Islands 4.57°N 127.07°E
	e	A	19 23	H = 12 01 57.3 h = 73 km MAG=5.3
	ePP	A	19 59	D = 102.2° Az = 324.0° (USCGS)
	LmV	B	13 05.4	LmH:20s 0.3 /um LmV:20s 0.4 /um
	LmH	B	06.8	MLH=4.8 MLV=5.0
8.	ePKIKP	A	13 41 43.5	Loyalty Islands Region 21.41°S 170.27°E
	-iPKHKP	A	41 45.0	H = 13 22 13.7 h = 90 km MAG=5.3 (USCGS)
	-iPKHKP	B	41 45.3	D = 146.4° h = ca. 90 km
	ipPKP	B	42 09	PV2(A):1.0s 104nm
	eSPP	C	57.6	LmH:30s 1.2 /um LmV:26s 1.4 /um
	eSS	C	14 04.0	MLH=5.4 MLV=5.6
	eSSP	C	04.7	e 41 57 e 42 04
	LmH	B	40	
	LmV	B	41	
8.	e	A	14 01 12	Explosion?
	i(Sg)	A	01 35	
	i	A	01 40	
8.	e(P)	A	20 14 37	Probably Crete (UPP)
	LmH	C	23.2	
	LmV	C	23.3	
8.	eSg	A	21 15 50.5	Yugoslavia 43.5°N 20.8°E
				H = 21 10 57 (BCIS)
				D = 9.5°
8.	e	A	21 29 58	Yugoslavia 46.0°N 14.8°E
	e(Sg)	A	30 00	H = 21 27 20 (BCIS)
	e	A	30 08	D = 5.1°

Day	Phase		h m s	Remarks
9.	ePKP	A	11 13 44	Fiji Islands Region 17.83°S 178.62°W
				H = 10 55 08.2 h = 576 km MAG=3.9
				D = 146.3° Az = 348.2° (USCGS)
9.	iPg	A	12 00 12.5	Explosion/ČSSR 50°25'N 13°50'E
	iSg	A	00 33.0	D = 1.4° Yield: 11.2 t
9.	iPKIKP	A	13 01 08.5	West of Macquarie Islands 52.36°S 143.48°E
	e	A	01 33.5	H = 12 41 22.0 h = normal MAG=5.1
	LmH	B	14 11.5	D = 150.4° Az = 286.2° (USCGS)
	LmV	B	11.5	LmH:22s 0.3 μm LmV:22s 0.5 μm
				MLH=4.9 MLV=5.3
9.	e	A	13 31 24.5	
	ei	A	31 28	
9.	+iPKHKP	A	17 24 46	Fiji Islands 20.62°S 178.64°W
	ePKP2	A	24 52	H = 17 05 58.9 h = 546 km MAG=4.5
				D = 149.0° Az = 347.3° (USCGS)
				PV1:1.4s 18.4nm PV2:1.0s 9.5nm
10.	eP	A	04 01 38.5	Fox Islands/Aleutian Is. 52.67°N 169.09°W
	e	A	01 55	H = 03 49 47.2 h = 32 km MAG=4.4 (USCGS)
	e	A	02 30	D = 77.1°
10.	ePg	A	04 44 52	Emilie/Italy 44.6°N 10.5°E
	e(Sn)	A	45 24.5	H = 04 42 49 (BCIS)
	eSg	A	46 07	D = 6.1°
	e	A	46 26	
	e	A	46 33	
10.	-eiPKIKP	A	05 45 28	Near Coast of Southern Chile
	e	A	45 53	41.31°S 73.64°W
	e	A	46 05	H = 05 26 44.4 h = 37 km MAG=5.7
	LmH	B	06 34.7	D = 117.9° Az = 45.9° (USCGS)
	LmV	B	34.7	PV:0.8s 11.8nm
				LmH:20s 0.8 μm LmV:19s 1.2 μm
				MLH=5.3 MLV=5.6

Day	Phase		h m s	Remarks
10.	eP	A	05 55 45	North of Ascension Island 3.61°S 12.14°W
	ei(pP)	A	55 51	H = 05 45 52.8 h = 12 km MAG=5.1
	eS	C	06 03 46	D = 57.8° Az = 17.6° (USCGS);
	eSS	C	07.6	(h = 24 km)
	LmH	B	14.8	PV:2.0s 72.6nm
	LmV	B	21.1	LmH:16s 0.9 μm LmV:15s 1.0 μm
				MPV=5.5 MLH=5.0 MLV=5.1
10.	eP	A	06 41 50	USSR-Mongolia Border Region
	e	A	41 54	49.49°N 97.29°E
	LmH	B	07 02.3	H = 06 32 43.0 h = normal MAG=4.5
	LmV	B	06.1	D = 52.0° Az = 306.3° (USCGS)
				PV:1.5s 10nm
				LmV:12s 0.4 μm
				MPV=4.7 MLV=4.8
10.	LmV	B	06 56.5	LmV:15s 0.4 μm
10.	ePKIKP	A	14 17 28.5	Fiji Islands Region 19.35°S 178.23°W
	eiPKHKP	A	17 32.5	H = 13 58 53.3 h = 596 km MAG=5.1
	iPKP2	A	17 37.5	D = 147.8° Az = 348.2° (USCGS);
	epPKP	A	19 52.5	h = ca. 600 km
				PV2:0.8s 28.4nm PV3:1.2s 40.8nm
10.	i(PKHKP)	A	14 26 01.0	Aftershock to Fiji Islands earthquake?
	e(PKP2)	A	26 05	
	e(pPKP)	A	28 20	
10.	eP	A	18 14 29	North Atlantic Ocean 16.40°N 46.64°W
	e(pP)	A	14 35	H = 18 04 39.6 h = normal MAG=4.9
	eS	C	22.5	D = 57.5° Az = 40.0° (USCGS);
	e(SS)	C	26.1	(h = 24 km)
	eLQ	C	28.5	LmH:18s 0.3 μm LmV:18s 0.5 μm
	eLR	C	31 16	MLH=4.5 MLV=4.7
	LmH	B	38.4	
	LmV	B	38.4	

Day	Phase		h m s	Remarks
11.	eP	A	05 38 38	Greece 38.07°N 22.87°E
	e	A	38 52	H = 05 35 03.7 h = 37 km MAG=4.4
	LmH	C	45.2	D = 14.9° Az = 331.1° (USCGS)
	LmV	C	45.2	
11.	e(P)	A	09 52 19	
11.	e	A	10 41 03.5	Vienna gives: Malé Karpaty/ČSSR?
	eiSg	A	41 08	
11.	e	A	11 39 40	
11.	e(P)	A	11 54 32	
	e	A	54 57	
11.	+eP	A	12 02 08.5	Kurile Islands 47.53°N 154.38°E
	e	A	02 14.5	H = 11 50 17.4 h = 36 km MAG=4.9
	LmH	B	41.5	D = 77.1° Az = 336.7° (USCGS)
	LmV	B	41.5	PV:1.1s 19.2nm
				LmH:15s 0.2 μm LmV:15s 0.3 μm
				MPV=5.1 MLH=4.6 MLV=4.8
12.	eP	A	00 14 52	North Atlantic Ridge 16.59°N 46.62°W
	e	A	14 59.5	H = 00 05 06.5 h = normal MAG=5.1
	e	A	15 04.5	D = 57.3° Az = 40.0° (USCGS)
	e	A	15 09	PV:2.5s 61.8nm
	LmH	B	38.5	LmH:18s 0.2 μm LmV:16s 0.2 μm
	LmV	B	38.5	MPV=5.3 MLH=4.3 MLV=4.4
12.	e(PKIKP)	A	00 23 20.5	Tonga Islands 20.97°S 174.55°W
	e(PKHKP)	A	23 26.5	H = 00 03 32.4 h = normal MAG=5.0
	e(PKP2)	A	23 34.5	D = 150.0° Az = 352.1° (USCGS)
	e	A	23 52	PV1:1.5s 20.2nm PV2:1.8s 30.6nm
				PV3:1.4s 30.6nm PV4:1.4s 18.4nm
12.	e	A	01 08 45.5	
12.	ePKP	A	01 08 50	Tonga Islands 21.06°S 174.38°W
	i	A	08 51.5	H = 00 48 59.2 h = 13 km MAG=5.1

Day	Phase		h m s	Remarks
cont.				
12.	ei	A	01 09 02.5	D = 150.1° Az = 352.3° (USCGS)
	ei	A	09 10.5	PV2:1.4s 24.5nm PV3:1.3s 38.9nm
				PV4:1.4s 43.0nm
12.	e(P)	A	01 32(44)	Greece 38.16°N 22.82°E
	e	A	32 47.5	H = 01 29 06.9 h = 27 km MAG=4.5
	e	A	32 54	D = 14.8° Az = 331.1° (USCGS)
	e	A	33 01	LmH:13s 0.5 /um LmV:14s 0.7 /um
	e	A	34 52	MLH=3.9 MLV=4.1
	LmH	B	39.5	The first onset of P must be about 8 s
	LmV	B	39.5	earlier than e(P).
12.	e(P)	A	02 09 14	North Atlantic Ridge 16.72°N 46.65°W
	e	A	09 20	H = 01 59 28.1 h = normal MAG=4.8 (USCGS)
	e	A	09 26	D = 57.2°
	LmH	B	26.0	LmH:16s 0.2 /um LmV:16s 0.2 /um
	LmV	B	26.0	MLH=4.2 MLV=4.4
				The first onset is very doubtful. The
				later onsets are bigger than the first
				one.
12.	eP	A	02 54 35	Greece 38.24°N 22.73°E
	e	A	54 39.5	H = 02 51 05.5 h = normal MAG=4.8
	eSS	B	57 34	D = 14.7° Az = 331.1° (USCGS)
	eLg1	B	58 58	PV1:1.5s 10.1nm PV2:1.4s 17.2nm
	eLg2	B	59 28	LmH:12s 3.2 /um LmV:14s 3.7 /um
	LmH	B	03 01.6	MLH=4.7 MLV=4.8
	LmV	B	01.6	e 54 44.5 e 58 10
12.	eP	A	03 20 45	Southern Sumatra 3.04°S 100.63°E
	epP	A	20 54	H = 03 07 38.8 h = normal MAG=5.1 (USCGS)
	e	A	21 08	D = 91.7° h = 32 km
	eS	C	31 44	LmH:20s 0.4 /um LmV:20s 0.5 /um
	LmH	B	04 08	MLH=4.9 MLV=5.0
	LmV	B	10	
12.	eP	A	05 34(45)	Prince Edward Islands 44.94°S 35.70°E
	ePP	B	38 36	H = 05 21 10.6 h = 36 km MAG=5.6

Day	Phase		h m s	Remarks
cont.				
12.	e(Pa)	B	05 42 16	D = 97.5° Az = 344.8° (USCGS)
	eSKS	C	45 20	PSH:18s 1.8 μm
	eS	C	46 08	LmH:21s 3.6 μm LmV:20s 3.6 μm
	iPS	B	47 35	MLH=5.8 MLV=5.9
	eSS	C	52 40	e 40 10 e 53 18
	LmV	C	06 21.2	
	LmH	C	21.5	
12.	e(P)	A	05 45 24.5	
12.	LmH	B	11 10.8	<u>Southern Greece</u> 37.25°N 23.09°E
	LmV	B	10.8	H = 11 00 15.2 h = normal MAG=4.5 (USCGS) D = 15.8° LmH:12s 0.4 μm LmV:12s 0.4 μm MLH=3.8 MLV=4.0
12.	e	A	11 29 30	Near earthquake?
12.	iPg	A	13 07 13.2	Explosion
	iSg	A	07 29.5	D = ca. 1.2°
12.	eP	A	15 15 38.5	<u>South of Skikotan Is.</u> 43.4°N 146.8°E H = 15 03 39 h = normal (AN USSR) D = 78.5°
12.	eP	A	18 15 55.5	<u>Greece</u> 39.21°N 21.43°E
	e	A	15 58	H = 18 12 47.4 h = 55 km MAG=4.5
	e	A	16 06.5	D = 13.4° Az = 332.0° (USCGS)
	LmH	B	21.8	PV1:1.0s 9.5nm PV2:1.0s 16.6nm
	LmV	B	21.8	PV3:0.7s 14.2nm LmH:10s 0.6 μm LmV:12s 0.7 μm MLH=3.9 AN USSR gives: Ionian Islands 38.2°N 20.9°E H = 18 12 32 h = normal D = 14.1°

Day	Phase		h m s	Remarks
12.	eP	A	21 30 55	Southern Sumatra 3.08°S 100.57°E
	epP	A	31 04.5	H = 21 17 48.9 h = normal MAG=5.4
	e	A	31 18	D = 91.7° Az = 320.4° (USCGS); h = 34 km
	eSKKS	C	41 36	LmH:20s 0.5 µm LmV:20s 0.6 µm
	eS	C	41 53	MLH=5.0 MLV=5.1
	ePS	C	43 06	
	LmH	B	22 18.3	
	LmV	B	20	
12.	+iP	A	23 34 34.5	Kurile Islands 47.42°N 154.32°E
	+iP	B	34 35.0	H = 23 22 45.3 h = 56 km MAG=5.4
	i	A	34 37.2	D = 77.2° Az = 336.7° (USCGS)
	eS	C	44 16	PV(A):1.0s 94.8nm PV(B):6s 1.4 µm
	ePS	C	44 56	PH(B):5s 0.5 µm SH:8s 0.8 µm
	eSS	C	49.3	LmH:16s 6.1 µm LmV:16s 6.3 µm
	LQ	C	55	MPV(A)=5.9 MPV(B)=6.2 MPH(B)=6.3
	LmH	B	00 15	MSH=5.9 MLH=6.0 MLV=6.1
	LmV	B	15	
13.	eP	A	01 23 11	Southern Sumatra 3.04°S 100.56°E
	epP	A	23 18	H = 01 10 05.9 h = normal MAG=5.0
	e	A	23 37	D = 91.7° Az = 320.4° (USCGS); h = 25 km
				PV:1.2s 7.7nm
				MPV=4.9
13.	eP	A	02 54 36	Kurile Islands 47.60°N 154.27°E
				H = 02 42 45.1 h = 32 km MAG=4.6
				D = 77.0° Az = 336.6° (USCGS)
13.	ePKP	A	03 31 36	Loyalty Islands Region 21.23°S 169.64°E
				H = 03 11 59.0 h = normal MAG=4.5
				D = 145.9° Az = 334.8° (USCGS)
13.	e(Sg)	A	06 02 10	Near earthquake?
	e	A	02 17	
13.	eP	A	09 51 13.5	Svalbard 78.65°N 8.20°E
				H = 09 45 22.4 h = normal MAG=4.6
				D = 28.1° Az = 175.4° (USCGS)
				PV:1.1s 9.6nm MPV=4.5

Day	Phase		h m s	Remarks
13.	ePg	A	12 00 51	Explosion/ČSSR-GDR Border Region
	e	A	00 59.5	D = 2.0° Yield: 10.2 t
	eSg	A	01 16.5	e 01 22.5 e 01 45.5 e 01 55
13.	e	A	12 11 34	Near earthquake?
	e(Sg)	A	12 15	
	e	A	12 26	
13.	e	A	15 22 12	Explosion?
	e	A	22 15	
	ei	A	22 21.5	
13.	eP	A	15 38 35.5	Molucca Passage 0.28°N 125.48°E
	e	A	38 46.5	H = 15 24 29.1 h = normal MAG=4.5
				D = 104.7° Az = 323.0° (USCGS)
13.	ePKIKP	A	15 58 00	New Britain Region 5.58°S 148.12°E
	epPKIKP	A	58 53	H = 15 39 29.7 h = 213 km MAG=5.4 (USCGS)
	ePP	A	59 38.5	D = 122.3° h = 213 km
	e(S)	C	16 07 12	LmH:20s 0.65 μm
	eSS	C	16.0	e 59 30 e 59 51 e 01 12.5 e 08 52
	e(sSS)	C	17.5	
	i	C	17 40	
	LmH	B	42.7	
13.	e	A	16 06 37.0	
13.	ei(Sn)	A	17 41 19.3	Austria 47.5°N 14.3°E
	i	A	41 30	H = 17 39 46 (BCIS)
	iSg	A	41 32.5	D = 3.6°
	i(Lg2)	A	41 42.1	
13.	eP	A	23 15 10	Eastern Caucasus 42.01°N 45.31°E
	e	A	15 25	H = 23 09 51.8 h = normal MAG=4.6
	e	A	15 55	D = 24.6° Az = 302.1° (USCGS)
	eS	C	19 40	LmH:16s 0.2 μm LmV:15s 0.3 μm
	LmH	B	25.2	MLH=3.7 MLV=4.1
	LmV	B	26.0	

Day	Phase		h m s	Remarks
14.	eP	A	03 27 45	Peru 14.86°S 73.42°W
	epP	A	28 13	H = 03 14 17.5 h = 99 km MAG=5.6
				D = 98.2° Az = 39.9° (USCGS); h = 109 km
				PV:1.8s 25.5nm
				MPV=5.6
14.	-eiP	A	03 57 14.5	Near E. Coast of Eastern Russia
				45.31°N 136.92°E
				H = 03 46 20.3 h = 360 km MAG=4.7
				D = 73.3° Az = 327.2° (USCGS)
				PV:0.9s 14.1nm
				MPV=4.7
				AN USSR gives: Sikhota Alin, h = 268 km
14.	-ePKP	A	05 25 53.5	Tonga Islands 15.24°S 173.61°W
	+i	A	25 55.8	H = 05 06 16.3 h = 11 km MAG=5.9
	e	A	26 06	D = 144.4° Az = 354.3° (USCGS)
	e	A	26 08	PV1:1.2s 20.4nm PV2:1.6s 91.6nm
	LmH	B	06 22.5	PV3:1.8s 112nm PV4:1.7s 96.5nm
	LmV	B	22.5	LmH:24s 1.0 µm LmV:24s 0.8 µm
				MLH=5.4 MLV=5.4
14.	e(Sg)	A	06 10 42	
14.	e(Sg)	A	06 25 55	
14.	ei(P)	A	07 12 13	(P)V:1.2s 10.2nm
14.	+eP	A	08 17 47.5	Kurile Islands 47.54°N 154.43°E
	+iP	B	17 48	H = 08 05 58.6 h = 55 km MAG=5.3
	eipP	A	18 01.5	D = 77.1° Az = 336.7° (USCGS); h = 50 km
	eisP	A	18 05	PV(A):2.2s 176nm PV(B):4s 0.65 µm
	eS	B	27 32	pPV:1.5s 53.6nm sPV:1.9s 106nm
	eSS	C	32 32	LmH:16s 2.8 µm LmV:16s 4.1 µm
	LQ	C	39.5	MPV(A)=5.8 MPV(B)=6.1 MLH=5.7 MLV=5.9
	LmH	B	56.8	e 18 20 (V:2.0s 87.5nm) e 27 40
	LmV	B	57.4	

Day	Phase		h m s	Remarks
14.	+iP	A	08 24 51.5	Kurile Islands 47.48°N 154.49°E
	+iP	B	24 52	H = 08 13 02.2 h = 53 km MAG=5.4
	ei	A	24 55.5	D = 77.2° Az = 336.8° (USCGS); h = 50 km
	eipP	A	25 05	PV(A):1.5s 121nm PV(B):4s 0.6 µm
	eisP	A	25 09.5	LmH:18s 2.3 µm LmV:18s 2.3 µm
	LmH	B	09 03.9	MPV(A)=5.8 MPV(B)=6.0 MLH=5.5 MLV=5.6
	LmV	B	04.5	
14.	eiPg	A	09 00 23.5	Explosion/GDR 51°01.1'N 13°10.2'E
	iSg	A	00 38.5	D = 1.0° Yield: 14.2 t
14.	e(Sg)	A	09 46 18	
14.	ei(Sg)	A	10 53 09.5	
14.	eP	A	15 47 42	Kurile Islands 46.50°N 153.17°E
				H = 15 35 46.3 h = normal MAG=4.5
				D = 77.7° Az = 336.1° (USCGS)
				PV:1.2s 12.8nm
				MPV=4.9
				AN USSR gives: H = 15 35 55 h = 110 km
14.	e(Sg)	A	16 51 16	Probably explosion.
	e	A	51 27	
15.	eiPKHKP	A	05 09 33	Tonga Islands 21.63°S 174.72°W
	ePKP2	A	09 43	H = 04 49 41.4 h = normal MAG=4.8
	e	A	09 54	D = 150.6° Az = 351.8° (USCGS)
15.	eP	A	07 29 56.5	Central Mid-Atlantic Ridge 1.02°N 29.59°W
	LmH	C	53.7	H = 07 19 45.3 h = 24 km MAG=4.6 (USCGS)
	LmV	B	54	D = 60.5°
				LmH:19s 0.2 µm LmV:18s 0.45 µm
				MLH=4.3 MLV=4.7
15.	iPg	A	12 59 06.7	Explosion/GDR
	eSg	A	59 19.5	D = ca. 1.0°
	i	A	59 20.2	
	e	A	59 33	

Day	Phase		h m s	Remarks
15.	eP	A	15 01 01	<u>Cyprus</u> 34.13°N 32.54°E
	ipP	A	01 14.8	H = 14 56 05.6 h = 62 km MAG=5.0 (USCGS)
	eisP	A	01 22	D = 22.5° h = 60 km
	eS	C	05 05	PV:1.1s 48.0nm pPV:1.0s 42.6nm
				sPV:1.2s 30.6nm
				MPV=4.9
15.	e(Sg)	A	15 23 12	(Sg)V:1.2s 12.8nm
15.	e	A	15 47 21.5	Probably near earthquake.
	eSg	A	47 31	
15.	LmH	C	18 28.5	<u>South Atlantic Ridge</u> 43.88°S 16.06°W
	LmV	C	28.5	H = 17 33 56.5 h = normal MAG=5.0 (USCGS)
				D = 97.2°
				LmH:20s 0.2 µm LmV:20s 0.2 µm
				MLH=4.6 MLV=4.7
15.	LmH	C	18 38.5	<u>South Atlantic Ridge</u> 43.87°S 16.00°W
	LmV	C	38.5	H = 17 44 14.9 h = normal (USCGS)
				D = 97.2°
				LmH:20s 0.15 µm LmV:20s 0.2 µm
				MLH=4.5 MLV=4.7
15.	eP	A	18 52 00	<u>Central Mid-Atlantic Ridge</u> 9.09°N 40.39°W
	epP	A	52 07	H = 18 41 57.7 h = normal MAG=4.8
	e	A	52 12.5	D = 59.5° Az = 35.6° (USCGS); h = 27 km
	LmH	C	19 11.5	PV:1.5s 25.0nm pPV:1.5s 40.0nm
	LmV	C	11.5	LmH:25s 0.4 µm LmV:25s 0.4 µm
				MPV=5.1 MLH=4.5 MLV=4.5
				AN USSR gives: 11.0°N 39.9°W
				H = 18 42 09 h = normal
				D = 57.6°
15.	LmH	C	21 59.9	<u>New Ireland Region</u> 5.43°S 153.03°E
	LmV	C	59.9	H = 20 44 47.9 h = 65 km MAG=5.0 (USCGS)
				D = 124.7°
				LmH:20s 0.2 µm LmV:20s 0.2 µm
				MLH=4.8 MLV=4.8

Day	Phase		h m s	Remarks
16.	ePKP	A	06 04 03	West of Macquarie Island 55.69°S 146.82°E
	e	A	04 21	H = 05 44 02.2 h = 27 km MAG=5.4
				D = 153.1° Az = 278.2° (USCGS)
16.	ePKP	A	06 23 09	West of Macquarie Is. 55.55°S 147.34°E
	e	A	23 17	H = 06 03 16.6 h = normal MAG=5.2
	i	A	23 22.0	D = 153.3° Az = 278.1° (USCGS)
	e	A	23 33.5	LmV:22s 0.3 µm
	LmH	C	07 37	MLV=5.0
	LmV	C	37.5	
16.	-ePKP	A	20 31 51.5	Tonga Islands 19.41°S 175.17°W
	e	A	32 14	H = 20 12 15.1 h = 90 km MAG=5.0
				D = 148.4° Az = 351.7° (USCGS)
				PV:1.2s 15.3nm
16.	LmH	C	21 06	Probably Panama-Columbia Border Region
	LmV	C	06	LmH:25s 0.3 µm LmV:25s 0.2 µm
				(MLH=4.5 MLV=4.5)
17.	eSg	A	23 01 45	Yugoslavia 44.0°N 20.0°E
				H = 22 56 57 (BCIS)
				D = 8.8°
17.	eP	A	00 28 29	Southern Sinkiang Prov./China
	LmH	B	51.9	40.73°N 89.56°E
	LmV	B	51.9	H = 00 19 07.9 h = 0 km MAG=4.6 (USCGS)
				D = 53.0°
				LmH:14s 1.4 µm LmV:14s 1.7 µm
				MLH=5.2 MLV=5.3
				Atmospheric nuclear explosion.
17.	LmH	B	02 01	Northern Easter Is. Cordillera
	LmV	B	01	4.50°S 104.73°W
				H = 00 56 29.4 h = normal MAG=4.8 (USCGS)
				D = 110.0°
				LmH:18s 0.6 µm LmV:18s 0.6 µm
				MLH=5.2 MLV=5.2

Day	Phase		h m s	Remarks
17.	e(pP)	B	05 15 19	South Sandwich Islands Region
	+iPKIKP	A	18 32.5	58.30°S 26.63°W
	eiPP	B	19 24.5	H = 05 00 11.8 h = 140 km MAG=6.1
	eiPP	A	19 25.5	D = 113.0° Az = 25.4° (USCGS);
	ipPP	B	20 00	h = ca. 140 km
	iSKS	B	25 00	PV2:1.1s 44.4nm SKSH:7s 3.3 µm
	isSKS	B	26 11	sSKSH:9s 4.1 µm SPV:10.5s 8.9 µm
	eiSP	B	28 46	sSPV:10s 7.1 µm
	eiPKKP1	A	29 20.5	LmH:19s 4.7 µm LmV:19s 9.4 µm
	iPKKP2	A	29 27.5	MLH=6.1 MLV=6.4
	eisSP	B	29 43	e 19 04.5 e 19 17 e 19 21 e 19 44
	ePcPPKP	A	33 21	e 28 24 ei 30 18.5 ei 39 05
	LmH	B	06 01.6	
	LmV	B	03.3	
17.	e	A	10 01 50	Probably to Eastern Caucasus
	e	A	03 25	41.75°N 45.27°E
	e	A	03 31	H = 09 56 10.4 h = normal MAG=4.4 (USCGS)
				D = 24.6°
				The first onset of P must be about 21 s earlier.
17.	ePn	A	15 45 14	Southern Italy 41.57°N 16.23°E
	e	A	45 20	H = 15 42 55.4 h = 24 km MAG=4.4
	ePg	A	46 12	D = 9.6° Az = 342.2° (USCGS)
	eLg2	A	48 26	e 47 00 e 45 27
17.	ePg	A	17 46 59.5	Carpathes/ČSSR 48.4°N 17.5°E
	e	A	47 38.5	H = 17 45 41 (BCIS)
	i(Sg)	A	47 54.5	D = 4.4°
	i	A	47 57.5	LmH:7s 0.8 µm LmV:5s 0.9 µm
	LmH	B	48.4	MLH=3.4
	LmV	B	48.7	ei 48 08.5
17.	eP	A	17 56 29	Guatemala 14.08°N 90.01°W
	epP	A	56 56.5	H = 17 43 55.2 h = 103 km MAG=4.8 (USCGS)
				D = 86.5° h = 105 km

Day	Phase		h m s	Remarks
17.	ePg	A	20 23 42	Valais/Switzerland 46.4°N 7.4°E
	iSn	A	24 13.0	H = 20 22 02 (BCIS)
	eiSg	A	24 47	D = 5.1°
18.	LmH	B	01 52.4	Tibet 35.24°N 87.64°E
	LmV	B	55.4	H = 01 20 21.6 h = normal MAG=3.9 (USCGS)
				D = 55.3°
				LmH:17s 1.4 μm LmV:16s 0.5 μm
				MLH=5.1 MLV=4.8
18.	eP	A	05 33 12.5	Turkey 36.78°N 29.50°E
	e	A	33 15	H = 05 28 54.1 h = 43 km MAG=4.8
	e	A	33 23	D = 18.9° Az = 322.8° (USCGS)
	e	A	33 39	PV1:1.1s 16.8nm PV2:1.2s 20.4nm
	LmH	B	39.6	LmH:14s 0.6 μm LmV:9s 0.5 μm
	LmV	B	40.5	MLH=4.0 MLV=4.3
18.	eSg	A	09 59 39.5	Yugoslavia 45.8°N 16.0°E
	e	A	59 42.5	H = 09 56 37 (BCIS)
				D = 5.7°
18.	LmV	B	15 04.5	Santa Cruz Islands 12.41°S 166.16°E
	LmH	B	08.4	H = 14 04 42.8 h = 38 km MAG=4.3 (USCGS)
				D = 136.7°
				LmH:18s 0.3 μm LmV:20s 0.4 μm
				MLH=5.0 MLV=5.2
18.	epP	A	16 47 37	Afghanistan-Hindukush Border Region
	e	A	47 56.5	37.43°N 71.89°E
				H = 16 39 11.7 h = normal MAG=5.0 (USCGS)
				D = 44.1° h = ca. 100 km
18.	ePKIKP	A	20 23 18.5	New Ireland Region 3.89°S 151.64°E
	ePP	A	25 59.5	H = 20 04 56.7 h = 301 km MAG=4.9
				D = 122.6° Az = 331.0° (USCGS)
				AN USSR gives: 4.1°S 152.2°E
				H = 20 04 25 h = normal

Day	Phase		h m s	Remarks
18.	e	A	21 50 00.5	Carpathes/ČSSR 48.4°N 17.5°E
	eiSg	A	50 04.5	H = 21 47 45 (BCIS)
	e	A	50 19.5	D = 4.4°
	e	A	50 28	e 50 06.5 e 50 23
19.	e	A	00 24 29.5	Carpathes/ČSSR 48.4°N 17.5°E
	e	A	25 06.5	H = 00 23 00 (BCIS)
	ei	A	25 15.5	D = 4.4°
	eSg	A	25 20	
19.	iPg	A	14 13 19.3	Explosion
	eiSg	A	13 35.3	D = ca. 1.2°
19.	+iP	A	14 42 24.9	Red Sea 20.65°N 38.39°E
	eS	B	48 02	H = 14 35 20.2 h = 35 km MAG=4.5
	LmH	B	15(00)	D = 36.6° Az = 331.3° (USCGS)
	LmV	B	(00)	PV:1.8s 71.5nm
				MPV=5.3
19.	+iP	C	17 19 36.0	Fox Islands/Aleutian Is. 52.74°N 166.92°W
	+iP1	A	19 36.4	H = 17 07 45.4 h = normal MAG=5.7
	+iP	B	19 36.8	D = 77.0° Az = 1.0° (USCGS); h = 34 km
	-iP2	A	19 38.2	PV1(A):0.8s 23.6nm PV2(A):1.6s 318nm
	ipP2	A	19 47.5	PV(B):8s 2.8 µm SH:15s 1.8 µm
	eS	B	29 23	LmH:16s 5.4 µm LmV:18s 7.7 µm
	iPPS	B	30 22	MPV1(A)=5.4 MPV2(A)=6.2 MPV(B)=6.4
	eSS	C	34 28	MSH=6.0 MLH=6.0 MLV=6.1
	e(SSS)	C	38.0	e 20 12 e 20(16) e 40.0
	e(Sa)	C	38 45	Multiple P-onset in the short-period
	eSa	B	39.0	records, the first much smaller than
	eLQ	C	39.5	the second one.
	eLR	C	44 25	
	LmH	B	18 04.3	
	LmV	B	04.6	
20.	é(Sg)	A	00 26 04	
20.	e(P)	A	02 21 54	Probably Aleutian Islands
	e	A	22 26	

Day	Phase		h m s	Remarks
20.	+eP	A	05 37 14	Fox Islands/Aleutian Is. 52.81°N 167.06°W
				H = 05 25 22.4 h = 31 km MAG=4.5
				D = 76.9° Az = 0.9° (USCGS)
				PV:1.6s 36.8nm
				MPV=5.3
20.	eP	A	06 32 44.5	Fox Islands/Aleutian Is. 52.74°N 166.94°W
	ei	A	32 49	H = 06 20 49.5 h = 9 km MAG=4.5
				D = 77.0° Az = 0.9° (USCGS)
				PV:1.3s 25.0nm
				MPV=5.2
20.	eP	A	07 47 36.5	Fox Islands/Aleutian Is. 52.88°N 166.90°W
				H = 07 35 45.2 h = 29 km MAG=4.0
				D = 76.8° Az = 1.0° (USCGS)
				PV:1.4s 23.8nm
				MPV=5.1
20.	+eP1	A	07 50 38	Fox Islands/Aleutian Is. 52.82°N 167.07°W
	+eP	B	50 39	H = 07 38 44.9 h = 11 km MAG=5.2
	eiP2	A	50 40.5	D = 76.9° Az = 0.9° (USCGS);
	eipP	A	50 51.5	h = ca. 45 km
	eiPP	A	53 40	PV(B):10s 1.2 µm PV2(A):1.1s 48.0nm
	eS	B	08 00 24	SH(B):15s 1.0 µm
	eiS	C	00 40	LmH:15s 2.7 µm LmV:16s 3.3 µm
	eSS	C	05 30	MPV(B)=6.0 MPV2(A)=5.5 MSH(B)=5.7
	e(SSS)	C	09 18	MLH=5.7 MLV=5.8
	eSa	B	10.1	e 53 50 e 10 40
	LQ	C	11	P is multiple in the short-period records.
	eLR	C	15 35	P1 is a small-amplitude precursor.
	LmV	B	35.7	
	LmH	B	36.8	
20.	eP	A	12 37 44.5	Fox Islands/Aleutian Is. 52.76°N 166.89°W
	e	A	37 51	H = 12 25 50.2 h = 11 km MAG=4.6
	e	A	37 55.5	D = 77.0° Az = 1.0° (USCGS)
				PV:1.0s 11.8nm
				MPV=5.0

Day	Phase		h m s	Remarks
20.	e	A	12 43 39	
20.	i(Pg)	A	12 51 27.0	Explosion
	i	A	51 32.5	
	iSg	A	51 35.8	
20.	eiPg	A	13 03 15.5	Explosion/GDR $51^{\circ}17.2'$N $12^{\circ}43.8'$E
	eiSg	A	03 29.5	D = 1.0° Yield: 1.0 t
20.	e(P)	A	16 40 43	Greece 38.47°N 20.85°E
	e	A	40 45	H = 16 37 24.8 h = normal MAG=4.8 (USCGS)
	e	A	40 53	D = 13.8°
	e	A	41 03.5	
20.	i(Sg)	A	21 53 25.9	
21.	eP	A	02 16 09.5	Central Mid-Atlantic Ridge 8.69°N 39.66°W
	e	A	16 14	H = 02 06 07.9 h = normal MAG=4.7
	LmH	C	40	D = 59.4° Az = 35.3° (USCGS)
	LmV	B	40	PV:1.8s 25.5nm
21.	eP	A	07 03(00)	Peru-Ecuador Border Region 2.17°S 77.57°W
	e	A	03 32	H = 06 49 56.6 h = 49 km MAG=5.3
	eSKS	C	13 36	D = 91.1° Az = 39.5° (USCGS)
	iS	C	13 57	PV:2.0s 92.7nm
	ePS	C	15(10)	LmH:19s 0.7 µm LmV:19s 1.4 µm
	eSS	C	20 05	MPV=5.7 MLH=5.2 MLV=5.4
	eLQ	C	28.0	The beginning of LQ is a very clear G-
	LmV	B	42.4	wave.
	LmH	B	46.5	
21.	LmH	B	12 22.5	LmH:18s 0.5 µm
	LmV	B	26	
21.	e(Sg)	A	14 18 19	(Sg)V:0.9s 9.4nm
21.	e(P)	A	15 36 02	(P)V:1.0s 9.5nm

Day	Phase		h m s	Remarks
21.	eP	A	15 58 38	Luzon/Philippine Islands 12.73°N 123.11°E
	epP	A	58 53.5	H = 15 45 28.3 h = 56 km MAG=5.2
	eS	C	16 09 44	D = 93.4° Az = 323.6° (USCGS); h = 52 km
	eiS	C	09 46	PV:0.6s 9.5nm
	ePS	C	10 46	LmH:17s 3.5 µm LmV:19s 2.9 µm
	eSS	C	15.9	MPV=5.3 MLH=5.9 MLV=5.7
	eSSS	C	19.6	e 02 58 e 03 13 e 03 58 e 05 58.5
	eiLQ	C	23.0	
	LmV	B	41.6	
	LmH	B	41.8	
21.	ePKIKP	A	16 05 52	West of Macquarie Is. 55.42°S 146.93°E
	ePKHKP	A	05 59	H = 15 46 01.7 h = normal MAG=5.0 (USCGS)
	e	A	06 02	D = 153.0°
	ePKP2	A	06 12	
21.	eP	A	18 15 21.5	Central Alaska 64.76°N 147.37°W
	epP	A	15 26.5	H = 18 04 49.5 h = 17 km MAG=5.4
	e	A	15 37	D = 63.8° Az = 14.7° (USCGS); h = 19 km
	eS	C	23 55	PV:1.8s 56.1nm pPV:1.2s 30.6nm
	eSSS	C	31 00	LmH:18.5s 2.5 µm LmV:19s 3.4 µm
	LmH	B	53.8	MPV=5.5 MLH=5.4 MLV=5.6
	LmV	B	54.4	e 15 42 e 15 51 e 16 00
21.	eP1	A	18 23 36	Central Alaska 64.76°N 147.37°W
	eiP	B	23 38.5	H = 18 13 02.9 h = 17 km MAG=5.6
	eP2	A	23 39	D = 63.8° Az = 14.7° (USCGS)
	e	A	23 43	PV:6s 0.5 µm PV2:2.0s 106nm
	ePP	B	26 00	MPV=5.9 MPV2=5.7
	eS	B	32 11	P1 is a small-amplitude precursor in the short-period records.
21.	eP	B	18 35 17	Central Alaska 64.76°N 147.37°W
	eP	A	35 20	H = 18 24 45.7 h = 17 km MAG=5.4
	ePP	B	37 38	D = 63.8° Az = 14.7° (USCGS)
				PV(A):2.0s 106nm
				MPV(A)=5.7
21.	e(P)	A	18 44 09	
	e	A	44 20	

Day	Phase		h m s	Remarks
21.	e(P)	A	18 52 26	
	i	A	52 33.3	
	e	A	52 38	
21.	eP	A	19 04 15.5	PV1:1.8s 15.3nm PV2:2.0s 39.8nm
	e	A	04 20	
21.	+iPKHKP	A	19 29 24	<u>South of Fiji Islands</u> 23.48°S 179.95°E
	+iPKP2	A	29 34.8	H = 19 10 31.1 h = 546 km MAG=5.0
	epPKP	A	31 36	D = 151.4° Az = 344.4° (USCGS)
				PV1:1.3s 19.5nm PV2:1.1s 43.3nm
				AN USSR gives: 24.8°S 179.8°E
				H = 19 09 31 h = normal
				D = 152.6°
21.	eP	A	20 23 32.5	<u>Near Coast of Northern Chile</u>
	ePP	A	27 50	25.19°S 70.55°W
	eSKS	C	34 12	H = 20 09 28.4 h = 23 km MAG=5.7 (USCGS)
	eSP	C	36 52	D = 104.3°
	eSS	C	42 48	LmH:19s 2.4 μm LmV:19s 3.6 μm
	LmV	B	21 09.8	MLH=5.8 MLV=5.9
	LmH	B	10	e 34 28
21.	ePKP	A	22 24 31.5	<u>Fiji Islands Region</u> 17.82°S 178.66°W
				H = 22 05 54.0 h = 574 km MAG=4.6
				D = 146.2° Az = 348.2° (USCGS)
22.	e	A	08 59 08	Probably explosion.
	eSg	A	59 28.5	
	e	A	59 46	
22.	iSg	A	10 01 55.5	Probably explosion.
22.	iP	A	11 01 46.0	<u>East Coast of Greece</u> 39.0°N 20.4°E
	ei	A	04 37	H = 10 58 36 (BCIS)
	LmH	B	06.3	D = 13.3°
	LmV	B	(07)	LmH:13s 0.6 μm
				MLH=3.8

Day	Phase		h m s	Remarks
22.	-ePKP	A	11 09 40	Samoa Islands 15.83°S 172.68°W
	e	A	10 11	H = 10 50 05.3 h = normal MAG=4.5
	e	A	10 19	D = 145.1° Az = 355.2° (USCGS)
				PV:1.3s 19.5nm
22.	eP	A	11 19 03	North of Aszension Is. 1.50°S 15.74°W
	e	A	19 11.5	H = 11 09 20.9 h = normal MAG=4.8 (USCGS)
	e	A	19 27	D = 57.0°
22.	eP	A	12 23 08	Turkey 40.90°N 33.44°E
				H = 12 18 53.0 h = 13 km MAG=4.7 (USCGS)
				D = 18.0°
22.	e(Sg)	A	14 13 16	Probably explosion.
	e	A	13 18	
	e	A	13 20.5	
22.	e(pP)	A	15 48 49.5	Probably Andreanof Islands/Aleutian Is.
	e	A	49 11	(USCGS)
	e	A	49 22	
22.	ePP	A	19 28 43	New Ireland Region 1.33°S 149.83°E
	eSP	C	38 30	H = 19 08 33.5 h = 34 km MAG=5.0 (USCGS)
	eSS	C	45.0	D = 119.5°
	LmH	C	20 21.7	LmH:19s 0.3/um LmV:18s 0.4/um
	LmV	B	22.4	MLH=4.9 MLV=5.1
23.	ePKP	A	00 45 02.5	Samoa Islands 15.02°S 172.31°W
	e	A	45 24	H = 00 25 29.8 h = normal MAG=5.1
	e	A	46 00	D = 144.3° Az = 355.7° (USCGS)
	LmH	B	01 55.6	PV:2.2s 68.1nm
	LmV	B	56.8	LmH:18s 0.4/um LmV:18s 0.6/um
				MLH=5.2 MLV=5.4
23.	ePKP	A	01 01 45	Samoa Islands 14.91°S 172.37°W
	+i	A	01 46.5	H = 00 42 13.4 h = normal MAG=5.1
	epPKP	A	01 58.5	D = 144.2° Az = 355.7° (USCGS); h = 47 km
	LmV	B	13.0	PV2:1.6s 45.4nm
	LmH	C	14.8	LmH:16s 0.3/um LmV:17.5s 0.5/um
				MLH=5.1 MLV=5.3

Day	Phase		h m s	Remarks
23.	ePKIKP	A	05 23 32	Banda Sea 5.81°S 130.46°E
	e(pPKIKP)	A	23 59.5	H = 05 05 04.8 h = 85 km MAG=5.9
	e(sPKIKP)	A	24 10	D = 112.5° Az = 322.9° (USCGS);
	ePP	A	24 20	h = ca. 100 km
	LmH	B	06 12	PV:1.1s 14.4nm
				e 25 15
23.	eP1	A	10 11 08	Turkey 40.80°N 33.61°E
	eiP2	A	11 10	H = 10 06 54.1 h = 14 km MAG=4.8
	eP	B	11 12	D = 18.2° Az = 310.2° (USCGS)
	eS	B	14 40	PV2:1.8s 56.1nm PV:7s 0.7 μm
	eLg1	B	17 10	PH:7s 0.6 μm
	e(Rg)	B	17 58	LmH:13s 0.9 μm LmV:15s 0.8 μm
	LmV	B	20.8	MLH=4.2 MLV=4.3
	LmH	B	21	ei 11 20 e 11 38
23.	eP	A	12 05 08	Central Alaska 64.82°N 147.45°W
	ePcP	A	05 44.5	H = 11 54 33.5 h = 9 km MAG=4.6
				D = 63.7° Az = 14.7° (USCGS)
23.	ePKP2	A	12 07 53	Balleny Islands 62.51°S 155.17°E
	e	A	08 02	H = 11 47 28.9 h = normal MAG=5.3
	e	A	08 16	D = 157.0° Az = 255.8° (USCGS)
	e	C	31.5	LmH:18s 0.6 μm LmV:18s 1.0 μm
	LmH	B	13 26	MLH=5.4 MLV=5.7
	LmV	B	26	
23.	iPg	A	14 09 11.5	Explosion/GDR 51°22.3'N 12°53.5'E
	iSg	A	09 27.5	D = 1.0° Yield: 3.4 t
23.	eiPn	A	14 45 30	Explosion/Adelebsen, GFR
	eiPg	A	45 32	51°36.48'N 9°44.80'E
	eSg	A	45 49.5	H = 14 45 00.8 Yield: 5 t
				D = 1.6°
23.	-ePKHKP	A	14 57 18	Fiji Islands 21.35°S 179.35°W
	ePKP2	A	57 25.5	H = 14 38 35.7 h = 605 km MAG=5.1
	e	A	57 30	D = 149.5° Az = 346.2° (USCGS)
	e	A	57 39.5	PV:1.2s 25.5nm
				AN USSR gives: H = 14 37 37 h = normal

Day	Phase		h m s	Remarks
23.	LmH	B	17 16.8	Probably Honshu/Japan (USCGS)
	LmV	B	21.5	LmH:18.5s 0.5 µm LmV:15s 0.2 µm
				(MLH=4.9 MLV=4.7)
23.	ePKP	A	21 49 41	New Hebrides Islands 19.24°S 167.68°E
	e(pPKP)	A	49 58.5	H = 21 30 11.5 h = 37 km MAG=5.3
	ei	A	50 23	D = 143.4° Az = 334.4° (USCGS);
	e	A	51 05	(h = 62 km)
	ePP	A	52 55	PV:1.8s 40.8nm PPV:2.0s 31nm
	LmH	C	23 05	MPPV=5.2
	LmV	C	05	
24.	i(Sg)	A	11 45 03.5	Probably explosion.
	e	A	45 19	
24.	+iPKHKP	A	13 47 19	Fiji Islands Region 21.35°S 179.28°W
	ePKP2	A	47 27	H = 13 28 35.9 h = 592 km MAG=4.7
				D = 149.5° Az = 346.3° (USCGS)
				PV:1.0s 19.0nm
24.	eP	A	21 14 24	South of Mariana Islands 12.49°N 141.59°E
	ePP	A	18 44	H = 21 00 23.9 h = 18 km MAG=5.5
	ePS	C	27 48	D = 103.5° Az = 329.9° (USCGS)
	ePPS	C	28 40	PPV:2.0s 53.0nm
	eSS	C	33 28	LmH:20s 2.3 µm LmV:20s 2.9 µm
	LmH	B	22 04.8	MPPV=5.7 MLH=5.7 MLV=5.8
	LmV	B	04.8	e 18 53.5
24.	eP	A	21 16(58)	Samar/Philippine Islands 11.43°N 125.67°E
	epP	A	17 14	H = 21 03 35.7 h = 53 km MAG=5.0
				D = 95.9° Az = 324.2° (USCGS)
				pPV:1.8s 15.3nm
24.	eP	A	21 36 13	Iceland 61.85°N 27.88°W
	LmV	B	47	H = 21 30 59.0 h = normal MAG=4.5
				D = 24.2° Az = 99.4° (USCGS)

Day	Phase		h m s	Remarks
25.	iP	A	01 06 23.5	Kurile Islands 46.68°N 152.55°E
				H = 00 54 30.7 h = normal MAG=4.7
				D = 77.4° Az = 335.7° (USCGS)
				PV:0.8s 9.4nm
				MPV=5.0
25.	e(pP)	A	20 04 49.5	Probably Fox Islands/Aleutian Is. (USCGS)
				P must be about 15 s earlier.
25.	eP	A	21 40 13	Off East Coast of Honshu/Japan
	epP	A	40 25	33.44°N 141.43°E
	e	A	40 36	H = 21 27 41.8 h = 59 km MAG=4.6
				D = 85.3° Az = 330.6° (USCGS); h = 45 km
25.	eiP	A	23 32 02	South of Mariana Islands 12.42°N 141.77°E
	epP	A	32 12	H = 23 18 04.3 h = 42 km MAG=5.6
	e	A	36 17	D = 103.7° Az = 329.9° (USCGS); h = 43 km
	ePP	A	36 21	LmH:20s 3.2 µm LmV:20s 4.3 µm
	epPP	A	36 31	MLH=5.9 MLV=6.0
	ePS	C	45 20	
	ePPS	C	46 20	
	eiSS	C	51 05	
	LmV	B	00 22.3	
	LmH	B	22.4	
26.	eP	A	02 35 41	Off East Coast of Jalisco/Mexico
	eSKS	C	46 14	18.40°N 105.25°W
	eS	C	46(40)	H = 02 22 34.8 h = 45 km MAG=5.0 (USCGS)
	ePS	C	47 54	D = 91.8°
	eSS	C	52 48	LmH:19s 1.5 µm LmV:19s 1.4 µm
	LmH	B	03 17.0	MLH=5.5 MLV=5.4
	LmV	B	17.0	
26.	iP	A	03 49 45.5	Fox Islands/Aleutian Is. 53.03°N 166.57°W
	epP	A	49 55.5	H = 03 37 55.7 h = normal MAG=4.3 (USCGS)
				D = 76.0°

Day	Phase		h m s	Remarks
26.	eiP	A	05 05 29	Kurile Islands 45.03°N 147.29°E H = 04 53 43.5 h = 88 km MAG=4.2 D = 77.2° Az = 332.9° (USCGS) PV:0.9s 9.4nm MPV=4.9
26.	e(P)	A	09 16 00	
26.	+iPKP	A	09 28 30	Fiji Islands Region 17.96°S 178.31°W
	e	A	29 07	H = 09 09 42.4 h = 477 km MAG=4.3 D = 146.4° Az = 348.5° (USCGS) PV:1.4s 15.3nm
26.	e(Sg)	A	11 38 32.5	Explosion?
26.	ePg	A	12 49 32.5	Explosion
	e(Sg)	A	49 38.5	
26.	LmH	C	15 08.6	East New Guinea Region 5.78°S 147.66°E H = 14 04 23.9 h = 33 km MAG=5.3 (USCGS) D = 122.2° LmH:24s 0.2 μm MLH=4.7
26.	e(P)	A	15 48 19	
26.	e(P)	A	17 49 57	Probably to Northern Chile (USCGS)
26.	LmH	C	23 58	South of Mariana Islands 12.49°N 141.76°E
	LmV	C	58.5	H = 22 54 43.1 h = normal MAG=4.6 (USCGS) D = 103.7° LmV:20s 0.2 μm MLV=4.7
27.	i	A	05 11 20.2	
	ei	A	11 33.5	

Day	Phase		h m s	Remarks
27.	eiP	A	20 44 56	Andreanof Is./Aleutian Is.
	LmH	C	21 22.7	51.33°N 179.99°W
				H = 20 32 59.3 h = 26 km MAG=5.1 (USCGS)
				D = 77.8°
				LmH:20s 0.2 μm
				MLH=4.5
27.	eSS	C	22 14 28	Southeast Indian Rise 46.39°S 96.01°E
	eSSSS	C	22 40	H = 21 37 48.1 h = normal MAG=5.4 (USCGS)
	LQ	C	28	D = 120.8°
	LmH	B	57.3	LmH:19s 0.7 μm LmV:20s 0.95 μm
	LmV	B	57.3	MLH=5.3 MLV=5.4
27.	eP	A	23 19 12.5	Taiwan 23.61°N 121.54°E
	epP	A	19 24.5	H = 23 06 47.0 h = 45 km MAG=4.8
	e	A	19 33	D = 83.8° Az = 323.0° (USCGS); h = 43 km
	LmH	C	24 01	PV:1.3s 27.9nm
	LmV	C	01	MPV=5.2
28.	+iP	A	01 21 57.9	Kurile Islands 46.04°N 151.47°E
	e	A	22 18	H = 01 10 03.9 h = 33 km MAG=5.4
	ei	A	22 26	D = 77.6° Az = 335.1° (USCGS)
	e	A	22 37	PV:1.4s 52.3nm
	LmH	B	59.5	LmH:20s 0.35 μm LmV:20s 0.5 μm
	LmV	B	59.5	MPV=5.5 MLH=4.7 MLV=4.9
28.	ePKP	A	05 53 34.5	Samoa 14.39°S 172.58°W
	epPKP	A	53 47.5	H = 05 34 06.4 h = 40 km MAG=4.8
	LmH	B	07 01.8	D = 143.7° Az = 355.5° (USCGS); h = 46 km
	LmV	B	02.0	PV1:2.3s 70.0nm PV2:2.0s 37.0nm
				LmH:18s 0.4 μm LmV:20s 0.4 μm
				MLH=5.2 MLV=5.2
28.	ePKIKP	A	14 54 02	Off W. Coast of S. Island N. Z.
	ePKP2	A	54 52	47.00°S 165.81°E
	eipPKP2	A	55 00.5	H = 14 34 04.5 h = 37 km MAG=5.6
	ePP	B	58.7	D = 162.7° Az = 291.6° (USCGS); h = 30 km
	ePPP	B	15 02.5	PV1:1.6s 26.3nm PV2:1.8s 28.2nm
	eSKKS	B	06 24	PV3:2.0s 66.7nm

Day	Phase		h m s	Remarks
cont.				
28.	eSKSP	C	15 09(00)	LmH:22s 0.9 µm LmV:20s 0.9 µm
	eSoSPKP	C	09 38	MLH=5.5 MLV=5.6
	ePPS	C	12 08	e 54 31 e 56 00 e 58 54
	eSS	C	19.0	
	ePSS	C	20.0	
	LmH	B	16 14.3	
	LmV	B	18.2	
28.	ei	A	17 57 14	Probably explosion.
	e(Sg)	A	57 25	
29.	+iP	A	03 04 46.0	Kazakh SSR 50.0°N 78.0°E
	ePP	A	06 22	H = 02 57 00 (BCIS)
				D = 41.1°
				PV:0.7s 71.1nm PH:0.7s 35.8nm
				MPV=5.7 MPH=5.7
				Probably underground explosion.
29.	eP	A	03 05 42	South of Panama 5.16°N 82.65°W
	eSKS	C	16.2	H = 02 52 50.1 h = normal MAG=4.8 (USCGS)
	ePS	C	17.5	D = 88.7°
	eSS	C	22.4	LmH:23s 0.7 µm LmV:22s 0.7 µm
	LQ	C	28.5	MLH=5.0 MLV=5.1
	LmH	B	38	
	LmV	B	38	
29.	eP	A	08 28(02)	Western Caucasus 41.59°N 43.98°E
	e	A	28 44	H = 08 22 47.9 h = 25 km MAG=4.9
	eS	C	32 17	D = 24.0° Az = 303.2° (USCGS)
	eS	B	32 28	PV:1.8s 31.3nm SH(B):12s 0.7 µm
	LmH	B	39.2	LmH:12s 0.2 µm LmV:12s 0.4 µm
	LmV	B	39.3	MPV=4.6 MSH(B)=5.0 MLH=3.9 MLV=4.3
29.	iPg	A	09 38 40.5	Explosion
	iSg	A	38 56	D = ca. 1.1°

Day	Phase		h m s	Remarks
29.	ePKP	A	09 45 23	Samoa Islands 15.78°S 172.42°W
	e	A	45 31	H = 09 25 47.4 h = 14 km MAG=4.8
	e	A	45 38.5	D = 145.1° Az = 355.5° (USCGS)
	e	A	45 52.5	
29.	ePKP	A	10 56 24	Tonga Islands 21.23°S 174.38°W
				H = 10 36 28.7 h = normal MAG=4.4 (USCGS)
				D = 150.2°
29.	e(Sg)	A	12 51 21	
	e	A	51 25	
29.	ei	A	13 03 09.5	
29.	iPKIKP	A	16 54 38.8	Banda Sea 7.23°S 128.60°E
	ePP	A	55 32	H = 16 36 15.7 h = 121 km MAG=5.4 (USCGS)
	ePS	C	17 04 52	D = 112.5°
	e(PPS)	C	06.0	PV:1.8s 25.0nm PPV:2.6s 250nm
	e(sPPS)	C	06(50)	MPPV=6.1
	eSS	C	11(10)	e 55 22 e 55 54
	LmH	B	46	
	LmV	B	46	
30.	eP	A	00 19 00	Jan Mayen Island 70.43°N 15.27°W
	e	A	19 03	H = 00 13 52.5 h = normal MAG=4.6
	e	A	19 09	D = 23.5° Az = 133.7° (USCGS)
	LmH	B	30.6	PV1:2.0s 44.5nm PV2:1.6s 47.4nm
	LmV	B	31.6	PV3:1.3s 32.6nm
				LmH:12s 0.3 µm LmV:13s 0.5 µm
				MPV1=4.7 MPV2=4.8 MLH=3.9 MLV=4.4
30.	iPg	A	09 30 25	Explosion/ČSSR 49°53.5'N 13°43'E
	i(Sg)	A	30 45	Yield: 13.4 t (PRUHONICE)
	e	A	30 57	D = 1.5°
30.	i	A	10 31 59.5	Explosion?
				V:1.1s 40.0nm
30.	e	A	11 21 43	

Day	Phase		h m s	Remarks
30.	LmH	B	11 46.2	
30.	eiSg	A	14 06 38.5	Explosion/Eschenlohe, GFR
30.	LmH	C	20 14.0	Rat Islands/Aleutian Is. (USCGS)
	LmV	C	14	

Day	Phase		h m s	Remarks
1.	ePx	A	02 57 39	Yugoslavia 44.15°N 19.14°E
	e	A	58 36	H = 02 55 33.3 h = normal MAG=4.5
	eSn	A	59 06.5	D = 8.3° Az = 324.5° (USCGS)
	e(Sg)	A	03 00 05	PxV:1.0s 13.1nm
	e(Rg)	A	00 31	LmH:11s 0.5 μm LmV:10.5s 0.7 μm
	LmH	B	00.5	MLH=3.5
	LmV	B	01.2	e 57 53.5 e 58 09.5
				Pn must be about 5 s earlier than Px.
1.	-iP	A	03 29 15.9	Turkey 36.09°N 31.42°E
	ei	A	29 19	H = 03 24 42.1 h = 67 km MAG=4.5
				D = 20.4° Az = 321.7° (USCGS)
				PV:1.0s 17.4nm
				MPV=4.2
1.	LmH	C	04 52	Traces
	LmV	C	53	
1.	+eiP	A	07 41 50.8	Southern Sumatra 0.82°S 98.67°E
	epP	A	41 58.5	H = 07 28 57.6 h = 26 km MAG=5.5 (USCGS)
	eX	A	42 09	D = 88.7° h = 28 km
	e	A	42 17	PV:1.4s 38.0nm pPV:1.3s 32.6nm
	e	A	42 25.5	XV:2.4s 91.0nm
	eiS	B	52 37	LmH:21s 0.9 μm LmV:21s 1.0 μm
	LmH	B	08 25.5	MPV=5.4 MLH=5.2 MLV=5.3
	LmV	B	27.5	eX is the greatest onset in our short-
				period broad-band registration.
1.	i(Sb)	A	08 38 32	Swabia/GFR 48.3°N 8.9°E
	i(Sg)	A	38 40	H = 08 37 02 (BCIS)
				D = 3.0°
1.	i(Sg)	A	11 07 53.5	
1.	e	A	12 03(00)	
	e	A	03 15	
1.	i	A	15 51 41.5	

Day	Phase		h m s	Remarks
1.	eiP	A	21 33 54.7	Alaska Peninsula 54.02°N 160.97°W
	eipP	A	34 06.2	H = 21 22 10.0 h = 19 km MAG=4.5 (USCGS)
	e	A	34 14	D = 75.4° h = 44 km
				PV:1.3s 28.0nm pPV:1.5s 30.0nm
				MPV=5.2
1.	-iP1	A	23 21 46.7	South of Alaska 54.36°N 158.05°W
	-iP	B	21 47.2	H = 23 10 07.2 h = normal MAG=6.2 (USCGS)
	iP2	A	21 48.0	D = 74.6° h = 42 km
	PL	B	21 48	PV1:1.2s 86.8nm PV2:1.7s 482nm
	ipP	A	21 57.7	pPV:1.7s 580nm PLV:15s 2.3 µm
	i(PP)	B	24 48	(PP)V:10s 1.5 µm SH:15s 13.9 µm
	i(Pa)	B	26 48	LmH:19s 15 µm LmV:18s 19.1 µm
	i!S	B	31 26	MPV1=5.8 MPV2=6.4 MSH=6.9 MLH=6.4
	iPS	C	31 58	MLV=6.5
	iSS	C	35 52	Multiple P in the short-period records.
	iLQ	C	41 16	The clear PL-waves (T = 15 s) in the long-
	ePKPPKP	A	49 06	period records are superimposed by relati-
	epPKPPKP	A	49 16	vely short-period P-waves (T ca. 2 - 3 s).
	e	A	49 42	(PP) is a very sharp onset in the long-
	LmH	B	58.0	period broad-band registrations, but about
	LmV	B	00 00.7	11 s later than expected for PP. SS is
				small compared to the very strong S-onset.
2.	eiPn	A	00 33 40	Yugoslavia 43.9°N 19.2°E
	eiPx	A	33 50	H = 00 31 37 (BCIS) MAG=4.2 (BEOGRAD)
	eSn	A	35 08	D = 8.5°
	i(Sg)	A	36 11	i 35 30.5
	i(Lg2)	A	36 32	
2.	ePn	A	01 16 08.2	Yugoslavia 43.70°N 19.20°E
	ePx	A	16 15.2	H = 01 14 03.3 h = normal MAG=4.2
	eiSn	A	17 42	D = 8.7° Az = 326.1° (USCGS)
	eiSg	A	18 43.5	e 17 08
	eiLg2	A	18 52	
2.	ei	A	01 21 03	Yugoslavia 43.9°N 19.2°E
	i	A	21 18	H = 01 17 13 (BCIS) MAG=3.9 (BEOGRAD)
	LmH	A	22 32	D = 8.5°

Day	Phase		h m s	Remarks
2.	eP	A	02 32 58.5	South of Alaska 54.5°N 158.0°W
	e	A	33 33	H = 02 36 19.8 h = 33 km MAG=4.7 (USCGS)
				D = 74.8°
2.	eP	A	03 13 12	Japan 45.7°N 142.9°E
				H = 03 02 03.0 h = 334 km MAG=4.5 (USCGS)
				D = 75.1°
2.	ePn	A	07 12 08.5	Yugoslavia 43.9°N 19.2°E
	e	A	12 18	H = 07 10 06 (BCIS) MAG=3.9 (BEOGRAD)
	e	A	14 10	D = 8.5°
	eSg	A	14 46	e 12 26 e 12 39
2.	eiP	B	07 15 47.5	Nicobar Islands Region 8.72°N 93.78°E
	eiP1	A	15 50	H = 07 03 52.9 h = normal MAG=5.7 (USCGS)
	eiP2	A	15 53.0	D = 78.3°
	iPP	A	18 52.5	PV:10s 1.1 µm PV2:1.6s 94.7nm
	iPP	B	18 54	PPV(B):9.5s 1.1 µm SH:13s 1.2 µm
	eS	B	25 42	LmH:17s 4.7 µm LmV:17.5s 9.7 µm
	ePPS	C	26 48	MPV=5.9 MPV2=5.7 MPPV=6.0 MSH=5.9
	eSS	C	31.1	MLH=5.9 MLV=6.2
	LmH	B	56.7	AN USSR gives: Andaman Sea 9.0°N 93.5°E
	LmV	B	56.7	H = 07 04 08 h = 172 km
				Multiple P in our short-period records, the first onset smaller than the second one.
2.	iP	A	07 50 52.2	Off East Coast of Honshu 33.0°N 141.6°E
	epP	A	51 05.5	H = 07 38 15.0 h = 39 km MAG=5.0 (USCGS)
				D = 85.7° h = 50 km
				pPV:1.8s 43.8nm
2.	eP	A	08 41 26	Eastern Kashmir 33.18°N 75.64°E
				H = 08 32 38.5 h = 33 km MAG=4.8 (USCGS)
				D = 49.2°
2.	LmH	C	14 21	Mariana Islands (USCGS)

Day	Phase		h m s	Remarks
2.	iP	A	14 21 37.2	<u>Nicobar Islands</u> 8.5°N 93.8°E
	eipP	A	21 46	H = 14 09 37.6 h = 36 km MAG=5.2 (USCGS)
				D = 78.5° h = 33 km
2.	eP	A	14 32 21.5	
	e	A	32 41	
2.	eP	A	16 28 28	<u>South of Honshu/Japan</u> 32.93°N 141.70°E
	e	A	28 43	H = 16 15 48.4 h = 19 km MAG=5.0 (USCGS)
	e	A	29 02	D = 85.8°
	e	A	29 30	LmH:15s 0.5 μm LmV:14s 0.7 μm
	e	A	30 19	MLH=5.1 MLV=5.2
	LmH	B	17 15.3	
	LmV	B	15.3	
2.	+iP	A	20 46 37.2	<u>Kyushu/Japan</u> 31.20°N 130.13°E
	i	A	47 07.5	H = 20 34 36.2 h = 181 km MAG=4.9 (USCGS)
				D = 82.1°
				PV:1.2s 40.8nm
				MPV=5.0
				AN USSR gives: Ryukyu Islands
				30.3°N 130.4°E
				H = 20 34 15 h = normal
				D = 83.0°
2.	eiP	A	22 13 46.5	<u>Volcano Islands</u> 23.0°N 142.7°E
	LmH	B	59.9	H = 22 00 38.1 h = 40 km MAG=4.7 (USCGS)
	LmV	B	23 00.2	D = 94.9°
				LmH:20s 0.2 μm LmV:20s 0.4 μm
				MLH=4.6 MLV=4.9
3.	eiPn	A	02 55 47	<u>Yugoslavia</u> 44.16°N 19.24°E
	eiPx	A	55 54	H = 02 53 47.9 h = 60 km MAG=4.3
	eiSn	A	57 16	D = 8.3° Az = 324.2° (USCGS)
	e	B	58 16	PnV:0.9s 20.7nm PxV:1.3s 53.5nm
	i(Lg2)	B	58 31.5	LmH:12.5s 3.1 μm LmV:10s 2.6 μm
	LmH	B	58.7	MLH=4.2
	LmV	B	59.5	i 56 11.5 i 57 33.5 i 57 45.5

Day	Phase		h m s	Remarks
3.	eiP	A	03 10 30.5	South of Alaska 54.7°N 157.7°W
				H = 02 58 52.9 h = normal MAG=4.2 (USCGS)
				D = 74.7°
3.	LmH	C	04 38.2	Mariana Islands (USCGS)
3.	eP	A	05 21 27	Kurile Islands 43.6°N 147.0°E
				H = 05 09 28.1 h = normal MAG=4.2 (USCGS)
				D = 78.3°
3.	iP	A	09 50 24.0	South of Alaska 54.6°N 157.7°W
				H = 06 54 43.4 h = normal MAG=4.6 (USCGS)
				D = 74.7°
3.	-eP	A	21 59 08.5	Ascension Island Region 7.46°S 13.45°W
	ipP	A	59 15.0	H = 21 48 50.9 h = normal MAG=4.8
	e	A	59 21.5	D = 61.8° Az = 17.8° (USCGS); h = 25 km
	LmH	C	22 25.5	PV:1.2s 35.7nm pPV:1.5s 57.1nm
				LmH:21s 0.3 μm
				MPV=5.4 MLH=4.5
3.	ePg	A	22 54 42	Apennin Toscan/Italy 44.0°N 12.0°E
	eSg	A	56 13.5	H = 22 52 27 (BCIS)
				D = 6.7°
4.	eSn	A	02 40 35.5	Probably Italy-Austria Border Region
	i	A	40 50.5	D = ca. 4.2°
	iSg	A	41 04.2	TRIESTE gives: D = ca. 110 km,
	eL	A	41 07	VIENNA gives: (D = 370 km)
				SgV:0.9s 25.7nm LV:1.4s 52.2nm
4.	e(P)	A	14 34 50.5	
4.	eiPKIKP	A	14 35 32	Near Coast of Central Chile 38.1°S 73.4°W
	e(PP)	A	36 29	H = 14 16 51.6 h = 28 km MAG=5.4 (USCGS)
	ePPP	B	38 57	D = 115.4°
	eSKS	C	42(16)	SH:23s 0.5 μm PSH:26s 1.5 μm
	eSKKS	C	43(30)	SSH:32s 0.5 μm
	eS	C	44 24	LmH:20s 2.6 μm LmV:20s 3.7 μm

Day	Phase		h m s	Remarks
cont.				
4.	eiPS	C	14 46 20	MLH=5.8 MLV=6.0
	eSPP	B	47 24	S is the diffracted S around the core.
	eiSPP	C	47 32	
	eSS	C	52 28	
	LmH	B	15 24.5	
	LmV	B	24.7	
4.	eSg	A	12 46 05	Explosion?
				SgV:0.8s 21.2nm
4.	eiSg	A	15 35 59	
	i	A	36 08.8	
4.	+iP	A	23 53 50.6	Hokkaido/Japan Region 43.24°N 142.50°E
	+iP	B	53 51	H = 23 42 13.7 h = 160 km MAG=5.6
	+iPcP	A	54 01.0	D = 77.2° Az = 330.4° (USCGS); h = 162 km
	eipP	B	54 30	PV(A):0.9s 94.3nm PV(B):7.5s 1.0 µm
	esP	B	54 46	SH:9s 0.9 µm
	ePP	B	56 47	LmH:15.5s 1.3 µm LmV:14s 1.4 µm
	eS	B	00 04(21)	MPV(A)=5.5 MPV(B)=5.6 MSH=6.0
	eiS	B	04 27	
	eScS	C	04 48	
	esS	C	05 32	
	eLR	C	18.5	
	LmH	B	27.0	
	LmV	B	34.4	
5.	eP	A	00 56 54	Southern Greece 36.76°N 21.32°E
	e	A	57 03.5	H = 00 53 14.2 h = 22 km MAG=4.8
	e	A	57 09	D = 15.5° Az = 336.4° (USCGS)
	e	A	57 14.5	PV1:1.0s 23.7nm PV2:1.2s 33.2nm
	eS	B	01 00(00)	PV3:1.2s 33.2nm PV4:1.4s 33.6nm
	e	C	00 32	LmH:10.5s 2.6 µm LmV:15.5s 3.5 µm
	eiLi	B	01 56	MLH=4.7 MLV=4.8
	eLg1	B	02 14	The first onset of P is followed by three
	LmH	B	04.2	clearly separated onsets of approximately
	LmV	B	04.2	constant amplitude.

Day	Phase		h m s	Remarks
5.	eP	A	04 14 45.5	South of Alaska 54.47°N 157.88°W
	epP	A	14 54	H = 04 03 06.8 h = 33 km MAG=4.8
	e	A	15 10	D = 74.9° Az = 6.9° (USCGS); h = 33 km
				PV:1.2s 17.8nm
				MPV=5.1
5.	ei(Pg)	A	06 56 14.6	Probably explosion.
	e(Sg)	A	56 31	
5.	eP	A	09 13 38.7	Ryukyu Islands 25.64°N 125.97°E
				H = 09 01 05.3 h = normal MAG=4.7 (USCGS)
				D = 84.5°
5.	e(Sg)	A	11 07 44	(Sg)V:0.8s 21.2nm
5.	iPg	A	11 24 43	Explosion/GDR
	eiSg	A	24 51	D = ca. 0.6°
	LmH	A	25 05	
	LmV	A	25 05	
5.	eiPg	A	12 00 02	Explosion/GDR
	iSg	A	00 16.5	D = ca. 1.1°
	LmH	A	00 32	
	LmV	A	00 32	
5.	e(Pg)	A	12 05 35.7	Explosion?
	ei(Sg)	A	05 51	(Sg)V:1.0s 28.4nm
	e	A	05 58	
	e	A	06 04	
5.	iSg	A	13 03 39.7	Explosion/GDR
5.	iSg	A	13 03 58.5	Explosion/GDR
	LmH	A	03 00	
5.	eiSg	A	13 04 31.5	Explosion/GDR
5.	e(Sg)	A	15 07 15	

Day	Phase		h m s	Remarks
5.	eP	A	16 53 14	Southern Greece 36.93°N 21.33°E
	e	A	53 23.5	H = 16 49 35.9 h = 41 km MAG=4.3
	LmH	B	17 00.5	D = 15.4° Az = 336.1° (USCGS)
	LmV	B	00.5	LmH:11s 0.6 µm LmV:11s 0.5 µm
				MLH=4.0 MLV=4.1
5.	i(Sg)	A	18 05 08.7	Explosion?
	e	A	05 10.5	
5.	e	A	18 42 42.5	
5.	LmH	C	22 07	Luzon/Philippine Islands (USCGS)
	LmV	C	07	
6.	LmH	C	01 52.5	South Pacific Cordillera (USCGS)
	LmV	B	58.2	
6.	-iP	A	05 16 55.7	Central Alaska 62.38°N 147.37°W
	ipP	A	17 10.7	H = 05 06 13.4 h = 59 km MAG=5.1
	e	A	17 20	D = 66.1° Az = 14.5° (USCGS); h = 59 km
	e	A	17 26.5	PV:0.8s 30.7nm
				MPV=5.5
6.	eP	A	08 25 29.5	Southern Greece 36.68°N 21.42°E
	e	A	25 43	H = 08 21 50.3 h = 35 km MAG=4.5 (USCGS)
	i	A	25 08.2	D = 15.7°
	LmH	B	32.7	LmH:11s 0.6 µm LmV:10s 0.9 µm
	LmV	B	32.7	MLH=4.0 MLV=4.4
				AN USSR gives: 35.5°N 20.8°E
				H = 08 21 39 h = normal
				D = 16.6°
6.	iSg	A	09 05 37	Explosion
				SgV:0.5s 53.0nm
6.	iSg	A	10 47 48.6	Probably explosion.
				SgV:1.1s 48.0nm

Day	Phase		h m s	Remarks
6.	iSg	A	11 08 27.8	Explosion
				SgV:0.8s 47.0nm
6.	e(Pg)	A	11 54 38.6	Explosion?
	eSg	A	54 54.5	
6.	eSg	A	12 04 49.6	Probably Upper Silesia/Poland
6.	+iP1	A	13 54 17.0	<u>Fox Islands/Aleutian Is.</u> 52.59°N 168.17°W
	+iP2	A	54 18.0	H = 13 42 22.5 h = 14 km MAG=5.9
	+iP	B	54 18.5	D = 77.1° Az = 0.1°
	+iP	C	54 20	PV2:1.7s 395nm PV(B):8s 2.3 µm
	eS	B	14 04 06	SH:14s 1.9 µm
	eSS	C	09 08	LmH:17s 4.8 µm LmV:15.5s 5.2 µm
	eSSS	C	12.4	MPV2=6.3 MPV(B)=6.4 MSH=6.0 MLH=5.9
	eSa	C	13 40	MLV=6.0
	eLR	C	19.4	ei 54 30 ei 55 00 e 04 17 e 04 21
	LmV	B	39.5	e 13 50 e 14(36) e 16(42)
	LmH	B	40.0	The first onset of P in the short-period records is much smaller than the second one.
6.	eiP	A	13 58 46	<u>Fox Islands/Aleutian Is.</u> 52.6°N 168.1°W
	e	A	59 03.5	H = 13 46 53.1 h = 28 km MAG=4.9 (USCGS)
				D = 77.1°
6.	eP	A	16 26 15	<u>Fox Islands/Aleutian Is.</u> 52.82°N 168.08°W
				H = 16 14 24.7 h = normal MAG=4.1
				D = 76.9° Az = 0.2° (USCGS)
6.	eP	A	18 42 53	<u>Leeward Islands</u> 18.88°N 61.92°W
	eS	B	51 35	H = 18 32 15.1 h = 57 km MAG=5.1
	LmH	B	19 07	D = 65.2° Az = 42.3° (USCGS)
	LmV	B	07.5	LmH:20s 0.8 µm LmV:18s 1.3 µm
				MLH=4.9 MLV=5.2
				AN USSR gives: 20.5°N 62.4°W
				H = 18 32 19 h = normal MAG=5¼
				D = 64.3°

Day	Phase		h m s	Remarks
6.	eP	A	19 07 22.5	Eastern Gulf of Aden 13.40°N 50.77°E
	e	A	07 28	H = 18 58 39.7 h = 44 km MAG=4.9
	eS	C	14 20	D = 48.8° Az = 327.7°
	e	C	18 16	
6.	eiP	A	19 29 47.5	Central Mid-Atlantic Ridge 8.09°N 38.51°W
	iPcP	A	30 35.5	H = 19 19 48.4 h = normal MAG=4.9
	eS	B	37 50	D = 59.2° Az = 34.7° (USCGS)
	iS	C	38 00	PV:1.2s 38.3nm SH(B):(16s) 1.1 µm
	eSoS	C	38 36	LmH:18s 2.9 µm LmV:19s 4.3 µm
	eiLQ	C	44 16	MPV=5.4 MSH=5.6 MLH=5.3 MLV=5.5
	iLR	C	47 00	e 29 50.5 e 30 21 e 30 52.5
	LmH	B	54.5	
	LmV	B	54.5	
6.	eiP	A	23 27 57.5	Kyushy/Japan 32.5°N 130.9°E
	e	A	28 24	H = 23 15 57.2 h = 159 km MAG=4.8 (USCGS)
				D = 81.3°
				AN USSR gives: 31.3°N 131.9°E
				H = 23 15 34 h = normal
				D = 82.8°
7.	eP	A	01 18 43	Eastern Gulf of Aden 13.46°N 50.77°E
	e	A	18 49	H = 01 09 59.0 h = 32 km MAG=4.8
	LmH	C	44.5	D = 48.8° Az = 327.1° (USCGS)
7.	e(P)	A	03 13 05	
7.	ePKIKP	A	10 00 51.5	Fiji Islands Region 20.31°S 177.68°W
	-iPKHKP	A	00 56	H = 09 42 08.0 h = 540 km MAG=4.6
	iPKP2	A	01 01.5	D = 148.8° Az = 348.5° (USCGS)
				PV2:1.0s 79.0nm
7.	eSg	A	10 09 08.4	Near earthquake?
7.	e(Sg)	A	10 59 59	Explosion
7.	e(Sg)	A	11 00 10	Explosion

Day	Phase		h m s	Remarks
7.	LmH	C	12 21	
7.	iPg	A	12 45 13.0	Explosion
	e	A	45 25.4	D = ca. 1.1°
	iSg	A	45 27.7	
7.	eSg	A	12 46 45.5	Explosion
7.	eiPg	A	12 56 28	Explosion/GDR
	iSg	A	56 38.5	D = ca. 0.8°
7.	eSg	A	13 44 29	Explosion
	e	A	44 32	
7.	iP	A	23 06 57.2	India-China Border Region 27.82°N 92.23°E
				H = 22 56 30.8 h = normal MAG=4.9
				D = 63.3° Az = 315.3° (USCGS)
8.	LmH	B	00 21.5	Tibet 35.5°N 87.8°E
	LmV	B	25.3	H = 23 49 23.6 h = normal MAG=4.4 (USCGS)
				D = 55.2°
				LmH:17s 1.2 μm LmV:14s 0.3 μm
				MLH=5.0 MLV=4.6
8.	eP	A	00 54 29	Near East Coast of Honshu/Japan
				38.24°N 141.69°E
				H = 00 42 18.1 h = 68 km MAG=4.2 (USCGS)
				D = 81.2°
8.	ePKHKP	A	01 18 04	New Hebrides Islands 15.36°S 167.50°E
	iPKIKP	A	18 08.0	H = 00 58 54.7 h = 137 km MAG=5.2
	eSKP	A	21 32	D = 139.8° Az = 336.2° (USCGS)
				PV2:1.5s 50.3nm SKPV:2.0s 73nm
				PKHKP is a small-amplitude precursor.
8.	ePKIKP	A	06 42 24	New Hebrides Islands 16.34°S 166.76°E
	ePP	A	45 24	H = 06 22 52.8 h = 9 km MAG=5.0
	LmH	C	07 43.4	D = 140.4° Az = 335.2° (USCGS)
	LmV	B	47.3	LmH:20.5s 0.6 μm LmV:22s 0.6 μm
				MLH=5.3 MLV=5.3

Day	Phase		h m s	Remarks
8.	iPg	A	08 40 20.4	Explosion
	iSg	A	40 37.7	D = ca. 1.3°
8.	e(Sg)	A	09 56 56	Near earthquake
	e	A	57 17.5	TRIESTE gives: D = 250 km
8.	e(Sg)	A	10 50 54.4	
8.	e(Sg)	A	11 37 28	
8.	-iPKHKP	A	13 32 17.0	Fiji Islands Region 19.95°S 178.07°W
	ePKP2	A	32 22	H = 13 13 29.1 h = 520 km MAG=4.3
				D = 148.4° Az = 348.2° (USCGS)
				PV:0.4s 20.0nm
8.	epP	A	19 30 57.5	Off East Coast of Honshu 37.7°N 143.7°E
	LmH	C	20 06	H = 19 18 22.3 h = 66 km MAG=4.3 (USCGS)
				D = 82.3°
8.	e(P)	A	22 07 21	Probably Crete (UPP)
8.	LmH	C	23 55	Sinkiang Prov./China 39.9°N 78.2°E
				H = 23 24 15 h = normal MAG=ca.4
				(AN USSR)
8.	eP	A	03 20 48	Hokkaido/Japan Region 43.95°N 144.71°E
	LmH	B	54.3	H = 03 09 03.2 h = 100 km MAG=4.6
	LmV	B	55	D = 77.3° Az = 331.5° (USCGS)
				LmV:17s 0.4 µm
				MLH=4.9
9.	e	A	19 24 25	Near earthquake
	e(Sg)	A	24 29.5	TRIESTE: D = 200 km
	e	A	24 33	
9.	eiP	A	21 40 40	North Atlantic Ridge 19.2°N 46.0°W
	e	A	40 54	H = 21 31 07.8 h = 18 km MAG=4.6 (USCGS)
				D = 55.0°

Day	Phase		h m s	Remarks
9.	LmH	C	22 40	South Pacific Cordillera (USCGS)
10.	eP	A	03 28 28.5	Kamchatka 59.80°N 161.23°E H = 03 17 36.5 h = normal MAG=4.4 (USCGS) D = 67.2°
10.	eP	A	03 47 32	Kamchatka 59.77°N 161.15°E H = 03 36 38.9 h = normal MAG=4.5 (USCGS) D = 67.1°
10.	LmH LmV	C C	07 02.7 02.7	Kamchatka (USCGS)
10.	e +ePKP -iPKP	A B A	06 48 05 48 11 48 11.5	Fiji Islands Region 17.60°S 178.83°W H = 06 29 30.5 h = 529 km MAG=4.8 D = 146.0° Az = 348.1° (USCGS) PV(B):4.5s 0.35 µm PV(A):1.5s 53.6nm e 06 48 05 is the beginning of anuniden-tified small-amplitude motion.
10.	iPKHKP	A	10 37 08	Fiji Islands Region 21.60°S 179.42°W H = 10 18 25.1 h = 621 km MAG=4.8 (USCGS) D = 149.8° PV:0.8s 37.4nm
10.	eiP	A	11 00 13.5	Fox Islands/Aleutian Is. 52.7°N 168.2°E H = 10 48 25.6 h = 26 km MAG=3.9 (USCGS) D = 75.2°
10.	LmH	C	11 59.1	Ceram Sea 3.2°S 130.0°E H = 10 56 26.2 h = normal MAG=5.1 (USCGS) D = 110.2° LmH:17.5s 0.4 µm MLH=5.0
10.	e e(Sg)	A A	11 43 13 43 19	Near earthquake?

Day	Phase		h m s	Remarks
10.	e(sP)	A	12 17 24	Java Sea 5.9°S 113.1°E
	eiSKS	B	24 06	H = 12 01 31.5 h = 591 km MAG=5.4 (USCGS)
	eSS	C	32.4	D = 101.8°
	e	C	37.2	
	eSSSS	C	40.0	
	eSa	C	43.2	
10.	eP	A	19 31 56	Talaud Islands 4.78°N 127.11°E
	e	A	32 10	H = 19 18 14.7 h = 118 km MAG=5.2
	ePP	A	36 12	D = 102.1° Az = 324.0° (USCGS)
	e	A	36 25	LmH:28.5s 3.0 /um LmV:20s 1.5 /um
	eSKS	C	42 30	(MLH=5.7) (MLV=5.5)
	eS	C	43 32	e 32 10 and e 36 25 are the greatest onsets
	eSS	C	50.9	in the short-period records.
	eLQ	C	20 00	AN USSR gives: h = normal
	LmH	C	11.8	
	LmV	C	22.4	
11.	ePKIKP	A	04 36 58	Solomon Islands 7.04°S 155.78°E
	e	A	36 11.5	H = 04 17 02.1 h = 88 km MAG=4.8
	e	A	36 25	D = 127.3° Az = 332.0° (USCGS)
				PV2:1.3s 14nm
				AN USSR gives: 7.6°S 150.4°E
				H = 04 18 17 h = 725 km
11.	iPg	A	10 36 56.2	Explosion
	iSg	A	37 14.0	D = ca. 1.3°
11.	ePn	A	12 43 05	Yugoslavia 44.5°N 17.3°E
	e	A	43 14.5	H = 12 41 19 (BCIS)
	eSn	A	44 22	D = 7.2°
	eLQ	B	45 13	LmH:9s 1.4 /um
	LmH	B	45.4	MLH=3.9
	LmV	B	46.0	ei 43 57 e 44 43 e 44 49
11.	e	A	13 35 42	Near earthquake
	e(Sg)	A	36 08	TRIESTE gives: e(Sg) 13 34 45.3
	e	A	36 16	VIENNA: e 13 33 52 e 34 06
				KHC: e 13 34 23 e 35 05.5

Day	Phase		h m s	Remarks
11.	e(Sg)	A	16 11 09	
	e	A	11 21	
11.	ePKP	A	17 50 23	Fiji Islands Region 19.44°S 177.73°W
				H = 17 31 22.8 h = 381 km MAG=4.2 (USCGS)
				D = 148.0°
12.	eP	A	10 43 40	Alaska 54.9°N 161.1°W
	e(PcP)	A	43 56	H = 10 32 01.6 h = 33 km MAG=5.0 (USCGS)
	e	A	44 08	D = 74.5°
12.	ePg	A	13 23 13	Explosion
	i(Sg)	A	23 22.0	
	i(Lg)	A	23 25.5	
12.	LmH	C	14 30.6	LmH:14s 0.3 µm
	LmV	C	32.4	
12.	e	A	15 26 15	
	e(Sg)	A	26 18.5	
12.	e(Sg)	A	17 27 54	Probably explosion.
12.	eP1	A	21 13 12.5	South of Panama 5.63°N 82.58°W
	eP	C	13 14	H = 21 00 20.9 h = normal MAG=5.0
	+iP2	A	13 20.7	D = 88.4° Az = 39.4° (USCGS)
	eiSKS	C	23 44	PV2:2.4s 190nm SSH:27s 8.6 µm
	eS	C	24 00	LmH:20.5s 6.2 µm LmV:23s 11.3 µm
	eSP	C	24 56	MPV2=5.9 MLH=6.0 MLV=6.2
	eSS	C	29 48	i 13 27.7 e 14 03 e 14 22 e 26 08
	eiLQ	C	36 24	Multiple P. The first onset is much smaller
	eLR	C	41 12	than the second one. Two shocks in the same
	LmH	B	47.2	focus?
	LmV	B	47.2	AN USSR gives: 7.2°N 83.6°W
				H = 21 00 30 h = normal MAG=6
				D = 87.7°
12.	ePKP	A	21 34 32.4	Fiji Islands 16.11°S 178.29°E
	e	A	35 48	H = 21 14 53.1 h = normal MAG=5.3
	e	A	36 26	D = 143.9° Az = 345.6° (USCGS)

Day	Phase		h m s	Remarks
13.	ePKP	A	01 11 53	South of Kermadec Islands 32.2°S 178.3°W
	epPKP2	A	12 04	H = 00 51 16.8 h = normal MAG=4.5 (USCGS)
				D = 160.3° h = 39 km
13.	eP	B	02 14 23.5	Algeria 35.53°N 0.11°W
	-eP1	A	14 23.5	H = 02 10 20.0 h = 13 km MAG=5.0
	eiP2	A	14 26.5	D = 17.3° Az = 25.8° (USCGS)
	ePP	A	14 38	PV:5.5s 0.5 μm PV1:1.7s 30nm
	e	A	14 43	PV2:1.5s 77.2nm
	iPPP	A	14 47.4	LmH:13s 1.2 μm LmV:13s 1.9 μm
	eS	B	17 38	MLH=4.3 MLV=4.7
	LmH	B	21.7	Successive P-onsets in our short-period
	LmV	B	21.7	records, the second bigger than the first
				one.
13.	e	A	02 34 53	Upper Silesia (KRAKOW)
	e(Sg)	A	37 04.5	e 36 45.5 e 37 08 e 37 17
13.	ePKP	A	07 55 36	New Hebrides Islands 16.2°S 178.1°E
	e(pPKP)	A	55 46	H = 07 36 07.2 h = 50 km MAG=5.4 (USCGS)
	eSS	C	08 17 40	D = 144.0° (h = 36 km)
	LQ	C	32	
	LmH	C	09 00	
	LmV	C	00	
13.	e(Sg)	A	10 01 23	Explosion
				(Sg)V:1.2s 28.0nm
13.	ePKP	A	10 23 51	New Hebrides Islands 20.40°S 169.27°E
	+iPKP	B	23 51.2	H = 10 04 19.0 h = 46 km MAG=5.0
	-iPKP	A	23 52.5	D = 145.0° Az = 335.0° (USCGS)
	-epPKP	B	23 02.5	LmH:21s 0.3 μm LmV:21s 0.6 μm
	+ipPKP	A	23 03.6	MLH=5.0 MLV=5.3
	LmH	B	11 31.5	i 23 10.0 ei 23 19
	LmV	B	31.5	
13.	eP	A	14 41 37	Albania 40.75°N 19.50°E
	e(PP)	A	41 45	H = 14 38 53.9 h = 23 km MAG=4.4
	e(Li)	B	44 18	D = 11.3° Az = 333.6°

Day	Phase		h m s	Remarks
cont.				
13.	eLg1	A	14 44 53	LmH:12s 1.4 μm LmV:11.5s 1.6 μm
	eLg2	B	45 18	MLH=4.1
	LmH	B	47.4	e 41 59.5 e 42 26.5
	LmV	B	47.4	Clear higher-mode surface waves in our records.
13.	e	A	22 10 24.5	Near earthquake.
	e(Sg)	A	10 41	TRIESTE gives: D = ca. 75 km
	e	A	10 49	
14.	epPKIKP	A	03 07 28	Santa Cruz Islands 11.44°S 166.23°E H = 02 47 53.0 h = 80 km MAG=5.2 (USCGS) D = 135.7°
14.	eP	A	03 18 41	Red Sea 19.81°N 38.93°E H = 03 11 28.2 h = normal MAG=4.7 D = 37.6° Az = 331.4° (USCGS) PV:1.8s 22nm MPV=4.8 AN USSR gives: Red Sea 18.1°N 38.6°E H = 03 11 16 h = normal D = 38.9°
14.	iSg	A	09 22 21.5	Explosion
14.	i	A	09 43 21.5	
14.	e	A	11 22 23	
14.	eSg	A	11 51 42	Explosion
14.	eP	A	11 17 06.5	Greece (UPP)
	e	A	17 10	D = ca. 15°
	e	A	22 49	LmH:12s 0.4 μm LmV:12s 0.5 μm
	LmH	B	24	MLH=3.8 MLV=4.0
	LmV	B	24	e 17 22 e 22 22
14.	e	A	11 52 17.5	

Day	Phase		h m s	Remarks
14.	LmH	C	12 06.4	Iran-Iraq Border Region (USCGS)
14.	ei(Pg)	A	12 48 49.5	Explosion
	iSg	A	49 10.5	(D = ca. 1.6°)
	i	A	49 14.0	
14.	eP	A	14 05 08	Unimak Island 54.0°N 164.3°W
				H = 13 53 23.8 h = normal MAG=4.7 (USCGS)
				D = 75.7°
14.	iPg	A	15 05 20.0	Explosion/Hilders, GFR 50°32.5'N 10°02.4'E
	eiSg	A	05 36.0	D = ca. 1.2° Yield: 8.75 t
	LmH	A	05 54	
14.	eP	A	18 48 06.5	Mid-Indian Rise 16.38°S 66.78°E
				H = 18 35 46.7 h = normal MAG=5.2
				D = 82.1° Az = 329.3° (USCGS)
14.	eSS	C	23 14.0	Bismarck Sea 3.6°S 149.4°E
	LmH	B	52	H = 22 36 59.9 h = normal MAG=4.6 (USCGS)
	LmV	B	52	D = 121.2°
				LmH:20s 0.3/um LmV:20s 0.4/um
				MLH=4.9 MLV=5.1
15.	ePg	A	02 24 31.5	Switzerland 46.8°N 8.8°E
	i	A	24 35.5	H = 02 23 12 (BCIS)
	ei	A	24 56	D = 4.3°
	eiSg	A	25 31.0	
	iSgSg	A	25 34.8	
15.	+iP	A	03 34 46.2	Eastern Kazakh SSR 49.83°N 78.11°E
	ePn	A	36 18	H = 03 26 57.4 h = 0 km MAG=5.4
				D = 41.2° Az = 297.7° (USCGS)
				PV:0.7s 100nm
				MPV=5.9
				Underground explosion.

Day	Phase		h m s	Remarks
15.	eP	A	08 26 54	Rat Islands/Aleutian Is. 51.5°N 176.8°E
	e	A	27 23	H = 08 14 59.3 h = 32 km MAG=4.9 (USCGS)
	LmH	B	09 04.5	D = 77.5°
	LmV	B	04.5	
15.	LmH	C	11 27	Luzon/Philippine Is. (USCGS)
15.	e(Sg)	A	12 01 17.5	Explosion?
	e	A	01 19	PRUHONICE gives: D = 1.5°
15.	iPg	A	12 10 56.5	Explosion
	iSg	A	10 16.0	D = ca. 1.5°
15.	LmV	C	12 50	
15.	LmH	B	14 34.0	LmH:16s 0.55/um LmV:16s 0.17/um
	LmV	B	38.2	
15.	eP	A	14 54 18	Mindanao/Philippine Islands
	+ipP	A	54 27.0	6.81°N 126.31°E
	e	A	55 16.5	H = 14 40 35.0 h = 37 km MAG=5.3
	ePP	A	58 30	D = 100.0° Az = 324.0° (USCGS); h = 33 km
	LmH	B	15 41.7	pPV:1.5s 30.0nm
	LmV	B	44.0	LmH:20s 0.5/um LmV:20s 0.5/um
15.	e(P)	A	16 35 24.5	
15.	iPKHKP	A	17 22 59.8	Fiji Islands Region 21.91°S 179.55°W
	eiPKP2	A	23 07.5	H = 17 04 13.0 h = 574 km MAG=4.1
				D = 150.0° Az = 345.7° (USCGS)
16.	e(P)	A	04 40 10	
16.	eP	C	13 49.0	West New Guinea Region 0.82°S 132.61°E
	eP	A	49 13.5	H = 13 34 29.9 h = normal MAG=5.2
	ePP	A	53 36.5	D = 109.8° Az = 324.5° (USCGS)
	iPP	B	53 37	PPV:15.5s 1.8/um PSH(B):17s 2.0/um
	ei(pPP)	A	53 43	PPSH(B):15s 2.3/um PPSV(B):15s 2.2/um
	eSKS	C	59 36	LmH:18s 18.5/um LmV:16s 7.9/um

Day	Phase		h m s	Remarks
cont.				
16.	eSKS	B	13 59 40	MPPV=6.6 MLH=6.7 MLV=6.3
	ei(SKKS)	B	14 00 36	e 49 53 e 52(10) ei52 41 e 59 00
	eS	C	01 16	(pPP) is the greatest onset in the short-
	ePS	C	02 58	period records. (SKKS) is greater in Z
	ePPS	C	03 54	than in the horizontal components. Another
	iPPS	B	04 05	possibility would be to interpret this
	eSS	C	09.1	phase as Pa (7.8 km/s).
	eSSS	C	13.9	
	iLQ	C	19 40	
	eLR	C	27.3	
	LmH	B	35.3	
	LmV	B	42.0	
16.	ei(Px)	A	14 05 36	France 47.3°N 5.4°E
	eiPb	A	05 48.5	H = 14 04 11 (BCIS); MAG=4.4 (BENSBERG)
	eiPg	A	05 55.0	D = 5.3°
	eSn	A	06 33	PbV:0.5s 33.7nm PgV:1.3s 27.8nm
	iSb	A	06 57.0	PgH:1.3s 49.8nm SbV:1.0s 52.1nm
	iSg	A	06 06.0	SgH:1.0s 230nm
	iSgSg	A	07 13.5	i 05 50.5 i 05 52.3 i 06 27.3 ei 06 56
				Pb and Sb are unusnally clear.
16.	e	A	19 03 49	Yugoslavia 44.0°N 17.0°E
	eSn	A	05 01	H = 19 01 49 (BCIS)
	eSg	A	05 55	D = 7.6°
	e(SgSg)	A	06 03	e 05 17.5 e 05 25 e 05 31 e 06 21
16.	iPKP	A	21 30 56.5	Tonga Islands 16.78°S 173.43°W
	epPKP	A	31 18.5	H = 21 11 16.5 h = 24 km MAG=4.9
	ei(sPKP)	A	31 28.5	D = 146.0° Az = 354.3° (USCGS); h = 78 km
				PV:1.0s 17.4nm
17.	e(P)	A	10 32 28.5	
17.	-eP	A	11 40 14.5	Fox Islands/Aleutian Is. 51.12°N 169.27°W
	e	A	40 24	H = 11 28 13.4 h = normal MAG=5.0
	e	A	40 45	D = 78.6° Az = 359.4° (USCGS)

Day	Phase		h m s	Remarks
17.	eP	A	12 48 22	Near East Coast of Honshu/Japan
	ipP	A	48 34.2	38.3°N 142.1°E
	LmH	B	13 27.5	H = 12 36 07.7 h = normal MAG=4.5
	LmV	B	27.5	D = 81.3° Az = 330.7° (USCGS); h = 44 km
				PV:1.2s 10.2nm
				MPV=4.8
17.	e(P)	A	12 54(37)	
17.	e(P)	A	14 04 57	
17.	e(P)	A	16 05 50.5	
17.	LmH	C	18 13	Taiwan Region (USCGS)
	LmV	C	13	
17.	LmH	C	19 53	West Chile Rise 42.7°S 83.4°W
	LmV	C	53.0	H = 18 45 02.5 h = normal MAG=4.9 (USCGS)
				D = 124.2°
				LmV:21s 0.13 µm
				MLV=4.6
17.	e	A	20 15 20	Probably to France 47.3°N 5.4°E
				H = 00 57 46 (BCIS)
				D = 5.3°
18.	e	A	02 59 52	
18.	eiPg	A	12 13 47.5	Explosion
	ei	A	13 56	
	iL	A	14 03.0	
18.	e	A	14 15 22	Explosion
	e(Sg)	A	15 33	
18.	eP	A	17 11 26	Near Coast of Honshu/Japan 40.13°N 142.43°E
	epP	A	11 39.5	H = 16 59 21.9 h = 52 km MAG=4.4
				D = 79.9° Az = 330.7° (USCGS); h = 50 km

Day	Phase		h m s	Remarks
18.	e	A	19 33 50	Explosion
	e(Sg)	A	33 52	SgV:1.2s 20.4nm
19.	eP	AB	09 10 27	Turkey 37.91°N 28.99°E
	è	A	10 32.5	H = 09 06 19.0 h = 28 km MAG=4.7
	eS	C	13(44)	D = 17.7° Az = 321.4° (USCGS)
	eLQ	C	13 52	PV(A):1.8s 51.1nm PV(B):7s 0.55 µm
	eiLg1	B	15 44	LmH:13.5s 3.5 µm
	e(Lg2)	B	16 22	MLH=4.8
	LmH	B	16.4	e 13 55 e 14 10
	LmV	B	18.1	BCIS gives: Turkey 37.3°N 30.0°E
				H = 09 06 06 MAG=4.9 (STRASBOURG),
				4.6 (ATHENS)
				D = 18.7°
19.	e	A	11 47 53	(Sg)V:1.2s 12.8nm
	e(Sg)	A	48 05	
19.	iPg	A	12 52 42.2	Explosion/GDR
	i(Sg)	A	52 48.0	
	iL	A	52 51.3	
19.	i(Sg)	A	12 53 13.0	Explosion/GDR
	iL	A	53 15.0	
19.	eiPKHKP	A	13 00 17	Fiji Islands Region 20.31°S 178.18°W
	iPKP2	A	00 23.0	H = 12 41 28.8 h = 518 km MAG=4.5
	e	A	00 47.5	D = 148.7° Az = 348.0° (USCGS)
	e	A	02 30	PV1:1.0s 47.3nm PV2:1.3s 22.2nm
	e	A	02 50	
19.	e	A	13 09 09.5	Explosion?
19.	e	A	13 37 51	Explosion
	e(Sg)	A	37 54	
	i	A	37 56.0	
19.	e(P)	A	15 11 10	
	e	A	11 23.5	

Day	Phase		h m s	Remarks
19.	eP	A	16 21 57.5	Greece 38.34°N 21.16°E
	e	A	22 04	H = 16 18 38.4 h = normal MAG=4.5 (USCGS)
	e	A	22 44.5	D = 14.1°
19.	e	A	16 46 04	Near earthquake?
	e	A	46 09.5	
19.	iP	A	17 36 17.0	Hindu Kush Region 36.50°N 70.30°E
				H = 17 28 32.2 h = 223 km MAG=4.7 (USCGS)
				D = 43.6°
				AN USSR gives: Hindu Kush Region
				35.8°N 70.8°E
				H = 17 28 06 h = normal
				D = 44.4°
20.	e	A	07 02 40.5	Explosion
	e	A	02 47	
20.	iP	A	07 09 29.8	Kurile Islands 44.37°N 149.32°E
				H = 06 57 30.8 h = normal MAG=4.4
				D = 78.5° Az = 334.1°
				PV:1.0s 7.1nm
				MPV=4.8
20.	iSg	A	09 08 05.5	Explosion
20.	iP	A	09 14 12.4	Kodiak Islands Region 56.52°N 153.27°W
				H = 09 02 47.3 h = 33 km MAG=4.5
				D = 72.5° Az = 10.0° (USCGS)
				PV:1.1s 9.6nm
				MPV=4.8
20.	e(Sg)	A	11 32 08.5	Explosion/ČSSR 50°05.3'N 16°20.8'E
	e	A	32 15	Yield: 19.8 t (PRUHONICE)
				D = 3.1°
20.	LmH	C	12 52.5	East New Guinea Region (USCGS)

Day	Phase		h m s	Remarks
20.	iPg	A	13 01 37.8	Explosion/GDR 51°17.2'N 12°43.8'E
	iSg	A	01 52.1	D = 0.9° Yield: 2.7 t
20.	eP	A	13 25 24	Catamarca Prov./Argentina 28.07°S 66.92°W
	ePP	AC	29 38	H = 13 11 35.0 h = 157 km MAG=5.3 (USCGS)
	iPP	B	29 40	D = 104.4°
	iSKS	B	35 52	PPV(B):4.5s 0.4 μm SKSH:8s 0.7 μm
	eSKKS	C	36 32	SH:10s 0.55 μm SPV:12s 0.5 μm
	eiSKKS	B	36 36	LmV:19s 0.4 μm LmH:18s 0.5 μm
	eS	C	37 00	MPPV=6.2
	eSP	B	38 36	
	eSS	C	44.3	
	LmH	B	14 14.0	
	LmV	B	14.0	
20.	iPg	A	14 12 06.5	Explosion
	iSg	A	12 22.5	D = ca. 1.1°
20.	eP	A	14 38 09	Rat Islands/Aleutian Is. 51.36°N 178.33°E
	e	A	38 16	H = 14 26 14.1 h = 33 km MAG=5.3
	e	A	38 34	D = 77.7° Az = 351.4° (USCGS)
				PV:1.1s 33.8nm
				MPV=5.4
20.	eP	A	15 50 26.5	West Caroline Islands 7.73°N 134.88°E
	e	A	50 34.5	H = 15 36 20.1 h = 8 km MAG=5.8
	e	A	50 44	D = 104.1° Az = 326.7°
	ePP	C	54 50	PV1:1.7s 35nm PV2:1.7s 35nm
	iPP	C	54 59	PV3:2.3s 80nm PPV(B):10s 1.3 μm
	eiSKS	B	16 00 56	SH(B):16s 2.1 μm SH(C):(30s) 3.4 μm
	eS	C	02 12	SSH:34s 8.8 μm
	eS	B	02 16	LmH:16.5s 17 μm LmV:17s 20.7 μm
	eSP	C	04.1	MPPV=6.4 MLH=6.8 MLV=6.9
	eiSS	C	09 38	LQ begins in the registrations of type C
	iSSS	C	13 54	with periods of about 72 s and great am-
	eLQ	C	20.8	plitudes. Very clear normal dispersion.
	LmH	B	41.5	
	LmV	B	41.5	

Day	Phase		h m s	Remarks
20.	e	A	16 21 30.5	Yugoslavia 46.18°N 15.15°E
	e(Pb)	A	21 35.5	H = 16 20 07.9 h = 22 km MAG=4.3
	e	A	21 42.5	D = 5.0° Az = 333.5° (USCGS)
	ePg	A	21 50	Istria 45 1/2°N 14 1/4°E
	iSn	A	22 21.5	H = 16 20 02 (BCIS)
	eiSg	A	22 52.5	D = 5.5°
	eSgSg	A	22 59.5	
	e	A	23 08	
20.	ePn	A	19 06 09.5	Albania 40.82°N 19.78°E
	ePg	A	07 17.5	H = 19 03 27.3 h = normal MAG=4.3
	e(Sb)	A	09 02	D = 11.4° Az = 332.7° (USCGS)
	eLg2	B	09 48	LmH:16s 0.9 µm LmV:16s 0.5 µm
	LmH	B	10.2	MLH=3.8
	LmV	B	11.2	e 06 31.5 e 08 22 e 09 43 e 10 09
20.	ePKIKP	A	23 31 39	South of Fiji Islands 26.46°S 178.47°E
	eiPKHKP	A	31 47.3	H = 23 12 54.4 h = 596 km MAG=5.2
	iPKP2	A	32 04.2	D = 153.8° Az = 340.8° (USCGS)
	e	A	32 17	PV2:1.3s 16.7nm PV3:1.4s 30.3nm
	e	A	34 16	AN USSR gives: Kermadec Islands
				28.1°S 179.6°W
				H = 23 11 45 h = normal
21.	eSg	A	01 22 45	Near earthquake.
				TRIESTE gives: D = 120 km
21.	eP	A	06 49 36	PV:0.8s 14.2nm
21.	LmH	C	08 03.4	Sumatra (USCGS)
	LmV	C	03.4	
21.	ei(P)	A	12 18 21	(P)V:0.5s 9.6nm
21.	i(Sg)	A	12 47 00	Explosion/GDR
	iL	A	47 03.1	
21.	i(Sg)	A	12 47 10	Explosion/GDR
	iL	A	47 12.5	

Day	Phase		h m s	Remarks
21.	iL	A	12 47 28.0	Explosion/GDR
21.	iPg	A	12 59 58.0	Explosion
	iSg	A	13 00 10.5	D = 0.9°
21.	ePKP	A	13 05 26.5	Fiji Islands Region 21.34°S 176.20°W
	e	A	05 28.5	H = 12 45 57.4 h = 199 km MAG=4.5
	e	A	05 31	D = 150.1° Az = 350.0° (USCGS)
	LmH	B	14 09.2	PV2:1.6s 30.4nm
	LmV	B	09.5	LmH:22s 0.3 µm LmV:24s 0.5 µm
21.	e(P)	A	14 07 03.5	(P)V:0.8s 11.8nm
21.	ePKIKP	A	19 47 36	New Hebrides Islands 19.17°S 168.60°E
	epPKIKP	A	47 47.5	H = 19 28 08.0 h = 53 km MAG=4.8
				D = 143.7° Az = 335.2° (USCGS); h = 41 km
				PV:1.2s 15.3nm
21.	eP	A	20 53 28	Cyprus 34.8°N 34.4°E
	e	A	53 45	H = 20 48 38.5 h = 114 km (USCGS)
				D = 22.8°
21.	ei(Sg)	A	21 51 02.5	Probably near earthquake.
22.	ePKIKP	A	04 17 56	South of Kermadec Islands 33.46°S 178.96°W
	ePKP2	A	18 42	H = 03 58 02.4 h = 39 km MAG=5.0
	epPKP2	A	18 53.5	D = 161.2° Az = 338.8° (USCGS); h = 41 km
	eSKKS	C	29 16	PV2:1.5s 26.8nm PV3:1.6s 45.5nm
	ePSKS	C	32 32	e 21 55 e 22 20
	ePKPScS	C	33 36	
	ePPS	C	36.0	
	eSS	C	42.5	
	eSSS	C	48.6	
22.	ePKP2	A	07 01 34.5	South of Kermadec Islands 33.69°S 178.71°W
	e(pPKP2)	A	01 44	H = 06 40 53.4 h = 26 km MAG=4.6 (USCGS)
				D = 161.5° (h = 34 km)
22.	e(Sg)	A	08 38 20	

Day	Phase		h m s	Remarks
22.	i(P)	A	08 19 57.5	Kodiak Island 58.3°N 151.6°W
	e	A	20 07	H = 08 08 35.1 h = normal MAG=4.1 (USCGS)
				D = 70.5°
22.	-eiPn	A	11 00 44.0	United Kingdom 51.41°N 1.32°E
	eiSn	A	01 54.5	H = 10 59 04.7 h = 0 km MAG=4.7
	ei(Sb)	A	02 24	D = 6.5° Az = 92.7° (USCGS)
	e(SgSg)	A	02 46	PnV:0.5s 33.7nm PnH:0.5s 32.5nm
				BCIS gives: 51°02'22"N 01°13'33"E
				H = 10 59 06 Yield: 500 t
				Chemical explosion.
22.	-iP	AB	17 00 49	Atapazari/Turkey 40.66°N 30.77°E
	eS	B	04(08)	H = 16 56 53.3 h = 4 km MAG=6.0
	LmH		(09)	D = 16.7° Az = 313.2° (USCGS)
				PH:(8s 32 μm)
				LmH:(10s 540 μm)
				MPH=ca.7 MLH=ca.7
				Multiple P with successively increasing amplitude. The amplitudes of surface waves are determined from the records of the Wiechert-seismograph.
22.	iP	A	17 34 02.5	Turkey
				H = 17 30.1 (UPP)
22.	eP1	A	17 52 00	Turkey 40.56°N 30.72°E
	iP2	A	52 01.5	H = 17 48 06.0 h = 26 km MAG=5.0
				D = 16.7° Az = 313.5° (USCGS)
				PV2:1.7s 94.5nm
22.	i	A	17 53 05	
22.	e	A	17 54 21	
22.	ei	A	17 54 35	
22.	eP	A	18 12 33	Turkey
				H = 18 08.6 (UPP)

Day	Phase		h m s	Remarks
22.	eiP	A	18 17 26	Turkey H = 18 13.6 (UPP)
22.	eP i	A A	18 13 45 13 48.2	Turkey 40.80°N 30.39°E H = 18 09 55.7 h = normal MAG=5.0 D = 16.3° Az = 313.3° (USCGS)
22.	eP	A	19 51 21	Turkey 40.81°N 30.93°E H = 19 47 25.6 h = normal MAG=4.6 D = 16.6° Az = 312.7° (USCGS)
22.	eP	A	20 39 36	Turkey 40.60°N 30.45°E H = 20 35 40.8 h = 16 km MAG=4.5 D = 16.5° Az = 313.7° (USCGS)
22.	eP	A	21 25 31	Turkey 40.46°N 30.51°E H = 21 21 34.1 h = 16 km MAG=4.4 (USCGS) D = 16.6°
22.	eP e	A A	22 12(27.5) 12 29.5	Turkey 40.70°N 30.70°E H = 22 08 29.5 h = 13 km MAG=4.5 (USCGS) D = 16.6°
22.	eP e	A A	23 45 51 45 57	Turkey 40.60°N 30.68°E H = 23 41 59.5 h = normal MAG=4.7 (USCGS) D = 16.7°
23.	i(P)	A	01 39 20.0	
23.	LmH	C	02 17	
23.	ePKIKP e ePP eSKSP LmH LmV	A A A A C C	03 28(12) 28 19 31 09 31 44.5 04 32.5 32.5	New Hebrides Islands 15.68°S 167.11°E H = 03 08 43.7 h = 33 km MAG=4.9 D = 139.9° Az = 335.8° (USCGS) LmH:23s 0.6 μm MLH=5.2

Day	Phase		h m s	Remarks
23.	eP	A	04 07 30	Turkey 40.58°N 30.56°E
				H = 04 03 38.5 h = 21 km MAG=4.5 (USCGS)
				D = 16.7°
23.	eP	A	04 53(45)	Turkey 40.6°N 30.7°E
	e	A	53 50	H = 04 48 53.3 h = normal MAG=4.5 (USCGS)
	LmH	C	58.2	D = 16.8°
23.	ePg	A	12 20 58	Explosion
	iSg	A	21 05.0	
	iL	A	21 06.5	
23.	LmH	B	07 54.2	Turkey 40.7°N 30.8°E
	LmV	B	54.2	H = 07 42 22 (BCIS)
				D = 16.9°
				LmH:9s 0.6 μm LmV:10s 0.6 μm
				MLH=4.1 MLV=4.3
23.	LmH	C	08 21.2	Ryukyu Islands 29.4°N 129.4°E
	LmV	C	21.2	H = 07 26 46.2 h = 45 km MAG=4.7 (USCGS)
				D = 83.3°
				LmH:16s 0.5 μm LmV:16s 0.4 μm
				MLH=5.0 MLV=4.9
23.	e	A	14 08 43	Macquarie Islands Region 56.18°S 158.33°E
	e(PKP2)	A	08 52	H = 13 48 05.8 h = normal MAG=5.1 (USCGS)
	e	A	09 11	D = 159.6°
	LmH	C	15 30.7	LmH:19s 1.1 μm
				MLH=5.6
23.	e	A	16 00 33	
23.	eP	A	16 01(02)	Turkey 40.7°N 30.6°E
	e	A	01 08	H = 15 57 08.8 h = normal MAG=4.4 (USCGS)
	e	A	01 56.5	D = 16.8°
23.	ePKP	A	19 03 48	Tonga Islands 19.96°S 175.48°W
	e	A	03 53	H = 18 44 13.6 h = 77 km MAG=4.9
				D = 148.0° Az = 351.2° (USCGS)
				PV2:1.5s 33.6nm

Day	Phase		h m s	Remarks
23.	e	A	20 06 03.5	Near earthquake?
	ei	A	06 19	
	ei	A	06 23	
23.	eP	A	23 23 07	Turkey 40.6°N 30.7°E H = 23 19 13.5 h = 15 km MAG=4.3 (USCGS) D = 16.9°
24.	eP	A	00 45 33	
24.	eP	A	03 44 16	Turkey 40.8°N 30.8°E
	i	A	44 20.5	H = 03 40 20.1 h = 4 km MAG=4.2 (USCGS) D = 16.6°
24.	LmH	C	08 17	Central California (USCGS)
24.	e	A	11 43 52.5	Explosion?
	e(Sg)	A	43 57	
	eL	A	44 05	
24.	e	A	12 25 47	Near earthquake.
	e	A	26 27.5	TRIESTE gives: D = 380 km
	ei	A	26 53	
	e	A	27 16	
24.	eP	A	13 56 58.5	
24.	eiP	A	15 40 24.5	Off East Coast of Honshu/Japan 33.11°N 142.07°E H = 15 27 45.4 h = 22 km MAG=4.6 D = 85.8° Az = 330.9° (USCGS)
24.	ei(Sg)	A	16 48 49	Explosion
25.	ipP	A	00 45 53.5	Southeast of Japan 33.1°N 142.0°E H = 00 33 02.8 h = 35 km MAG=4.5 (USCGS) D = 85.8°

Day	Phase		h m s	Remarks
25.	eP	A	08 40(28)	Greece-Bulgaria Border Region
	e	A	40 48	41.86°N 24.58°E
	eLg1	A	43 56	H = 08 37 25.7 h = normal MAG=4.2
	eLg2	A	44 24	D = 12.5° Az = 318.9° (USCGS)
	LmH	B	44.7	LmH:13.5s 0.5/um LmV:11s 0.5/um
	LmV	B	46	MLH=3.6
25.	iSn	A	11 25 08.7	Yugoslavia 45.6°N 14.5°E
	e	A	25 37.5	H = 11 22 47 (BCIS)
	eSg	A	25 40	D = 5.4°
	e	A	25 49	
25.	iPg	A	12 00 16.2	Explosion/GDR
25.	e(Sg)	A	12 40 58	Probably explosion.
25.	LmH	B	12 49.4	LmH:13s 0.2/um
	LmV	B	51	
25.	LmH	C	13 23.0	Southern Iran 28.9°N 54.5°E
				H = 13 00 38.8 h = 34 km MAG=4.5 (USCGS)
				D = 38.1°
				LmH:26s 0.2/um
				MLH=3.8
26.	eP	A	03 41 45	PV:1.3s 8.3nm
26.	e(P)	A	06 03(00)	Turkey
	i	A	03 06.0	H = 06 59.1 (UPP)
26.	LmV	B	04 24.4	LmH:11s 0.2/um LmV:11s 0.2/um
	LmH	B	24.6	
26.	ePKIKP	A	06 51 04.5	Kermadec Islands 31.79°S 178.68°W
	e	A	51 35	H = 06 31 10.6 h = 37 km MAG=5.1 (USCGS)
	e(PKP2)	A	51 43.5	D = 159.7°
	ei	A	51 58	

Day	Phase		h m s	Remarks
26.	ei(P)	A	08 06 50.5	(P)V:0.7s 11.9nm
	ei	A	07 06	
	e	A	07 21	
26.	+ePKIKP	A	08 34 36	Loyalty Islands Region 22.01°S 170.11°E
	epPKIKP	A	34 48	H = 08 14 56.3 h = 30 km MAG=5.0
	LmH	B	09 43	D = 146.8° Az = 334.8° (USCGS); h = 43 km
	LmV	B	43.5	PV(B):8s 1.1 /um
				LmH:19s 0.4 /um LmV:20s 0.5 /um
				MLH=5.2 MLV=5.3
26.	iP	A	09 19 59.6	Turkey 40.31°N 30.57°E
	ei	A	20 04.3	H = 09 16 06.4 h = normal MAG=4.5 (USCGS)
	e	A	20 09	D = 16.8°
	LmH	B	28.0	PV2:1.6s 22.8nm
	LmV	B	28.0	LmH:10s 0.7 /um LmV:9.5s 0.6 /um
				MLH=4.2 MLV=4.3
26.	eP	A	09 35 44.5	Greenland Sea 73.43°N 7.16°E
	e	A	35 51	H = 09 30 42.6 h = normal MAG=4.4
				D = 23.0° Az = 172.7° (USCGS)
				PV:1.5s 26.8nm
				MPV=4.6
26.	ePKP	A	09 58 19.5	New Hebrides Islands 20.54°S 169.89°E
				H = 09 38 44.1 h = 34 km MAG=4.6
				D = 145.4° Az = 335.5° (USCGS)
26.	e(Sg)	A	12 08 17.5	(Sg)V:1.0s 14.2nm
26.	iPg	A	14 11 45.5	Explosion/GDR
	eiSg	A	12 01.5	D = ca. 1.2°
26.	ei	A	18 01 19.5	Explosion?
	i	A	01 25	
	eL	A	01 30	
26.	-eiP1	A	18 58 04	Turkey 39.48°N 40.36°E
	-iP	B	58 04	H = 18 53 01.3 h = normal MAG=5.6
	-iPL	C	58 05	D = 23.0° Az = 308.5° (USCGS)

Day	Phase		h m s	Remarks
cont.				
26.	+iP2	A	18 58 06.0	PLV:26s 2.7 /um PLH:26s 2.6 /um
	iS	B	19 02 15	SH:12.5s 19.8 /um
	i	C	02 22	LmH:12s 39 /um LmV:11s 29.2 /um
	eLR	C	03.8	MSH=6.4 MLH=6.1 MLV=6.1
	LmH	B	10.8	Multiple P in the short-period records,
	LmV	B	10.8	P1 much smaller than P2. Clear PL-waves
				in the long-period registrations are su-
				perimposed by short-period P-waves (T ca.
				5 s).
26.	ePKP	A	19 12 03	Tonga Islands 17.41°S 174.00°W
	e	A	12 35	H = 18 52 21.2 h = 15 km MAG=5.0
				D = 146.5° Az = 353.2° (USCGS)
				PV: 1.6 s 49.3 nm
27.	i(PKP)	A	00 27 52.9	Probably to Solomon Islands
				6.76°S 155.44°E
				H = 00 08 40.1 h = 54 km MAG=5.2
				D = 126.9° Az = 332.0° (USCGS)
				PKIKP must be 13 s earlier than (PKP).
27.	LmH	B	00 57.5	Revilla Gigedo Islands 19.9°N 109.4°W
	LmV	B	57.5	H = 00 00 47.9 h = 31 km MAG=5.1 (USCGS)
				D = 92.7°
				LmH:15s 0.4 /um LmV:16s 0.8 /um
				MLH=5.0 MLV=5.3
27.	ei	A	01 50 43.5	Explosion?
	e(Sg)	A	50 45.5	(Sg)V:0.8s 14.1nm
27.	e	A	01 48 07.5	Iran 31.7°N 50.8°E
	e	A	48 24	H = 01 40 53.6 h = 65 km MAG=5.0 (USCGS)
	e	A	50 24	D = 34.5°
	LmH	C	02 01.5	
27.	eP	A	04 59 20	Crete 34.28°N 26.66°E
	ei	A	59 24	H = 04 54 52.0 h = 53 km MAG=4.2
	e	A	59 57	D = 19.7° Az = 330.7° (USCGS)

Day	Phase		h m s	Remarks
27.	+eP	A	05 22 43.5	Iceland 64.02°N 20.73°W
	e	A	22 53.5	H = 05 17 54.0 h = normal MAG=5.0
	e	A	23 01	D = 21.7° Az = 112.8° (USCGS)
	e	A	23 08.5	PV:1.9s 165nm PH:2.0s 157nm
	eiS	B	26 50	SH:7s 1.1/um
	LmH	B	33.1	LmH:12s 1.3/um LmV:12s 1.8/um
	LmV	B	33.3	MPV=5.1 MPH=5.2 MSH=5.2 MLH=4.6 MLV=4.8
27.	eP	A	11 48 47	South Indian Ocean 35.12°S 54.01°E
	eS	C	59(40)	H = 11 35 33.8 h = normal MAG=5.0
	ePPS	C	12 01 08	D = 93.2° Az = 334.5° (USCGS)
	LmH	C	30.6	LmH:24s 0.3/um LmV:20s 0.4/um
	LmV	C	32.0	MLH=4.7 MLV=4.9
27.	ePg	A	13 15 21.5	Explosion/Zeulenroda, GDR
	iSg	A	15 26.8	H = 13 15 16.5
				D = ca. 0.3°
28.	LmH	C	03 00.1	LmH:18s 0.3/um LmV:16s 0.2/um
	LmV	C	01	
28.	LmV	C	04(35)	Oaxaca/Mexico (USCGS)
	LmH	C	36	
28.	eiPKP	A	05 36 08	Loyalty Islands Region 22.05°S 169.91°E
	e	A	36 38	H = 05 16 28.2 h = 33 km MAG=4.1
				D = 146.8° Az = 334.5°
				PV:1.2s 15.3nm
28.	eiP	A	07 14 05	Kurile Islands 47.4°N 153.1°E
				H = 07 02 17.2 h = 60 km MAG=4.5 (USCGS)
				D = 76.9°
28.	iPg	A	09 13 06.0	Explosion
	iSg	A	13 07.5	
28.	e(P)	A	09 56 14.5	Hokkaido/Japan Region 42.59°N 145.38°E
				H = 09 44 21.1 h = 142 km MAG=3.9 (USCGS)
				D = 78.7°
				P must be about 6 s earlier than e(P).

Day	Phase		h m s	Remarks
28.	ePKIKP	A	10 06 34.5	South of Australia 49.68°S 117.04°E
				H = 09 47 19.3 h = normal MAG=5.0
				D = 134.1° Az = 301.2° (USCGS)
28.	eSg	A	10 32 22.5	Explosion?
	eL	A	32 38	
28.	eiSg	A	12 00 44	
28.	ei	A	12 47 26.5	Explosion/GDR
	eiSg	A	47 29	
28.	eSg	A	12 47 39	Explosion/GDR
28.	eiSg	A	12 47 46	Explosion/GDR
28.	eSg	A	12 47 55.5	Explosion/GDR
28.	ei(Sg)	A	14 07 17	
28.	ePKIKP	A	14 44 31	Fiji Islands Region 20.73°S 178.53°W
	iPKHKP	A	44 36.5	H = 14 25 50.1 h = 555 km MAG=4.7
	iPKP2	A	44 43.2	D = 149.1° Az = 347.4° (USCGS)
	epPKP	A	46 58	PV2:1.5s 53.7nm PV3:1.3s 22.2nm
28.	iPn	A	15 05 35.0	Explosion/Bransrode, GFR
	i	A	05 35.8	51°14.0'N 9°51.6'E
	iPg	A	05 36.9	D = 1.2° Yield: 21 t
	i	A	05 52.2	
	iSg	A	05 53.2	
28.	eP	A	15 39 53	Iceland 63.94°N 20.53°W
	e	A	39 56	H = 15 35 03.4 h = 31 km MAG=4.6
	eS	B	44 00	D = 21.6° Az = 112.9° (USCGS)
	LmH	B	48.3	PV1:1.6s 44.0nm PV2:1.4s 40.0nm
	LmV	B	50.4	LmH:18s 0.7 μm LmV:12s 0.9 μm
				MPV=4.8 MLH=4.1 MLV=4.6

Day	Phase		h m s	Remarks
28.	eP	A	17 40 14.5	Northern Sumatra 2.11°N 98.01°E
	e(pP)	A	40 29	H = 17 27 35.7 h = 32 km MAG=5.1
	e(sP)	A	40 34	D = 86.1° Az = 320.4° (USCGS); (h = 56 km)
				PV:1.5s 13.4nm
				MPV=4.9
28.	eP	A	22 57 46.5	Southwestern Ryukyu Islands
	LmH	B	23 39	23.89°N 125.44°E
	LmV	B	41	H = 22 45 07.1 h = 14 km MAG=4.9
				D = 85.7° Az = 324.3° (USCGS)
				LmV:18s 0.3 μm
				MLV=4.7
29.	e(Sg)	A	00 18 56	Near earthquake.
				TRIESTE gives: D = 115 km
29.	eiP	A	02 25 59	Iceland 64.05°N 20.56°W
	ei	A	26 03.5	H = 02 21 09.5 h = normal MAG=4.7
	e	A	26 18.5	D = 21.6° Az = 113.1° (USCGS)
	eS	C	30.0	PV:2.0s 51.8nm SH(B):9s 0.4 μm
	eiS	B	30 09	LmH:12s 0.5 μm LmV:12s 0.6 μm
	LmH	B	35.5	MPV=4.7 MSH=4.8 MLH=4.1 MLV=4.3
	LmV	B	36.5	
29.	LmH	C	03 49	Off Coast of Hokkaido/Japan
	LmV	C	49	42.75°N 146.73°E
				H = 02 57 18.1 h = normal MAG=4.6 (USCGS)
				D = 79.1°
				LmH:18s 0.1 μm LmV:16s 0.2 μm
				MLH=4.3 MLV=4.5
29.	e(P)	A	10 36 23.5	Northern Colombia 6.81°N 72.99°W
	+eiP	BC	36 24.5	H = 10 24 24.6 h = 161 km MAG=6.0
	+iP	A	36 25.2	D = 81.4° Az = 39.9° (USCGS);
	epP	B	37 05	h = ca. 165 km
	epP	C	37 06	PV(A):2.5s 735nm PV(B):11s 6.2 μm
	esP	C	37 20	PH(B):11s 2.2 μm SH(B):17s 11.6 μm
	e(sP)	B	37 24	sSH:15s 18.7 μm PKPPKPV:2.2s 71.5nm
	ei(pPP)	B	40 19	LmH:15.5s 5.4 μm LmV:17s 9.8 μm

Day	Phase		h m s	Remarks
cont.				
29.	eS	AB	10 46 22	MPV(A)=6.0 MPV(B)=6.2 MPH(B)=6.1
	eSP	C	47 00	MSH(B)=6.6
	epS	B	47 10	e 37 34 e 37 48.5 i 47 56 i 48 13
	isS	B	47 29	P has an exceptionally long period in the
	isSS	B	52 46	short-period registrations (T = 2.5 s).
	iSa	B	58.6	(P) is a small-amplitude precursor. The
	iR	C	11 02 16	long-period Rayleigh-wave (R; T ca. 50 s)
	ePKPPKP	A	03 02	shows a sharp beginning and a great ampli-
	epPKPPKP	A	03 44	tude.
	eSKPPKP	A	06 16	
	esSKPPKP	A	06 33	
	LmV	B	10.9	
	LmH	B	11.1	
29.	eP	A	11 38 56	Kurile Islands 44.53°N 149.15°E
				H = 11 26 57.4 h = normal MAG=4.0 (USCGS)
				D = 78.2°
				PV:0.8s 7.1nm
				MPV=4.8
29.	eP	A	15 10 28	
29.	e	A	21 28 27.5	
29.	LmH	B	22 04.1	LmH:13.5s 0.3/um LmV:12s 0.3/um
	LmV	B	04.9	
29.	ePKP	A	22 23 50.5	Fiji Islands Region 17.07°S 177.15°W
	e	A	24 07	H = 22 04 27.2 h = 187 km MAG=4.2
				D = 145.8° Az = 350.1° (USCGS)
30.	eP1	B	00 11 41	Near Coast of Venecuela 10.56°N 67.26°W
	iP2	B	11 48.5	H = 23 59 58.7 h = 10 km MAG=5.6
	iP3	B	12 02	D = 74.9° Az = 40.3° (USCGS)
	iP3	A	12 03.5	PV2:8s 1.5/um PV3(B):12s 3.0/um
	eX	B	13 08	PH3(B):12s 1.1/um XV:14s 2.2/um
	e(PP)	B	14 20	(PP)V:16s 2.3/um SH2:12s 1.2/um
	ePa	B	17.3	SH3:14s 5.5/um

Day	Phase		h	m	s	Remarks
cont.						
30.	eS1	C	00	21	12	LmH:21s 24 µm LmV:21.5s 44.6 µm
	eS2	B		21	18	MPV2=6.2 MPV3=6.3 MPH3=6.2 MSH2=5.9
	iS3	B		21	31	MSH3=6.5 MLH=6.5 MLV=6.8
	iSS	C		26	20	e 15 14 i 22 46 e 23 33
	iG	C		31	00	Multiple P- and S-onsets with increasing
	iR	C		35	00	amplitudes. Probably multiple shocks in
	LmH	B		41.3		the same focus Exceptionally clear G-
	LmV	B		41.3		(T ca. 50 s) and R- (T ca. 40 s) waves
						with abnormally great amplitudes in the
						seismographs of the type C.
30.	ePKIKP	A	01	21	10	South of Kermadec Islands 33.04°S 179.11°W
						H = 01 01 12.2 h = normal (USCGS)
						D = 160.8°
30.	eP1	A	01	34	52.5	Turkey 40.68°N 30.45°E
	iP2	AB		34	55.0	H = 01 31 01.7 h = 16 km MAG=5.6
	eiS	C		37	50	D = 16.5° Az = 313.5° (USCGS)
	iSS	C		38	14	PV1:1.5s 23.5nm PV2(A):1.8s 316nm
	eLg1	B		39	56	PV2(B):12s 2.9 µm
	LmH	B		42.8		LmH:11s 14.3 µm LmV:11s 14.0 µm
	LmV	B		42.8		MLH=5.4 MLV=5.6
						P is multiple in the short-period records.
30.	e(P)	A	02	01	12	Probably Turkey
						H = 01 57.3 (UPP)
30.	LmH	B	09	24		South Sandwich Island Region 60.1°S 28.5°W
	LmV	B		24		H = 08 19 28.3 h = 33 km MAG=5.2 (USCGS)
						D = 114.8°
						LmH:19s 0.4 µm LmV:19s 0.6 µm
						MLH=5.1 MLV=5.2
30.	e(P)	A	10	28	09	Probably Turkey
	e	A		28	37.5	H = 10 25 09 (UPP)
30.	ePKIKP	A	11	09	23	West of Macquarie Island 56.16°S 146.95°E
	ei	A		09	29.5	H = 10 49 32.8 h = normal MAG=5.1
	eiPKHKP	A		09	31.5	D = 153.2° Az = 277.2° (USCGS)

Day	Phase		h m s	Remarks
cont.				
30.	ePKP2	A	11 09 43.5	LmH:20s 1.3 µm LmV:20s 2.1 µm
	eSS	C	32 50	MLH=5.6 MLV=5.9
	eSSP	C	33 38	ei 09 29.5 e 09 53 e 47 30
	eSPSP	C	34 12	SS2 and SSP2 are travelled to the station
	eSSS	C	38 30	along the farther path (D_2 = 360° - D).
	eSS2	C	41.0	
	eSSP2	C	43.2	
	eLQ	C	49.7	
	LmH	B	12 23.5	
	LmV	B	24.1	
30.	ePKIKP	A	13 54 13.5	<u>New Ireland Region</u> 5.27°S 153.55°E
	ePP	B	55 58	H = 13 35 14.4 h = 50 km MAG=5.2
	ePP	A	55 59.5	D = 124.7° Az = 331.5° (USCGS)
	ePS	C	14 05 50	PPV(B):13s 0.75 µm
	ePPS	C	07 26	LmH:17s 1.7 µm LmV:18s 2.9 µm
	eSS	C	13.1	MPPV=5.2 MLH=5.8 MLV=6.0
	e(PSPS)	C	13 46	e 54 25 e 54 33
	eSSS	C	17.9	
	eLQ	C	28	
	LmV	B	52.9	
	LmH	B	53	
30.	ePKIKP	A	17 43 19.5	<u>Fiji Islands Region</u> 17.77°S 178.76°W
	iPKHKP	A	43 21.5	H = 17 24 43.1 h = 564 km MAG=5.1
	e	A	46 08.5	D = 146.2° Az = 348.1° (USCGS)
	ePP	A	46 54	PV2:1.4s 55.2nm
				AN USSR gives: 17.6°S 178.1°W
				H = 17 23 50 h = 66 km
30.	eP	A	19 02 35	<u>Turkey</u> 40.71°N 30.69°E
	LmH	B	10.5	H = 18 58 45.3 h = normal MAG=4.5
	LmV	C	10.5	D = 16.6° Az = 313.2° (USCGS)
30.	LmH	C	21 17.5	<u>Luzon/Philippine Islands</u> (USCGS)
	LmV	C	17.5	
30.	LQ	C	23 07	<u>South Sandwich Island Region</u> (USCGS)
	LR	C	15	

Day	Phase		h m s	Remarks
30.	eP	A	23 15 12.5	Kurile Islands 46.04°N 153.11°E
	e(pP)	A	15 36	H = 23 03 15.8 h = normal MAG=4.5
				D = 78.1° Az = 336.1° (USCGS); (h = 93 km)
				(pP)V:1.5s 13.4nm
31.	eiPn	A	02 19 30	GFR 50.4°N 7.5°E
	e(Pg)	A	19 36	H = 02 18 48 (BCIS) MAG=2.5 (BENSBERG)
	eiSn	A	19 58.5	D = 2.6°
	eSb	A	20 05.5	
	ei	A	20 08	
	eiSg	A	20 11.0	
31.	eiP	A	07 15 41	Turkey 40.6°N 27.6°E
	iS	A	18 30	H = 07 12 05.3 h = 14 km MAG=4.2 (USCGS)
	eLg1	B	20 12	D = 15.0°
	eRg	B	21 00	LmH:16s 0.7 μm LmV:14s 0.4 μm
	LmH	B	22	MLH=3.9 MLV=3.9
	LmV	B	22	e 15 59.5 e 17 10 e 17 47.5 i 17 55
				ei 19 04.5
31.	eiPKP	A	10 24 22	Tonga Islands 15.98°S 174.32°W
				H = 10 05 01.3 h = 151 km MAG=3.8
				D = 145.1° Az = 353.4° (USCGS)
				PV:0.8s 9.4nm
31.	eiSg	A	11 31 51	Near earthquake?
				TRIESTE gives: e(Sn) 31 15.7 e(Sg) 31 29.3
31.	iPg	A	13 06 11.5	Explosion/GDR 51°17.2'N 12°43.8'E
	iSg	A	06 26.5	D = 0.9° Yield: 1.1 t
31.	e(P)	A	13 06 38.5	
31.	eP	A	17 12 37	Near East Coast of Kamchatka
				53.60°N 160.36°E
				H = 17 01 10.3 h = normal MAG=4.5
				D = 72.9° Az = 339.8° (USCGS)
				PV:0.8s 9.4nm
				MPV=5.0

Day	Phase		h m s	Remarks
31.	ePKP2	A	23 09 20	Macquarie Island Region 60.0°S 159.1°E
	e(SS)	C	33.0	H = 22 48 35.6 h = normal MAG=5.2 (USCGS)
	eSS	C	33.8	D = 159.8°
	eSSS	C	38.6	LmH:20s 0.6 µm LmV:20s 0.7 µm
	LmH	B	00 33.2	MLH=5.3 MLV=5.5
	LmV	B	33.5	

Day	Phase		h m s	Remarks
1.	eP	A	00 17 27	Turkey 40.8°N 30.4°E
				H = 00 13 34.5 h = normal MAG=4.5 (USCGS)
				D = 16.4°
1.	eSKS	C	01 37 52	Near Coast of Peru 13.00°S 76.78°W
	eSS	C	45.5	H = 01 13 42.6 h = 66 km MAG=5.5 (USCGS)
	LmH	C	02 03.0	D = 98.9°
	LmV	C	03.0	LmH:30s 0.9 μm LmV:30s 0.8 μm
				MLH=5.1 MLV=5.1
1.	e(Sg)	A	04 14 38.5	
1.	LmH	C	04 31	Molucca Passage (USCGS)
1.	e	A	06 46 48.5	
1.	e	A	09 26 26.5	South of New Zealand 59.95°S 159.17°E
	ePKP2	A	26 28	H = 09 05 49.3 h = normal MAG=5.5 (USCGS)
	epPKP2	A	26 39	D = 159.6° h = 39 km
	ePP	B	30(00)	PV2:1.9s 34.5nm PV3:2.0s 40.8nm
	eSS	C	50.2	LmH:19s 0.8 μm LmV:20s 1.0 μm
	eSSS	C	56.5	MLH=5.5 MLV=5.6
	LQ	C	10 11	AN USSR gives: South of Macquarie Island
	LmH	B	46.0	60.3°S 158.2°E
	LmV	B	50.3	H = 09 05 58 h = 64 km
1.	e(Sg)	A	11 36 17.5	
1.	iP	A	14 11 55.7	Kurile Islands 43.85°N 147.87°E
				H = 13 59 58.6 h = 52 km MAG=4.7
				D = 78.5° Az = 333.3° (USCGS)
1.	iP	A	17 00 30.7	Turkmen SSR 39.73°N 53.33°E
				H = 16 54 14.6 h = normal MAG=4.5
				D = 31.0° Az = 304.6° (USCGS)
				PV:0.9s 9.5 μm
				MPV=4.7

Day	Phase		h m s	Remarks
1.	ePKP	A	19 37 47.5	South of Kermadec Islands 33.36°S 179.42°E
	ePKP2	A	38 21	H = 19 18 06.2 h = 228 km (USCGS)
				D = 161.0°
2.	-eP	A	00 56 21	Kurile Islands 44.56°N 146.42°E
	i	A	56 22.2	H = 00 44 41.4 h = 149 km MAG=5.0
				D = 77.4° Az = 332.4° (USCGS)
				PV:1.3s 55.5nm
				MPV=5.1
2.	ei(P)	A	04 22 54.5	
	e	A	22 57	
2.	LmH	B	07 28.2	Taiwan 23.58°N 121.45°E
	LmV	B	28.2	H = 06 34 19.3 h = 40 km MAG=4.4 (USCGS)
				D = 83.7°
				LmH:13.5s 0.6/um LmV:14s 0.8/um
				MLH=5.1 MLV=5.2
2.	ePKIKP	A	09 56 06.5	Fiji Islands Region 20.80°S 179.14°W
	iPKHKP	A	56 12.0	H = 09 37 29.5 h = 592 km MAG=4.7
	iPKP2	A	56 19.0	D = 149.0° Az = 346.6° (USCGS)
				PV2:1.3s 11.1nm
2.	eiPL	C	11 11 36	Jan Mayen Island Region 71.22°N 7.98°W
	+iP1	A	11 36.5	H = 11 06 38.7 h = normal MAG=5.0
	iP2	A	11 38.0	D = 22.5° Az = 146.1° (USCGS)
	eS	C	15 35	PLV:26s 3.1/um PLH:25s 3.8/um
	iSE	BC	15 44	PV1:1.3s 128nm PV2:1.5s 266nm
	iSN	BC	15 47	SN(B):13s 5.2/um SE(B):(24s 8.7/um)
	iSS	C	16 21	SN(C):20s 7.3/um SE(C):32s 13.3/um
	eiLR	C	17 00	LmH:15s 9.3/um LmV:16.5s 8.0/um
	LmH	B	22.1	MPV1=5.2 MPV2=5.5 MLH=5.3 MLV=5.4
	LmV	B	22.1	Compared to the Jan Mayen earthquake about 3 hours later, the first shock has clearly greater PL, S and surface waves, whereas short-period P-waves have larger amplitudes in the second shock. This remarkable difference could be due to a greater focus depth of the second shock or possibly to a

Day	Phase		h m s	Remarks
cont. 2.				different time function of the source (compare Bath's remarks in the Bulletin of Seismological Institute Uppsala). S has clearly different onset times and periods in the long-period NS- and EW-components. The exceptionally long periods in the EW-components can be due to the superposition of S and LQ.
2.	iPg	A	13 06 26.6	Explosion/GDR
	iSg	A	06 32.5	D = ca. 0.4°
2.	iSg	A	13 06 41	Explosion/GDR
2.	iSg	A	13 06 55	Explosion/GDR
2.	+iP	A	14 11 16.5	Jan Mayen Island Region 71.16°N 8.51°W
	eiPL	C	11 17	H = 14 06 17.8 h = normal MAG=5.3
	i	A	11 21.8	D = 22.6° Az = 145.2° (USCGS)
	ei	A	11 39	PV:1.3s 455nm PH:1.4s 262nm
	eS	C	15 14	PLH:21s 1.3 µm SN(B):11s 2.6 µm
	iSE	BC	15 24	SE(B):(16s) 1.7 µm SN(C):17s 2.9 µm
	iSN	BC	15 27	SE(C):28s 4.2 µm
	iSS	C	16 00	LmH:13.5s 4.9 µm LmV:15s 3.8 µm
	eiLR	C	16 33	MPV=5.8 MPH=5.7 MLH=5.1 MLV=5.1
	LmH	B	21.9	See remarks given to the preceding Jan
	LmV	B	21.9	Mayen earthquake (August 2., H = 11 06 38.7).
2.	eP	A	15 37 16	Turkey 40.71°N 30.57°E
	e	A	37 20.5	H = 15 33 22.5 h = normal MAG=4.5
				D = 16.5° Az = 313.3° (USCGS)
				PV:1.5s 10.0nm
2.	eP	A	18 30 46	Southern Sumatra 4.59°S 103.25°E
	e	A	31 23	H = 18 17 32.0 h = 83 km MAG=5.1
				D = 94.6° Az = 320.3° (USCGS)
				PV:1.0s 9.5nm
				MPV=5.2

Day	Phase		h m s	Remarks
3.	e(PKIKP)	A	00 28 00	Tonga Islands 20.93^{o}S 174.32^{o}W
	e(PKHKP)	A	28 04	H = 00 08 12.9 h = 43 km MAG=4.4
	ei(PKP2)	A	28 11	D = 150.0^{o} Az = 352.4^{o} (USCGS)
	e	A	28 16	PV2:1.8s 51.0nm PV3:1.5s 40.3nm
	LmH	B	01 18.4	LmH:16s 0.2 µm LmV:16s 0.4 µm
	LmV	B	18.4	MLH=5.0 MLV=5.2
3.	LmH	B	00 43.8	Southeast Spain 37¾^{o}N 1¼^{o}W
	LmV	B	44.5	H = 00 34 12 (BCIS)
				D = 15.9^{o}
				LmH:12s 0.2 µm LmV:10s 0.3 µm
				MLH=3.6
3.	iPg	A	07 59 40.5	Explosion/GDR 51°15.2'N 12°39.6'E
	i	A	59 53.8	D = 0.8^{o} Yield: 14 t
	iSg	A	59 54.5	
3.	LmH	B	08 02	Santa Cruz Islands 11.8^{o}S 165.5^{o}E
	LmV	B	08.6	H = 06 43 24.7 h = 57 km MAG=4.7 (USCGS)
				D = 135.8^{o}
				LmH:17s 0.5 µm LmV:16s 0.6 µm
				MLH=5.3 MLV=5.4
3.	e(Sg)	A	10 55 23	Probably explosion.
	e	A	55 28.5	KAŠPERSKÉ HORY gives: D = 1.5^{o}
	eL	A	55 48	
3.	eiPg	A	13 00 23.5	Explosion
	ei	A	00 38.5	D = ca. 1.2^{o}
	iSg	A	00 40.0	LmH:1.3s 59.2nm LmV:1.2s 56.0nm
	LmH	A	00 50	
	LmV	A	00 55	
3.	e	A	13 05 04.5	
3.	e(P)	A	16 34 54	
3.	e(Sg)	A	18 18 02	Austria 47.7^{o}N 16.0^{o}E
				H = 18 15 52 (VIENNA)
				D = 4.2^{o}

Day	Phase		h m s	Remarks
3.	LmH	C	20 02.1	Ryukyu Islands 27.8°N 128.0°E
	LmV	C	09.0	H = 19 14 38.5 h = 93 km MAG=4.7 (USCGS)
				D = 83.8°
				LmH:17.5s 0.8 μm LmV:16s 0.7 μm
				MLH=5.2 MLV=5.1
3.	iP	A	21 49 18.0	Fox Islands/Aleutian Is. 53.0°N 166.7°W
	LmH	B	22 27.5	H = 21 37 26.7 h = 29 km MAG=4.6 (USCGS)
	LmV	B	27.5	D = 76.6°
				PV:0.9s 14.2nm
				LmH:18s 0.4 μm LmV:17s 0.3 μm
				MPV=5.1 MLH=4.7 MLV=4.7
3.	epP	A	23 29 24.5	Fox Islands/Aleutian Is. 53.78°N 170.05°W
				H = 23 17 08.4 h = 194 km MAG=4.9 (USCGS)
				D = 75.7°
4.	eP	A	01 56 45	PV:1.2s 8.7nm
4.	eP	A	06 11 04	Central Mid-Atlantic Ridge 7.37°N 36.28°W
	e(PoS)	C	15 48	H = 06 01 09.9 h = normal MAG=5.0 (USCGS)
	eS	C	19 08	D = 58.6°
	eSS	C	22 40	PV:1.9s 17.7nm
	eLQ	C	25 34	LmH:16s 0.5 μm
	LmV	C	28	MPV=4.9 MLH=4.7
	LmH	B	28.6	
4.	+eP	A	07 05 46.5	Eastern Kazakh SSR 50.0°N 78.0°E
	ePn	A	07 19.5	H = 06 58 00 (BCIS)
	ePP	A	07 24	D = 41.2°
				PV:0.7s 51.1nm PH:0.7s 30.2nm
				MPV=5.6 MPH=5.6
				Underground explosion.
4.	e(Sg)	A	09 54 16	(Sg)V:0.9s 9.4nm Explosion?
4.	eSg	A	09 56 43	Explosion/probably ČSSR
				KAŠPERSKÉ HORY gives: D = 1.1°

Day	Phase		h m s	Remarks
4.	eSg	A	10 44 17.5	Explosion
				KAŠPERSKÉ HORY gives: D = 1.1°
4.	eiSg	A	12 44 37	Explosion/GDR
4.	e(Pg)	A	12 45 18.5	Explosion/GDR
	ei(Sg)	A	45 27	D = ca. 0.6°
4.	ei	A	12 45 48.5	Explosion/GDR
	ei(Sg)	A	45 50	
4.	i(Sg)	A	12 46 35	Explosion/GDR
	ei	A	46 37.5	
4.	ePn	A	14 56 39.5	Adriatic Sea 42.91°N 17.73°E
	eiSn	A	58 18.5	H = 14 54 33.3 h = normal MAG=4.4
	eLg1	B	59 08	D = 8.8° Az = 333.7° (USCGS)
	e(Sg)	A	59(17.5)	LmH:10s 2.4/um LmV:9s 1.1/um
	LmH	B	59.5	MLH=4.3
	LmV	C	15 00.1	i 56 41.2 ei 56 44
4.	ePKP	A	22 54 27.5	Tonga Islands 17.69°S 173.16°W
	e(pPKP)	A	54 37.5	H = 22 34 47.7 h = normal MAG=4.8
	ei(sPKP)	A	54 43.5	D = 146.9° Az = 354.4° (USCGS); h = 36 km
	LmH	C	00 10.5	PV1:1.8s 46.0nm PV3:1.6s 53.1nm
	LmV	C	10.5	LmH:16s 0.2/um LmV:16s 0.2/um
				MLH=5.0 MLV=5.1
5.	eP	A	01 56 45.0	Kurile Islands 43.32°N 147.53°E
	LmH	C	02 29.0	H = 01 44 43.2 h = normal MAG=4.4 (USCGS)
	LmV	C	35	D = 78.8°
5.	eiP	A	05 41 24	Kurile Islands 43.33°N 147.62°E
	e(pP)	A	41 42.5	H = 05 29 21.8 h = normal MAG=4.8
	e(sP)	A	41 52	D = 78.9° Az = 333.2° (USCGS); (h = 74 km)
	LmH	C	06 14.0	PV:1.1s 10.7nm
				LmH:21s 0.5/um
				MPV=4.9 MLH=4.8

Day	Phase		h m s	Remarks
5.	e(P)	A	07 44 22	
5.	e(Sg)	A	13 11 05	Explosion?
	e	A	11 14	
5.	ei	A	12 02 21	
5.	e	A	18 10 37	KAŠPERSKÉ HORY gives: ei Sg 09 53 PRUHONICE gives: eiPg 18 08 53
5.	e(P)	A	22 59 49	
6.	ePKP	A	05 05 16	<u>Fiji Islands Region</u> 20.73°S 178.39°W H = 04 46 29.4 h = 536 km MAG=4.0 D = 149.1° Az = 347.6° (USCGS)
6.	eP	A	10 39 04.5	<u>Tadzhik-Sinkiang Border Region</u> 37.99°N 74.52°E H = 10 31 06.3 h = 215 km MAG=4.8 D = 45.4° Az = 307.3° (USCGS) PV:1.1s 7.2nm MPV=4.6
6.	-eiPKP	A	13 32 54.5	<u>Fiji Islands Region</u> 21.37°S 179.54°W H = 13 14 08.7 h = 561 km MAG=4.2 D = 149.5° Az = 346.0° (USCGS) PV:0.7s 9.5nm
6.	eP	A	22 57 58.5	<u>Fox Islands/Aleutian Is.</u> 52.7°N 168.4°W H = 22 46 08.1 h = 44 km MAG=4.3 (USCGS) D = 77.0°
7.	e(Sg)	A	04 32 44	Near earthquake. (Sg)V:1.0s 9.5nm TRIESTE gives: e 04 30 40 i(Sg) 30 51.2
	e	A	32 51.5	
7.	eiP	A	05 57 46	<u>Afghanistan-USSR Border Region</u> 36.48°N 71.24°E H = 05 49 57.5 h = 229 km MAG=5.0 D = 44.3° Az = 308.1° (USCGS)
	e	A	59(33)	

Day	Phase		h m s	Remarks
7.	iPg	A	06 53 21	Explosion
	e(Sg)	A	53 38.5	
7.	iPg	A	09 05 42.3	Explosion/GDR
	iSg	A	05 44.3	D = 16 km
				SgV:0.5s 268nm
7.	e(P)	A	10 06 24	(P)V:1.8s 20.4nm
7.	ei(Sg)	A	11 24 49	Explosion
	e	A	24 51	
	eL	A	24 54	
7.	eP	A	11 25 54	Alaska Peninsula $58.7^{o}N$ $154.6^{o}W$
	LmV	C	12 01.5	H = 11 14 42.7 h = 37 km MAG=5.1 (USCGS)
	LmH	C	04.5	D = 70.3^{o}
				LmH:16s 0.3 µm LmV:16s 0.2 µm
				MLH=4.7 MLV=4.5
7.	i(Pg)	A	11 28 37.5	Explosion
	eiL	A	28 49	LmH:1.6s 49.2nm LmV:1.3s 30.6nm
	LmH	A	28 52	
	LmV	A	28 54	
7.	eP	A	11 47 13	PV:1.3s 13.9nm
7.	LmV	C	12 01.5	Alaska Peninsula $58.7^{o}N$ $154.6^{o}W$
	LmH	C	04.5	H = 11 14 42.7 h = 37 km MAG=5.1 (USCGS)
				D = 70.4^{o}
				LmH:16s 0.3 µm LmV:16s 0.2 µm
				MLH=4.7 MLV=4.5
7.	i(Sg)	A	14 56 17.5	Explosion
	eL	A	56 21	
7.	eiPKP2	A	17 27 32	Kermadec Islands Region $29.36^{o}S$ $177.38^{o}W$
				H = 17 07 20.1 h = 147 km MAG=4.8 (USCGS)
				D = 158.0^{o}

Day	Phase		h m s	Remarks
8.	ePKP	A	07 32 32	Fiji Islands Region 17.47°S 179.04°W
				H = 07 13 50.8 h = 523 km MAG=4.4
				D = 145.8° Az = 347.9° (USCGS)
8.	LmH	C	11 58	Traces.
8.	ePg	A	14 24 21	Explosion
	iSg	A	24 30	D = ca. 0.6°
	eL	A	24 35	LmH:1.4s 104nm LmV:1.4s 101nm
	LmH	A	24 45	
	LmV	A	24 45	
8.	LmH	C	15 41	Traces; Off Coast of Mexico (USCGS)
	LmV	C	41	
8.	eP	A	16 18 13.5	Near East Coast of Honshu/Japan
	e	A	18 34	37.16°N 141.06°E
				H = 16 05 59.1 h = 54 km MAG=3.9
				D = 81.9° Az = 330.2° (USCGS)
9.	eP	A	00 37 25	Turkey 37.06°N 28.80°E
	e	A	37 30.5	H = 00 33 11.8 h = normal MAG=4.2
				D = 18.3° Az = 323.2° (USCGS)
				PV1:2.0s 29.8nm PV2:1.8s 25.0nm
9.	ePKIKP	A	08 38 32.5	Banda Sea 6.40°S 130.41°E
	epPKIKP	A	38 53	H = 08 20 03.7 h = 89 km MAG=5.7
	ePP	A	39 23	D = 113.0° Az = 322.7°
	e(S)	C	46 48	PPV:1.0s 9.5nm
	e(SPP)	C	49 16	MPPV=5.2
	e(SSS)	C	59.0	
	eLQ	C	09 05.8	
9.	i	A	12 58 00	Explosion/GDR
	eiSg	A	58 04.7	
	iL	A	58 07.5	
9.	iSg	A	12 58 54	Explosion/GDR
	iL	A	58 57.0	

Day	Phase		h m s	Remarks
9.	eiP	A	13 36 45	Colorado 39.88°N 104.73°W
	LmV	C	14 07.5	H = 13 25 06.2 h = 5 km MAG=5.3
	LmH	C	08.4	D = 74.0° Az = 36.4° (USCGS)
				PV:1.5s 23.4nm
				LmH:19s 0.3 µm LmV:19s 0.2 µm
				MPV=5.1 MLH=4.6 MLV=4.5
9.	eSg	A	21 01 17	Vallis/Switzerland 46.4°N 7.4°E
	e	A	01 21	H = 20 58 34 (BCIS)
				D = 5.1°
10.	LmH	C	05 22.5	Off Coast of Southern Chile
	LmV	C	22.5	45.0°S 79.5°W
				H = 04 20 27.8 h = 33 km MAG=5.0 (USCGS)
				D = 123.4°
				LmH:35s 0.3 µm LmV:35s 0.2 µm
				MLH=4.6 MLV=4.5
10.	iPg	A	09 40 28.5	Explosion
	i	A	40 43.5	D = ca. 1.2°
	iSg	A	40 44.5	
10.	ei(Sg)	A	11 14 30	Explosion?
				(Sg)V:1.0s 16.6nm
10.	+iP	A	11 33 17.8	Kurile Islands 45.37°N 150.30°E
	eiX	A	33 22	H = 11 21 22.3 h = 37 km MAG=5.7
	ePcP	A	33 28.5	D = 77.9° Az = 334.5° (USCGS); h = 53 km
	epP	A	33 32	PV:1.3s 94.5nm XV:1.4s 49.1nm
	ePP	A	36 23.5	PcPV:1.2s 30.6nm
	eS	C	43 00	LmH:22s 3.1 µm LmV:22s 3.0 µm
	ePS	C	43(40)	MPV=5.8 MLH=5.6 MLV=5.6
	LmH	C	12 10.5	i 33 38.8 e 48(56)
	LmV	C	10.6	X is a very clear unidentified phase. It seems possible to interpret this phase as P converted in S at the base of the crust.
10.	e	A	11 49 27	Explosion?
	e	A	49 59	e 49 43 e 49 46
	ei	A	50 09.5	

Day	Phase		h m s	Remarks
10.	e(P)	A	13 01 05.5	(P)V:0.8s 9.4nm
10.	iSg	A	13 39 13.0	Explosion KAŠPERSKE HORY gives: D = 1.1^{o} ePg 13 38 47 iSg 39 02
10.	e(P)	A	16 01 29.5	(P)V:0.8s 11.8nm
11.	ei(P)	A	01 20 15.5	(P)V:1.0s 14.4nm
11.	e(Sg)	A	06 57 11	Explosion?
11.	e i(Sg) i	A A A	12 49 53 49 56 50 03.7	Explosion
11.	LmH LmV	C C	13 16.4 25.0	Nicaragua 11.81^{o}N 85.88^{o}W H = 12 26 18.3 h = 21 km MAG=4.7 (USCGS) D = 85.7^{o} LmH:18s 0.5 /um LmV:18s 0.3 /um MLH=5.0 MLV=4.8
11.	i(Sg)	A	17 31 29	Explosion (Sg)V:0.6s 64.3nm
11.	i e i	A A A	19 08 29.0 08 58 09 19	
11.	iPP epPP	A A	19 11 39.8 12 11	Volcano Islands 22.1^{o}N 144.0^{o}E H = 18 54 28.8 h = 125 km MAG=5.3 (USCGS) D = 96.3^{o} h = 130 km
12.	eP e e ePP	A A A A	04 42 49 42 56 45 46 45 52	Near East Coast of Honshu/Japan 38.53^{o}N 141.91^{o}E H = 04 30 38.5 h = 53 km MAG=5.4 D = 81.1^{o} Az = 330.6^{o} (USCGS) PV:1.2s 25.5nm MPV=5.2

Day	Phase		h m s	Remarks
12.	eiSg	A	08 59 55.5	Near earthquake?
12.	+iPKIKP	AB	09 59 19	South of Fiji Islands 24.71°S 177.47°W
	+iPKHKP	B	59 26.5	H = 09 39 44.3 h = 134 km MAG=5.8
	-iPKP2	B	59 40	D = 153.2° Az = 347.1° (USCGS)
	e(pPKP)	B	59 54	PV1(B):6s 2.0 µm PV2:5s 1.8 µm
	i(sPKP)	B	10 00 07	PV3:5.5s 2.5 µm (sPKP)V:8s 3.8 µm
	iSKP	B	02 37.5	PPV:9.5s 5.2 µm (pPP)V:10s 4.7 µm
	-iPP	B	03 12	SKKSH:10s 5.5 µm SKSPV:14.5s 5.4 µm
	ei(pPP)	B	03 39	PSKSH:14s 2.8 µm RH:55s 32.8 µm
	iPP2	B	07 31	RV:55s 32.4 µm
	eiSKKS	B	09 50	MPPV=6.6
	eSKKKS	B	10 32	ei 00 04 e 00 12 e 03 34 i 06 30
	iSKSP	B	13 06	PP2 is the PP travelled along the farther
	iPSKS	B	13 26	path to our station (D2 = 360° - D). Very
	iSPP	B	16 00	clear R-waves are registered. The G-waves
	eiSSP	B	23 28	(LQ) have amplitudes 3 times smaller than
	eSSS	C	28 22	R-waves and do not show such a sharp be-
	eiLQ	C	43.0	ginning. PKP-phases in the short-period
	iR	C	52.0	registrations are very complex and not ty-
				pical for PKP-registrations of deep earth-
				quakes of this region at our station.
12.	+iP	A	10 52 12.0	Near East Coast of Kamchatka
	e	A	52 24.5	53.71°N 160.36°E
	eiPcP	A	52 30	H = 10 40 43.9 h = 25 km MAG=5.0
				D = 72.7° Az = 339.8° (USCGS)
				PV:1.0s 28.5nm
				MPV=5.4
12.	e(Sg)	A	11 15 49	Explosion/CSSR 50°47.4'N 14°31.5'E
	ei	A	15 52.5	(PRUHONICE)
	e	A	15 57	D = 1.8° Yield: 8.9 t (PRUHONICE)
12.	ePKIKP	A	12 50 23	New Hebrides Islands 14.85°S 166.67°E
	LmH	C	13 55.8	H = 12 30 56.1 h = 23 km MAG=5.2 (USCGS)
	LmV	C	55.8	D = 139.1°
				PV:1.6s 22.8nm
				LmH:21s 0.9 µm LmV:21s 0.8 µm
				MLH=5.5 MLV=5.4

Day	Phase		h m s	Remarks
12.	eP	A	17 03 35.5	Turkey 41.03°N 34.26°E
	LmH	C	11.5	H = 16 59 21.1 h = normal MAG=4.4
				D = 18.4° Az = 309.1° (USCGS)
				PV:1.5s 20.1nm
				LmH:19s 0.2 µm
				MLH=3.4
12.	iP	A	23 02 35.3	Afghanistan-USSR Border Region
	ei	A	02 39	37.04°N 71.37°E
	epP	A	02 57	H = 22 54 38.6 h = 121 km MAG=5.1
	e	A	05 10	D = 44.0° Az = 307.7° (USCGS); h = 100 km
				pPV:1.3s 16.7nm
13.	eiPKP	A	05 34 41	Tonga Islands 18.32°S 174.01°W
	e	A	35 02.5	H = 05 14 57.9 h = 25 km MAG=4.5
				D = 147.4° Az = 352.3° (USCGS)
				PV:1.0s 4.7nm
13.	ePP	C	16 51 10	South of Africa 50.88°S 29.07°E
	eSKS	C	57 38	H = 16 33 04.0 h = normal MAG=5.4 (USCGS)
	ePS	C	17 00 22	D = 102.3°
	e	C	02 20	LmH:16.5s 1.8 µm LmV:18s 1.5 µm
	eSS	C	05 40	MLH=5.7 MLV=5.6
	LQ	C	17	
	LR	C	20.8	
	LmH	C	39.2	
	LmV	C	39.3	
13.	epPKP	A	17 13 51	New Britain Region 4.33°S 152.47°E
	ePP	A	15(25)	H = 16 54 45.7 h = 25 km MAG=5.0 (USCGS)
	e	A	15 30	D = 123.4°
	e	A	15 42	
13.	+iP	A	20 18 28.0	Southern Honshu/Japan 35.29°N 135.31°E
	+e(pP)	B	19 50	H = 20 06 50.6 h = 357 km MAG=6.0
	+ipP	A	19 51.7	D = 81.1° Az = 327.6° (USCGS); h = 370 km
	eiPP	A	21 38	PV:2.0s 880nm PH:2.0s 335nm
	epPP	B	22(46)	PPV:2.7s 1680nm PPH:2.8s 896nm
	ePa	B	24 50	SH:9s 4.5 µm

Day	Phase		h m s	Remarks
cont.				
13.	eiS	C	20 28 00	MPV=6.2 MPH=6.1 MPPV=6.7 MPPH=6.6
	iS	B	28 06	MSH=6.0
	iSP	B	28 50	i 19 54.5 i 31 30 e 39 12 e 44 50
	i(pS)	C	29 50	e 47 50
	esS	C	30 22	
	eiSS	C	33 28	
	iSSS	C	36 58	
	eiSa	C	39 28	
	eiPKPPKP	A	44 59	
	eSKPPKP	A	47 54	
13.	-eP1	A	22 10 29	<u>Pyrenees</u> 43.17°N 0.54°W
	eP2	A	10 39	H = 22 07 47.5 h = 15 km MAG=5.3
	eS	B	12 40	D = 11.2° Az = 43.8° (USCGS)
	i(S)	A	12 47.0	LmH:7.5s 25.4 µm LmV:10s 18.6 µm
	eSS	B	12 54	MLH=5.6
	iLi(3.71)	A	13 20.5	i 13 58 i 15 06
	iLg2(3.38)	B	13 52	The clear phase, interpreted as P2 is
	LmH	C	15.2	following P1 after 10s (also at the Swe-
	LmV	C	15.2	dish stations. See bulletin of Seismologi-
				cal Institute Uppsala). It is probably the
				P of a second shock in the same focus.
13.	ePKP	A	22 34 12	<u>New Britain Region</u> 4.37°S 152.51°E
	ePP	A	36 00	H = 22 15 09.6 h = 29 km MAG=5.3
	e	A	36 05.5	D = 123.4° Az = 331.2° (USCGS)
	ePS	C	45 40	LmH:18s 3.2 µm LmV:18s 3.8 µm
	ePPS	C	47 12	MLH=6.0 MLV=6.1
	ei(PKKS)	C	48 00	
	eSS	C	52 38	
	LQ	C	06	
	LR	C	12.3	
	LmH	B	37.8	
	LmV	B	37.8	
13.	eP	A	23 54 23.5	<u>Ascension Island Region</u> 6.96°S 12.57°W
	epP	A	54 31.5	H = 23 44 11.1 h = 28 km MAG=5.0 (USCGS)
	e	A	54 39.5	D = 61.2° h = 30 km

Day	Phase		h m s	Remarks
cont.				
13.	e	A	23 54 45	PV:1.2s 17.9nm
	LmH	B	00 15.5	LmH:26s 2.8 µm
	LmV	B	17	MPV=5.1 MLH=5.3
				Multiple P-phases.
14.	iPn	A	10 17 15.2	Northern Italy 46.87°N 10.41°E
	iPb	A	17 22.5	H = 10 16 18.1 h = 20 km MAG=4.3
	ePg	A	17 31	D = 3.9° Az = 11.5° (USCGS)
	iSn	A	18 01	PnV:0.6s 76.1nm
	iSg	A	18 22	i 17 50 i 18 11
	LmH	B	18.5	
14.	iP	A	12 56 37.0	Mexico 17.3°N 94.6°W
				H = 12 44 04.7 h = 120 km MAG=4.5 (USCGS)
				D = 86.6°
14.	e(Sg)	A	16 05 19	Probably explosion.
	e	A	05 22	
	e	A	05 24	
14.	e	A	16 49 05	Near earthquake?
	e	A	49 36	
	e(Sg)	A	49 56.5	
	eL	A	50 04	
14.	eP	A	20 13 16	Turkey 40.70°N 30.53°E
	LmH	B	21.1	H = 20 09 25.8 h = normal MAG=4.7
	LmV	B	21.3	D = 16.5° Az = 313.4° (USCGS)
15.	LmH	C	04 00.0	North Atlantic Ocean 19.19°N 68.50°W
				H = 03 23 52.3 h = 39 km MAG=4.9 (USCGS)
				D = 69.2°
				LmH:23s 0.4 µm
				MLH=4.5
15.	eP1	A	04 39 23	Mediterranean Sea 36.52°N 19.35°E
	eP2	A	39 32	H = 04 35 53.3 h = normal MAG=4.5
	+iP2	A	39 32.6	D = 15.2° Az = 340.9° (USCGS)

Day	Phase		h m s	Remarks
cont.				
15.	LmH	C	04 45.6	PV1:0.8s 18.9nm PV2:1.1s 77.0nm Multiple P, the second onset much greater than the first one. Probably a double shock.
15.	eP	A	07 09 26.5	<u>Sicily</u> 38.85°N 15.00°E
	e	A	09 41.5	H = 07 06 35.5 h = normal MAG=4.5
	e	A	09 48	D = 12.0° Az = 349.6° (USCGS)
	e	A	09 58.5	PV:0.9s 23.6nm
	e(SS)	C	12 14	LmH:16s 3.2 /um
	LmH	C	13.9	MLH=4.4
15.	eP	A	07 48 18	<u>Hindu Kush Region</u> 36.29°N 70.16°E H = 07 40 28.7 h = 189 km MAG=4.7 (USCGS) D = 43.7°
15.	eiP1	A	09 31 20.4	<u>Tibet</u> 31.09°N 93.66°E
	eP2	A	31 23.5	H = 09 21 02.3 h = normal MAG=5.7
	ePcP	A	32 05	D = 61.8° Az = 314.3° (USCGS)
	ePP	A	33 37.5	PV1:1.8s 51.0nm PV2:1.9s 153.0nm
	eS	B	39 44	LmH:20s 1.7 /um LmV:15s 1.7 /um
	LQ	C	50.5	MPV1=5.4 MPV2=5.8 MLH=5.2 MLV=5.4
	LmH	C	55.7	e 33 53 e 34 04
	LmV	B	10 00.2	P is multiple. Probably a double shock, the second is stronger than the first one.
15.	e(P)	A	11 29 54	Near earthquake.
	LmH	C	34.4	TRIESTE gives: D = 785 km LmH:16s 0.4 /um
15.	iPg	A	11 34 42.5	Explosion
	iSg	A	34 43.5	D = 8 km SgV:0.4s 462nm
15.	iPg	A	12 49 44.5	Explosion
	iSg	A	49 50	
	iL	A	49 53	

Day	Phase		h m s	Remarks
15.	eP	A	15 47 29	E. Russia - N. E. China Border Region
	e	A	47 43	44.76°N 132.38°E
	LmH	B	16 17.3	H = 15 36 06.6 h = normal MAG=5.3
	LmV	B	22	D = 72.0° Az = 324.9° (USCGS)
				PV:1.4s 30.7nm
				LmH:14s 0.85 μm LmV:12s 0.5 μm
				MPV=5.2 MLH=5.2 MLV=5.0
16.	ePn	A	00 04 47.5	Near earthquake.
	ePg	A	05 06	Probably Julian Alps 46 1/2°N 13 1/3°E (MOXA)
	eSn	A	05 35	D = 4.2°
	eSg	A	05 59	
16.	e(Sg)	A	07 29 23	Near earthquake.
	e	A	29 31	TRIESTE gives: D = 20 km
16.	e	A	09 04 57	Traces
16.	e(Sg)	A	11 41 46.5	Probably explosion.
	e	A	41 53	
16.	iSg	A	12 51 58.0	Explosion/GDR
	i	A	51 59.5	
16.	eL	A	12 52 16.5	Explosion/GDR
16.	eSg	A	13 15 17	Explosion?
	e	A	15 29	
17.	e(Sg)	A	11 23 42	Near earthquake?
	e	A	23 48	
	e	A	24 00	
	e	A	24 09	
17.	iSg	A	12 03 10.0	Probably explosion.
	eL	A	03 13	
17.		A	12 59 03	Central Mid-Atlantic Ridge 0.78°S 21.15°W
	e	A	59 40	H = 12 49 08.9 h = 40 km MAG=4.5
	e(PcP)	A	59 57.5	D = 58.3° Az = 23.9° (USCGS)

Day	Phase		h m s	Remarks
cont.				
17.	eS	B	13 07 00	LmH:14s 0.5 /um LmV:16s 0.6 /um
	eS	C	07 10	MLH=4.8 MLV=4.8
	eSS	C	11.0	
	eLQ	C	14.2	
	LmH	B	24.5	
	LmV	B	27.0	
17.	e(Sg)	A	14 00 40	Explosion?
	eL	A	00 56.5	e 00 48 e 01 06
17.	e	A	14 44 28.5	South Sandwich Islands Region
	LmH	C	21 32	60.3°S 27.0°W
	LmV	C	33	H = 20 28 33.7 h = 98 km MAG=5.2 (USCGS)
				D = 114.6°
				LmH:20s 0.2 /um
				MLH=4.7
17.	iP	A	22 53 13.5	Kenai Peninsula/Alaska 59.41°N 151.43°W
	epP	A	53 26	H = 22 42 09.3 h = 55 km MAG=5.0
				D = 69.4° Az = 11.4° (USCGS); h = 50 km
				PV:1.2s 23nm pPV:1.5s 36.9nm
				MPV=5.1
18.	+iP	A	03 48 00.2	Ryukyu Islands 27.77°N 127.65°E
	LmH	C	04 22.0	H = 03 35 40.5 h = 94 km MAG=5.4
				D = 83.7° Az = 324.9° (USCGS)
				LmH:24s 0.4 /um
				MLH=4.7
				AN USSR gives: h = normal
18.	eiP	A	06 01 24	Southern Alaska 61.53°N 151.01°W
	ipP	A	01 30.0	H = 05 50 29.0 h = 19 km MAG=4.5
	esP	A	01 32.5	D = 67.3° Az = 11.9° (USCGS); h = 23 km
	ei	A	01 44	PV:1.0s 21.3nm pPV:1.1s 43.2nm
				MPV=5.3
18.	ei(Sg)	A	07 33 31	Probably explosion.
	eL	A	33 43	

Day	Phase		h m s	Remarks
18.	ei(Sg)	A	10 33 01.5	Near earthquake?
	i	A	33 04.0	e 33 19
	eL	A	33 13	
18.	eiSn	A	12 04 30	Austria-Italy-Switzerland Border Region
	eSb	A	04 46	46.8°N 10.5°E
	eSg	A	04 55	H = 12 02 45 (BCIS)
				D = 3.9°
18.	ei(Pg)	A	12 54 05.5	Explosion/GDR
	iSg	A	54 11.0	
	eiL	A	54 13	
18.	iSg	A	12 54 19	Explosion/GDR
	eL	A	54 21	
18.	ei	A	12 54 50	Explosion/GDR
	eL	A	54 52.5	
18.	e	A	12 55 21.5	Explosion/GDR
	eiL	A	55 24.5	
18.	ei	A	14 11 23.5	
18.	(LQ)	C	15 28.7	Traces
	LmH	C	36.5	LmH:14s 0.1 µm LmV:14s 0.2 µm
	LmV	C	36.5	
18.	(LQ)	C	16 08.3	Traces
	LmH	C	17.0	LmH:13s 0.4 µm LmV:13s 0.5 µm
	LmV	C	17.0	
19.	eP	A	01 42 41	Afghanistan-USSR Border Region
				36.90°N 71.51°E
				H = 01 34 43.5 h = 127 km MAG=4.9 (USCGS)
				D = 44.1°
19.	eP	A	07 47 06.5	
	i	A	47 30.0	

Day	Phase		h m s	Remarks
19.	ei(P)	A	09 12 47.5	
19.	e(P)	A	10 22 08	
19.	iP	A	11 54 50.2	
19.	eP	A	12 26 25	Off East Coast of Honshu/Japan
	epP	A	26 37.5	40.77°N 143.49°E
	e	A	26 53.5	H = 12 14 21.7 h = 45 km MAG=4.8 (USCGS)
	LmH	C	13 05.0	D = 79.7° h = 45 km
	LmV	C	05.0	LmH:19s 0.3 µm LmV:20s 0.3 µm
				MLH=4.7 MLV=4.7
19.	+eP	A	15 41 34	Philippine Islands Region 10.41°N 125.97°E
	+iP	B	41 35	H = 15 28 08.5 h = 58 km MAG=5.6
	+iP	C	41 37	D = 96.2° Az = 324.3° (USCGS); h = 50 km
	epP	A	41 47	LmV:19s 5.0 µm
	ePP	B	45 31	MLV=6.0
	ePP	A	45 33.5	
	eSKS	C	52 16	
	eS	C	52 54	
	eSP	C	54 08	
	eSPP	C	54 44	
	eSS	C	59 30	
	LmH	C	16 21.8	
	LmV	B	30.5	
19.	eiPKIKP	A	16 01 06	Santa Crúz Islands 12.41°S 166.64°E
	-iPKIKP	A	01 07.2	H = 15 41 53.3 h = 86 km MAG=5.4
	epPKIKP	A	01 30	D = 136.8° Az = 336.9° (USCGS); h = 87 km
	ePP	A	03 51.5	ei 01 15.5 i 01 46.5
	epPP	A	04 13	
	eSKP	A	04 33	
20.	LmH	C	00 09.3	Traces. Szechwan Prov./China (USCGS)
	LmV	C	(13)	

Day	Phase		h m s	Remarks
20.	eP	A	00 18 52.5	Alaska Peninsula 58.07°N 156.48°W
				H = 00 07 46.0 h = 127 km MAG=4.8
				D = 71.2° Az = 8.0° (USCGS)
				AN USSR gives: Kodiak Island Region
				56.5°N 154.8°W
				H = 00 07 24 h = normal
				D = 72.6°
20.	iP1	A	02 10 20.0	Kazakh-Sinkiang Border Region
	-eiP2	A	10 21.2	45.32°N 80.14°E
	ipP	A	10 29.3	H = 02 02 05.2 h = normal MAG=5.1
	ePP	A	12 21.5	D = 44.7° Az = 302.6° (USCGS); h = 36 km
	eScP	A	15 50.5	PV2:1.3s 72.2nm
	eS	C	16 52	LmH:18s 5.3 µm LmV:12s 3.5 µm
	eScS	C	20 16	MPV2=5.5 MLH=5.5 MLV=5.6
	eiL(3.87)	B	23 30	i 10 37.5 i 10 46.0 e 11 57 e 12 11.5
	ei(Li)	B	24 32	e 14 37 e 15 29 e 22 32
	eL(3.57)	B	25 18	Multiple P. P1 is much smaller than P2.
	eLg1(3.49)	B	25 50	The first onset of P, which must be 3.5 s
	LmH	C	26.2	earlier than our P1, is not detectable in
	LmV	B	29.5	our records. Båth gives the first P-onset
				of Swedish stations 4 - 5 s before a se-
				cond and much stronger phase, which would
				be correspond to our P2-onset. Probably a
				double shock in the same focus. Clearly
				developed higher-mode surface waves.
20.	i(P)	A	02 17 17.0	
20.	i(P)	A	02 20 17.5	
20.	i(P)	A	02 21 56.0	
20.	e(P)	A	06 41 43	
20.	eP	A	07 36 47.5	Komandorsky Islands Region 55.81°N 164.20°E
				H = 07 25 29.9 h = 46 km MAG=4.8 (USCGS)
				D = 71.3°
				PV:1.0s 7.1nm
				MPV=4.8

Day	Phase		h m s	Remarks
20.	e	A	10 09 52	Near earthquake.
	e	A	10 20	TRIESTE gives: D = 370 km ePn 10 06 49
	e	A	10 24	eiSn 07 28 iSg 07 46
	iL	A	10 40.2	
20.	e	A	12 38 36	Explosion/GDR
	e(Sg)	A	38 40	
	iL	A	38 44	
20.	e(Sg)	A	12 39 05	Explosion/GDR
	iL	A	39 08	
20.	e	A	14 55 53	Near earthquake.
	e	A	57 33	TRIESTE gives: D = 380 km eiPn 14 53 43.8
	eL	A	58 35	iSn 55 56.5 iSg 55 41.5
20.	iP	A	15 17 25	Chile 25.2°S 69.0°W
	ePP	A	21 48	H = 15 03 36.2 h = 109 km MAG=5.6 (USCGS)
	epPP	A	22 13	D = 103.5°
	eSKS	C	28 00	e 21 36 e 22 04
	eSS	C	36.3	
	LmH	C	16 02	
	LmV	C	02	
20.	eP	A	17 29 33	Tsinghai Prov./China 37.13°N 95.75°E
	e(pP)	A	29 38	H = 17 19 35.0 h = normal MAG=4.5 (USCGS)
				D = 58.9° (h = 20 km)
20.	ePS	C	20 27 46	Northern Easter I. Cordillera
	eSS	C	34.2	8.85°S 108.3°W
	LmH	C	21 06.2	H = 19 58 22.1 h = normal MAG=4.9 (USCGS)
	LmV	C	06.2	D = 115.5°
				LmH:19s 0.5 μm LmV:19s 0.4 μm
				MLH=5.2 MLV=5.0
21.	+iP	AB	07 45 26.2	Off W. Coast of Northern Sumatra
	i(pP)	A	45 39.5	3.64°N 95.76°E
	i(sP)	A	45 45	H = 07 33 00.6 h = normal MAG=5.9
	eiPP	A	48 40.5	D = 83.5° Az = 320.4° (USCGS); (h = 50 km)

Day	Phase		h m s	Remarks
cont.				
21.	+iPP	B	07 48 42.5	PV(A):2.0s 380nm PV(B):10s 4.9 µm
	iS	B	55 44	PPV(B):10s 1.7 µm SH(B):12s 7.2 µm
	iS	C	55 45	LmH:18s 14.5 µm LmV:18s 20 µm
	i(sS)	B	56 00	MPV(A)=6.2 MPV(B)=6.6 MPPV=6.4 MSH=6.7
	i(sS)	C	56 04	MLH=6.4 MLV=6.6
	iPS	C	56 44	
	eSS	C	08 01 15	
	eSSS	C	04 30	
	eiLQ	C	08.1	
	LmH	B	29.3	
	LmV	B	29.3	
21.	iSg	A	12 44 39.5	Explosion
	eL	A	44 43	PRUHONICE gives: D = 1.2°
21.	ePn	A	13 43 38	<u>North Sea</u> 56.99°N 4.94°E
	eiPx	A	43 42	H = 13 41 48.8 h = normal MAG=4.2
	eSn	A	45 02	D = 7.5° Az = 145.5° (USCGS)
	i(Sb)	A	45 28.0	LmH:12s 0.2 µm LmV:14s 0.2 µm
	iLi	B	45 36	MLH=3.0
	LmH	C	47.9	i 45 11.5 i 45 14.0
	LmV	C	47.9	
21.	e(Sg)	A	16 37 16	Explosion?
	e	A	37 18	
	eL	A	37 27	
22.	ePP	A	09 13 38.5	<u>Near Coast of Northern Chile</u>
	eSKS	C	20(20)	19.70°S 70.70°W
	ePS	C	22(45)	H = 08 55 54.6 h = 46 km MAG=4.8 (USCGS)
	LmH	C	51.3	D = 100.3°
	LmV	C	51.3	PPV:1.8s 20.4nm
				LmH:20s 0.7 µm LmV:18s 0.7 µm
				MPPV=5.4 MLH=5.4 MLV=5.2
22.	e	A	11 25 38	
22.	e	A	11 39 18	Explosion?
	e	A	39 33.5	

Day	Phase		h m s	Remarks
22.	eP	C	13 16 56	South Sandwich Islands Region
	ePKIKP	A	20 43.5	60.83°S 24.58°W
	eiPP	B	21 38	H = 13 02 06.8 h = normal MAG=6.1
	eiPP	A	21 40	D = 114.9° Az = 24.5° (USCGS); h = 35 km
	i(pPP)	A	21 50.2	PPV(B):7.5s 2.7 µm PPH(B):8s 1.0 µm
	eSKS	C	27 32	SPV(B):16s 3.6 µm SSH:36s 20 µm
	eSKKS	B	28 39	LmH:18s 25.6 µm LmV:17.5s 31.5 µm
	eS	C	29(32)	MPPV=7.1 MPPH=6.8 MLH=6.9 MLV=7.0
	iS	C	29 40	i 31 25.8 i 32 48
	eSP	C	31 24	
	iSP	B	31 29	
	iPKKP	A	31 31.0	
	eipPKKP	A	31 41	
	eiPPS	C	32 40	
	eSKKP	A	35 19	
	e(SS)	C	37.4	
	iSS	C	37 40	
	eiSSS	C	41.6	
	LmH	B	14 07.3	
	LmV	B	07.3	
22.	ePKIKP	A	13 35 37	South Sandwich Islands 60.9°S 23.2°W
	ePP	A	36 35	H = 13 17 02.5 h = 19 km MAG=5.9 (USCGS)
				D = 114.5°
22.	eiSg	A	14 58 21.5	Explosion
	eL	A	58 24.5	
22.	eiP	A	23 21 51	Lake Baikal Region 56.24°N 112.57°E
	epP	A	21 57	H = 23 12 18.9 h = 22 km MAG=5.0
	eS	C	29.6	D = 55.1° Az = 310.3° (USCGS); h = 24 km
	e	C	34.0	PV:1.0s 30.8nm
	LmH	B	47.6	LmH:14s 0.9 µm LmV:15s 0.9 µm
	LmV	B	47.6	MPV=5.4 MLH=5.0 MLV=5.0
23.	LmH	C	01 05	Off Coast of Northern Chile 19.69°S 70.96°W
	LmV	C	05	H = 00 09 26.0 h = 29 km MAG=4.6 (USCGS)
				D = 100.4°
				LmH:20s 0.2 µm LmV:20s 0.2 µm
				MLH=4.6 MLV=4.6

Day	Phase		h m s	Remarks
23.	LmH	C	02 42	Traces
23.	LmH	C	05 34.7	South Sandwich Islands Region
	LmV	C	34.7	54.44°S 22.41°W
				H = 04 19 32.8 h = normal MAG=4.5 (USCGS)
				D = 108.4°
				LmH:14s 0.2 µm LmV:15s 0.2 µm
				MLH=4.8 MLV=4.7
23.	eP	A	09 35 22.5	Near Coast of Northern Peru 4.29°S 81.46°W
	e	A	38 37	H = 09 21 59.4 h = normal MAG=5.0
	eSKS	C	46.0	D = 95.2° Az = 39.7° (USCGS)
	eS	C	46.7	LmH:20s 0.4 µm LmV:22s 0.4 µm
	ePS	C	48.0	MLH=4.9 MLV=4.8
	eSS	C	53.1	
	LmH	C	10 16.5	
	LmV	C	16.5	
23.	eiPg	A	10 15 27.5	Explosion
	eSg	A	15 45	D = ca. 1.3°
23.	eSg	A	10 32 57	Probably explosion.
	eL	A	33 00	
23.	iPg	A	13 04 50.2	Explosion/GDR
	iSg	A	04 57.0	
	eiL	A	04 58.5	
23.	eiL	A	13 05 12	Explosion/GDR
23.	ei	A	13 05 20.5	Explosion/GDR
	eiL	A	05 23.5	
23.	ePg	A	14 00 26	Explosion/ČSSR 50°37'N 14°21'E
	eiSg	A	00 49.5	Yield: 9 t (PRUHONICE)
	ei	A	00 53	D = 1.7°
	LmH	A	00 56	
23.	e	A	14 44 41.5	Explosion?
	e	A	44 45.5	

Day	Phase		h m s	Remarks
23.	e(Pg)	A	14 45 17.5	Explosion?
	eSg	A	45 37.5	(D = ca. 1.5°)
23.	LmH	C	15 21	South Sandwich Islands Region (USCGS)
	LmV	C	21	Traces
23.	iPg	A	16 33 21	Explosion
	iSg	A	33 40.5	D = ca. 1.5°
23.	ePKP	A	17 27 32	South of Fiji Islands 23.4°S 179.7°E
				H = 17 08 46.7 h = 640 km MAG=4.6 (USCGS)
				D = 151.2°
23.	LmH	C	18 56.3	
	LmV	C	57.5	
24.	LmH	C	01 48.4	Tibet 35.29°N 87.95°E
	LmV	C	52.5	H = 01 17 09.6 h = 31 km MAG=4.5 (USCGS)
				D = 55.5°
				LmH:20s 0.2 µm
				MLH=4.3
24.	+iP	A	03 33 13.4	Kurile Islands 43.48°N 147.51°E
	epP	A	33 29.5	H = 03 21 17.6 h = 70 km MAG=5.4
	eisP	A	33 38	D = 78.7° Az = 333.1° (USCGS);
	LmV	C	04 04	h = 64 ± 3 km
	LmH	C	04.6	PV:1.2s 40.8nm pPV:1.4s 27.6nm
				sPV:1.4s 40.0nm
				LmH:25s 0.25 µm
				MPV=5.4 MLH=4.5
24.	+iPKHKP	A	05 48 42.7	Fiji Islands Region 21.05°S 179.36°W
	eiPKP2	A	48 49.6	H = 05 30 05.8 h = 672 km MAG=4.7
				D = 149.2° Az = 346.3° (USCGS)
				PV1:1.0s 26.1nm
24.	iPg	A	09 02 44.5	Explosion
	iSg	A	02 46.5	D = 16 km
				SgV:0.5s 294nm

Day	Phase		h m s	Remarks
24.	iSg	A	10 29 28.5	Explosion/ČSSR 49°30.3'N 12°46'E
	eL	A	29 35	Yield: 8 t (PRUHONICE)
	LmH	A	29 41	D = 1.4°
24.	ePKP	C	10 52 10	New Hebrides Islands 14.91°S 166.85°E
	ePKIKP	A	52 14	H = 10 32 52.6 h = 23 km MAG=5.3
	e	A	52 19	D = 139.1° Az = 336.0° (USCGS)
	ePP	C	55 11.5	PPV(A):2.0s 66.6nm PPV(B):10.5s 0.5 µm
	iPP	B	55 14.5	LmH:20s 1.0 µm LmV:20s 1.4 µm
	eiPKS	AB	55 56	MPPV(A)=5.4 MPPV(B)=5.7 MLH=5.5 MLV=5.7
	eSSS	C	11 19.5	
	LmH	B	56	
	LmV	B	56	
24.	eP	A	10 54 44	Mozambique Channel 17.11°S 40.33°E
	e	A	54 51.5	H = 10 43 26.2 h = normal MAG=5.1 (USCGS)
	i	A	54 58.5	D = 72.1°
				Successive P-onsets with increasing amplitudes.
24.	e	A	10 59 23	Probably explosion.
	eSg	A	59 39	
24.	iSg	A	11 15 33.3	Explosion/ČSSR
	eL	A	15 44	KAŠPERSKÉ HORY gives: D = 85 km
				PRUHONICE: D = 1.1°
24.	+iP	A	12 10 49.3	PV:0.9s 14.2nm
24.	i(Pn)	A	12 31 17.5	Explosion? Probably ČSSR.
	i(Sn)	A	31 48.5	(D = ca. 2.7°)
	e(Sg)	A	31 58.5	PRUHONICE: D = 1.5°
	iL	A	32 06.5	KAŠPERSKÉ HORY: D = 1.3°
				VIENNA: D = 165 km
24.	eiPKP	A	13 53 24	South of Fiji Islands 22.29°S 178.08°W
	e	A	53 27.5	H = 13 34 10.5 h = 330 km MAG=4.6
	epPKP	A	54 55	D = 150.7° Az = 347.3° (USCGS); h = 375 km
				PV:1.1s 12.0nm

Day	Phase		h m s	Remarks
24.	e	A	14 07 30	Explosion
	eL	A	07 34	Traces
24.	e	A	15 58 39	Explosion?
	e	A	58 43	Traces
	e	A	58 51	
24.	+iP	A	23 24 09.5	Congo 10.5°S 27.3°E
	e	A	24 21.5	H = 23 14 45.0 h = 21 km MAG=5.0 (USCGS)
				D = 62.5°
25.	eiPg	A	09 03 43.0	Explosion
	eiSg	A	03 57.7	D = ca. 1.1°
25.	iPg	A	13 01 25.5	Explosion
	eiSg	A	01 39	D = ca. 1.0°
25.	eiPg	A	15 05 20.5	Explosion/Hilders, GFR 50°32.5'N 10°02.4'E
	eiSg	A	05 35.5	Yield: 4 t (HANNOVER)
				D = 1.0°
25.	eP	A	15 15 17.5	Rat Islands/Aleutian Is. 51.74°N 177.22°E
	LmH	C	50	H = 15 03 25.1 h = 37 km MAG=4.8
				D = 77.3° Az = 350.7° (USCGS)
				PV:1.8s 30.6nm
				MPV=5.1
26.	-eP	A	00 50 40	West Caroline Islands 12.23°N 140.73°E
	-eiP	B	50 41	H = 00 36 42.1 h = 33 km MAG=6.1
	e(pP)	A	50 54	D = 103.3° Az = 329.5° (USCGS)
	eX	B	53 46	PV(A):1.5s 67.2nm PV(B):7.5s 1.1 µm
	eiPP	B	54 50	PH(B):7.5s 0.3 µm PPV:12s 3.5 µm
	eSKS	C	01 01 14	PPH:12s 1.6 µm
	ei(S)	C	02 11	LmV:14.5s 10.4 µm
	iPS	B	03 54	MPPV=6.8 MPPH=6.7 MLV=6.5
	iPPS	B	04 47	i 50 43.3 e 50 46.5 e 50 59 e 51 24
	eiSPP	B	04 52	e 51 45 e 53 38 e 53 51.5 e 54 13
	eiPKKP	A	06 35	e 55 02.5 e 01 21 i 01 36 e 02 25
	eSS	C	09 24	i 04 40 e 06 55.5

Day	Phase		h m s	Remarks
cont.				
26.	ePSPS	B	01 09 40	X is a clear unidentified phase in our
	iLQ	C	22 32	long-period records preceding PP.
	iLR	C	26.0	
	LmV	B	41.2	
26.	e	A	01 59 37	Traces
26.	eP	A	02 21 07.5	West Caroline Islands 12.19°N 140.79°E
				H = 02 07 08.9 h = 30 km MAG=5.3
				D = 103.4° Az = 329.5° (USCGS)
26.	LmV	C	06 27.5	West Caroline Islands 12.11°N 140.74°E
	LmH	C	28	H = 05 25 17.4 h = normal MAG=4.7 (USCGS)
				D = 103.3°
				LmH:20s 0.2 µm LmV:22s 0.2 µm
				MLH=4.6 MLV=4.6
26.	ei(Sg)	A	13 41 01	Explosion/GDR
	iL	A	41 04.0	
26.	i(Sg)	A	13 41 14.0	Explosion/GDR
	iL	A	41 17.0	
26.	eiL	A	13 41 25.5	Explosion/GDR
26.	ei(Sg)	A	13 41 40.5	Explosion/GDR
	iL	A	41 42.5	
26.	eiL	A	13 41 52.5	Explosion/GDR
26.	eP	A	14 22 20	Turkey 37.35°N 30.78°E
	eS	C	25 56	H = 14 17 57.4 h = normal MAG=4.5
	LmH	C	28.5	D = 19.1° Az = 320.3° (USCGS)
	LmV	C	30.5	LmV:15s 0.3 µm
				MLV=3.8
26.	eP	A	15 37 57	Mid Indian Rise 20.17°S 67.15°E
	e	A	38 08	H = 15 25 20.0 h = normal MAG=5.0
				D = 85.7° Az = 328.2° (USCGS)

Day	Phase		h m s	Remarks
26.	-ePKP	A	18 39 32.5	Samoa Islands Region 15.39°S 172.73°W
	+i(pPKP)	A	39 43.8	H = 18 19 58.2 h = 37 km MAG=5.0
	e	A	40 50	D = 144.7° Az = 355.2° (USCGS);
	eLQ	C	19 20.0	(h = 40 km)
	LmH	C	51.0	PV1:1.6s 37.9nm PV2:1.6s 121nm
	LmV	C	51.0	LmH:18s 0.5 µm LmV:18s 0.8 µm
				MLH=5.3 MLV=5.5
26.	eP	A	21 56 31	Alaska Peninsula 55.27°N 160.57°W
	epP	A	56 49	H = 21 44 59.2 h = 65 km MAG=4.7
				D = 74.3° Az = 5.2° (USCGS); h = 70 km
				PV:1.0s 14.2nm
				MPV=5.1
27.	eP	A	02 29 51	Aleutian Islands 52.4°N 168.7°W
				H = 02 17 57.2 h = normal MAG=3.9 (USCGS)
				D = 77.5°
27.	+eP	A	13 21 13.5	Nicaragua 12.27°N 86.18°W
	+iP	B	21 14	H = 13 08 55.9 h = 183 km MAG=5.2
	+iP	C	21 15	D = 85.5° Az = 39.3° (USCGS); h = 180 km
	eX	A	21 36	PV(A):1.6s 37.9nm PV(B):8s 0.8 µm
	eipP	A	21 57.5	PPV(A):2.6s 272nm PPH(A):2.6s 175nm
	+ipP	B	21 58.5	PPV(B):5s 0.8 µm PPH(B):3.5s 0.4 µm
	isP	AB	22 17.5	pPPV(B):5s 0.4 µm SH(B):18s 1.3 µm
	-eiPP	B	24 41	(SP)H:18.5s 1.7 µm
	eiPP	A	24 42.5	MPV(A)=4.9 MPV(B)=5.6 MPPV(A)=5.9
	epPP	AB	25 20	MPPH(A)=5.9 MPPV(B)=6.2 MPPH(B)=6.1
	esPP	B	25 42	MSH=5.6
	eS	B	31 20	X is a clear unidentified phase between P
	eS	C	31 22	and pP in our short-period records. After
	e(SP)	C	32 22	Båth (Bulletin of the Seismological In-
	esS	C	32 36	stitute Uppsala) this phase is the P of
	ePPS	C	33 18	another shock in the same place.
	eSS	C	37.1	
	eLQ	C	43.6	
	eLR	C	48.7	

Day	Phase		h m s	Remarks
27.	eP	A	13 46 28.5	Vancouver Island Region 50.22°N 129.97°W
				H = 13 34 52.6 h = 24 km MAG=5.1
				D = 74.3° Az = 24.3° (USCGS)
27.	ePP	A	14 35 20	Molucca Passage 0.50°N 126.12°E
	e	C	55.0	H = 14 16 56.1 h = 62 km MAG=5.4 (USCGS)
	LQ	C	15 00	D = 104.9°
	LmV	B	21	LmV:20s 0.6 µm
				MLV=5.1
27.	LmH	C	17(32)	Luzon/Philippine Islands (USCGS)
	LmV	C	(32)	Traces
27.	iP	A	18 40 43.5	Vancouver Island 50.2°N 129.7°W
	ei	A	40 45.5	H = 18 29 07.4 h = normal MAG=4.5 (USCGS)
	e	A	40 55.5	D = 74.1°
27.	ePg	A	21 27 02	Valais/Switzerland 46.5°N 7.4°E
	iSg	A	28 07.0	H = 21 25 26 (BCIS)
	e	A	28 10.5	D = 5.0°
	eLg2	A	28 14.5	e 27 20 e 28 04 e 28 08.5
27.	eiPKHKP	A	22 28 58	Fiji Islands Region 20.38°S 178.12°W
	ePKP2	A	29 03	H = 22 10 11.6 h = 545 km MAG=4.3
				D = 148.8° Az = 348.0° (USCGS)
				PV:1.4s 27.6nm
28.	LmH	C	03 47.8	Aegean Sea (USCGS)
	LmV	C	50.1	
28.	ei(Sg)	A	12 17 52	Explosion
	eL	A	17 59	
28.	eiSg	A	13 24 41	Explosion
	e	A	24 44	PRUHONICE gives: D = 1.2°
28.	eP	A	15 37 27.5	Vancouver Island Region 50.36°N 129.91°W
	e	A	37 30.5	H = 15 25 51.8 h = normal MAG=5.2
	eS	C	47 08	D = 74.2° Az = 24.3° (USCGS)

Day	Phase		h m s	Remarks
cont.				
28.	ePS	C	15 47 40	LmH:25s 0.6 /um LmV:25s 0.4 /um
	eSS	C	52(20)	MLH=4.8 MLV=4.7
	LmH	C	16 08	e 37 33 e 47 40
	LmV	C	08	
28.	eP	A	16 31 44	<u>Vancouver Island Region</u> 50.40°N 129.84°W
	e	A	31 50	H = 16 20 06.6 h = normal MAG=5.1
	eS	C	41 20	D = 74.1° Az = 24.4° (USCGS)
	LmH	C	17 02.0	SH:15s 0.5 /um
	LmV	C	02.2	LmH:23s 0.6 /um LmV:24s 0.7 /um
				MSH=5.4 MLH=4.9 MLV=4.9
				AN USSR gives: South of Queen Charlotte
				Island 52.3°N 131.6°W
				H = 16 20 19 h = normal MAG=5
				D = 72.9°
28.	-iP	A	17 40 38.7	<u>Dodecanese Islands</u> 36.70°N 26.82°E
				H = 17 36 41.0 h = 174 km MAG=4.3
				D = 17.7° Az = 326.7° (USCGS)
				PV:1.2s 15.3nm PH:1.1s 19.8nm
28.	eP	A	18 43 20	<u>West Pakistan</u> 30.3°N 69.9°E
	e	A	43 58.5	H = 18 34 44 h = normal (AN USSR)
				D = 47.5°
28.	e(P)	A	21 14 48	<u>Southern Sinkiang Prov./China</u>
	e	A	14 53	36.51°N 80.14°E
				H = 21 05 51.7 h = normal MAG=4.7
				D = 49.9° Az = 309.2° (USCGS)
				The first onset of P must be about 4 s
				earlier.
28.	eP	A	21 20 41	<u>Morocco</u> 31.49°N 6.06°W
	e	A	20 51	H = 21 15 35.7 h = normal MAG=4.6
	e	A	20 58.5	D = 23.2° Az = 29.4° (USCGS)
	eS	B	24 53	PV:2.8s 143nm SH:8s 0.6 /um
	LmH	C	30.4	LmH:12s 0.5 /um LmV:14s 0.4 /um
	LmV	C	30.8	MPV=5.1 MSH=5.1 MLH=4.2 MLV=4.2

Day	Phase		h m s	Remarks
29.	e(PP)	A	07 46 20	Banda Sea 6.8°S 123.5°E
	ePP	C	46 34	H = 07 27 37 h = normal MAG=6 (AN USSR)
	eSP	C	55 56	D = 109.1°
	eSPP	C	57 00	LmH:30s 3.1 /um LmV:25s 1.2 /um
	eSS	C	08 01 44	MLH=5.7 MLV=5.4
	eLQ	C	12.2	e 46 45 e 47 13
	LR	C	18	
	LmH	C	27.1	
	LmV	C	33.2	
29.	i	A	09 21 57.5	
29.	iPg	A	10 42 52.6	Explosion
	ei	A	43 06.5	D = ca. 1.2°
	iSg	A	43 08.6	
29.	ePP	C	11 10 08	New Guinea 3.25°S 141.50°E
	ePP	A	10 11	H = 10 50 09.4 h = 41 km MAG=5.1 (USCGS)
	ePS	C	19 40	D = 116.8°
	eSS	C	26 30	LmH:21s 0.8 /um LmV:21s 0.5 /um
	LmH	C	55.7	MLH=5.3 MLV=5.1
	LmV	C	59.5	
29.	ei(Sg)	A	17 14 44	Probably explosion.
	eL	A	14 55	
	LmH	A	15 04	
29.	LmH	C	22 08.6	China 41.1°N 79.0°E
	LmV	C	08.6	H = 21 46 39 h = normal MAG=4½ (AN USSR)
				D = 46.4°
				LmH:9s 0.2 /um LmV:14s 0.3 /um
				MLH=4.5 MLV=4.4
30.	eP	A	02 17 23	Japan 35.6°N 140.0°E
	e	A	17 27.5	H = 02 06 11.1 h = 72 km MAG=4.7 (USCGS)
	e	A	17 58	D = 82.3°

Day	Phase		h m s	Remarks
30.	-iP1	A	04 32 47.2	Szechwan Prov./China 31.66°N 100.30°E
	-iP	B	32 48.0	H = 04 22 01.5 h = 3 km MAG=6.1
	-iP2	A	32 49.7	D = 65.5° Az = 315.6° (USCGS)
	+iPP	C	35 08	PV:7s 14.2 /um PH:7s 6.6 /um
	-iPP	B	35 15	PPV(B):8s 5.8 /um PPPV:12s 3.3 /um
	+iPPP	B	36 49.5	PaV:22s 2.8 /um PaH:24s 5.6 /um
	iPa	C	37 04	SH:13s 8.4 /um
	iS	B	41 32	LmH:15s 37 /um LmV:17s 45.5 /um
	iSS	C	46 12	MPV=7.2 MPH=7.3 MPPV=7.2 MSH=6.7
	iSa	C	48 41	MLH=6.7 MLV=6.8
	LmH	B	05 03.3	i 41 38 i 46 28
	LmV	B	03.3	Multiple P in our short-period records, the first onset smaller than the second one.
30.	iP	A	05 08 24.5	Szechwan Prov./China 31.67°N 100.32°E
				H = 04 57 42.6 h = normal MAG=4.9
				D = 65.5° Az = 315.6° (USCGS)
30.	e(Pg)	A	07 21 27	Near earthquake.
	e(Sg)	A	22 37	(D = 5.3°)
				TRIESTE gives: D = 160 km
30.	eP	A	08 21 55	Near East Coast of Honshu/Japan
				36.18°N 140.02°E
				H = 08 09 40.8 h = 77 km MAG=4.7
				D = 82.3° Az = 329.8° (USCGS)
				AN USSR gives: Japan 37.0°N 139.2°E
				H = 08 09 59 h = 190 km
				D = 81.3°
30.	eSg	A	10 46 41.0	Probably explosion.
	eL	A	46 50	
	LmH	A	46 54	
30.	eP	AB	11 19 31	Szechwan Prov./China 31.61°N 100.29°E
	e	A	19 37	H = 11 08 49.6 h = normal MAG=5.1
	ePP	A	21 50	D = 65.5° Az = 315.6° (USCGS)
	eS	B	28 16	PV(A):1.5s 33.5nm PV(B):5.5s 0.5 /um

Day	Phase		h m s	Remarks
cont.				
30.	eiS	B	11 28 19	SH:9.0s 0.4 /um
	eSS	C	32(30)	LmH:14s 2.2 /um LmV:14s 2.7 /um
	eSSS	C	35.5	MPV(A)=5.3 MPV(B)=5.9 MSH=5.6 MLH=5.5
	LmH	B	50.2	MLV=5.6
	LmV	B	50.2	
30.	ePKIKP	A	12 15 28	Kermadec Islands Region 30.37°S 178.65°W
	e	A	15 30	H = 11 55 50.5 h = 161 km MAG=4.8
	iPKP2	A	16 04.0	D = 158.3° Az = 342.1° (USCGS)
				PV3:1.3s 36.2nm
				AN USSR gives: Kermadec Islands Region
				30.3°S 178.9°W
				H = 11 55 34 h = normal MAG=5½
				D = 158.2°
30.	ei	A	12 58 40	Explosion/GDR
	ei(Sg)	A	58 44.5	
	iL	A	58 48.5	
30.	ei	A	12 59 09.5	Explosion/GDR
	eiL	A	59 11.5	
30.	eiSg	A	12 59 18	Explosion/GDR
	eiL	A	59 20	
30.	iPg	A	13 32 09.1	Explosion
	iSg	A	32 23.5	D = ca. 1.1°
30.	+iP	A	13 45 23.7	Kurile Islands 45.40°N 151.50°E
	+i(pP)	A	45 30.2	H = 13 33 26.4 h = normal MAG=5.5
	iPcP	A	45 34.6	D = 78.2° Az = 335.2° (USCGS)
	eS	C	55 14	PV:1.4s 221nm
	eSS	C	14 00.2	LmH:17.5s 5.9 /um LmV:20s 4.5 /um
	LQ	C	07.5	MPV=6.1 MLH=6.0 MLV=5.8
	eLR	C	10.6	ei 45 40 i 55 47
	LmH	B	19.5	
	LmV	B	23.3	

Day	Phase		h m s	Remarks
30.	eP	A	13 46 45	Kurile Islands H = 13 34 49 (UPP) PV:1.5s MPV=5.5
30.	eP	A	20 15 29	Kurile Islands 45.27°N 151.39°E
	LmH	C	49.4	H = 20 03 31.7 h = normal MAG=4.5
	LmV	C	53	D = 78.3° Az = 335.2° (USCGS) PV:1.4s 15.3nm LmH:15.5s 0.2 µm LmV:20s 0.2 µm MPV=4.9 MLH=4.6 MLV=4.4
31.	-iPKP	A	11 06 36	New Hebrides Islands 18.74°S 169.14°E H = 10 47 30.4 h = 234 km MAG=4.9 D = 143.5° Az = 335.9° (USCGS) PV:1.3s 27.8nm
31.	e	A	11 30 46	Traces
	e	A	30 54	
31.	eP	A	13 51 30.5	Luzon/Philippine Islands 18.30°N 121.27°E H = 13 38 50.7 h = 99 km MAG=4.8 D = 87.9° Az = 323.1° (USCGS)
31.	ei(Pg)	A	14 47 41	Explosion
	ei	A	47 47.5	
	eiSg	A	48 02	
31.	ePKP	A	19 12 34.5	Tonga Islands 17.46°S 175.25°W
	-iPKP	A	12 35.4	H = 18 53 25.2 h = 277 km MAG=5.4
	i	A	12 37.5	D = 146.4° Az = 352.1° (USCGS); h = 270 km
	ipPKP	A	13 43.0	PV2:1.0s 90.0nm
	esPKP	B	14 08	e 13 46.5 e 15 51

Day	Phase		h m s	Remarks
1.	-eiPKIKP	A	03 49 44	East New Guinea Region 5.57°S 147.18°E
	e	A	51 13	H = 03 31 10.5 h = 182 km MAG=5.6
	ePP	A	51 28.5	D = 121.8° Az = 328.4° (USCGS)
	e	B	52 06	PPV:2.0s 33.1nm
	eSKKS	C	04 07(40)	LmH:35s 1.0 /um
	LmH	C	30.7	MPPV=5.2
1.	ei(Sn)	A	05 26 46.0	
	ei	A	26 48.3	
	i(Sg)	A	26 51.6	
1.	e(PKP2)	A	07 26 57.5	South of Kermadec Islands 34.4°S 179.0°E
				H = 07 06 21.9 h = 33 km MAG=4.7 (USCGS)
				D = 161.3°
1.	e	A	11 42 30	
1.	LmH	C	16 06.5	West Chile Rise 44.1°S 82.1°W
	LmV	C	06.5	H = 14 53 55 h = normal MAG=5.2 (USCGS)
				D = 124.3°
				LmH:19s 0.3 /um LmV:19s 0.2 /um
				MLH=5.9 MLV=5.9
1.	iPg	A	17 09 22.5	Explosion
	iSg	A	09 25.0	D = 0.2°
1.	+iP	A	22 53 42.9	Kurile Islands 44.88°N 146.96°E
	ePcP	A	53 55	H = 22 42 01.8 h = 134 km MAG=5.4
	epP	A	54 13	D = 77.3° Az = 332.7° (USCGS); h = 122 km
	eisP	A	54 28	PV:1.3s 139nm
	ePP	A	56 34	LmH:20s 1.2 /um LmV:25s 1.0 /um
	eS	C	23 03 20	MPV=5.9 MLH=5.3 MLV=4.2
	eScS	C	03 50	e 53 48
	LmH	C	25.3	AN USSR gives: Kurile Islands
	LmV	C	29.5	44.9°N 147.3°E
				H = 22 41 50 h = normal MAG=5.4

Day	Phase		h m s	Remarks
2.	ePKP2	A	01 44(54)	South of Kermadec Islands 33.7°S 178.8°W
	ei	A	44 58	H = 01 24 22.4 h = 129 km MAG=4.7 (USCGS)
	ei	A	45 23.5	D = 161.4°
	e	A	45 38	ei 45 21
2.	eiP1	A	03 51 14	Jan Mayen Island 71.56°N 8.23°W
	iP2	A	51 17.0	H = 03 46 13.9 h = normal MAG=4.4
	iP3	A	51 22.5	D = 22.8° Az = 146.2° (USCGS)
	eS	B	55 24	PV2:1.3s 27.9nm PV3:1.7s 52.9nm
	LmH	C	04 02	MPV2=4.8 MPV3=4.9
	LmV	C	03	Successive P-onsets with increasing amplitude.
2.	eP	A	05 34 20	
	e	A	34 42.5	
2.	ePKP2	A	05 57 40	Kermadec Islands 29.07°S 178.98°W
				H = 05 37 50.0 h = 307 km MAG=4.6 (USCGS)
				D = 157.0°
				AN USSR gives: Fiji Islands Region
				22.1°S 175.4°E
				H = 05 38 00 h = normal
				D = 148.8°
3.	i(P)	A	00 14 05.0	
	ei	A	14 49	
3.	e	A	00 49 30	
3.	eSS	C	02 00.4	New Guinea 7.8°S 147.1°E
	eSSS	C	05.0	H = 01 23 19.6 h = 139 km MAG=5.4 (USCGS)
	eSSSS	C	08.3	D = 123.6°
	LQ	C	14	AN USSR gives: h = normal
	LmH	C	26.5	LmH:26s 0.5 μm
3.	eP	A	04 58 01	Japan 31.0°N 129.8°E
	epP	A	58(40)	H = 04 45 57.0 h = 165 km MAG=4.6 (USCGS)
				D = 82.1°

Day	Phase		h	m	s	Remarks
3.	ePKP	A	06	41	30.5	Fiji Islands 18.96°S 178.00°W
	e	A		41	33	H = 06 22 45.8 h = 547 km MAG=4.1
						D = 147.5° Az = 348.5° (USCGS)
						PV:1.1s 14.4nm
3.	eiP	A	07	49	50	Southern Greece 37.51°N 21.85°E
						H = 07 46 19.1 h = 48 km MAG=4.4
						D = 15.0° Az = 334.1° (USCGS)
3.	e(P)	A	09	23	17	Greece 38.5°N 22.1°E
						H = 09 19 47.6 h = 35 km MAG=4.3 (USCGS)
						D = 14.3°
						The first onset of P must be about 7 s earlier.
3.	eiP	A	11	41	47.5	Alaska 60.5°N 151.6°W
	ipP	A		42	06	H = 11 30 51.7 h = 79 km MAG=4.7 (USCGS)
						D = 68.3° h = 75 km
3.	eP	A	21	21	08	Off Coast of Peru 10.58°S 79.75°W
	+iP	B		21	08.1	H = 21 07 30.8 h = 38 km MAG=6.5
	+iP	C		21	09.5	D = 99.0° Az = 40.1° (USCGS)
	eX	C		23	24	PV(B):13s 7.9 μm PH(B):14s 2.6 μm
	eY	C		24	38	PPV(B):(16s) 8.5 μm SKSH(B):(18s) 11 μm
	ePP	A		25	14	SKSH(C):(24s) 15 μm SH(C):32s 13.1 μm
	+iPP	B		25	17.5	LmH:23s 60 μm LmV:23s 91 μm
	iPPP	C		27	24	MPV(B)=7.1 MPH(B)=7.3 MPPV(B)=6.9
	ei(Pa)	C		30	48	MLH=7.0 MLV=7.2
	iSKS	BC		31	51	ei 21 12.5 i 24 44 e 27 04
	iS	C		32	40	PP has an unusually long period (T = 6 s)
	iSP	C		34	03	in the registrations of our short-period
	iPPS	C		34	54	vertical component. X and Y are two clear
	ePKKP	A		37	36.5	unidentified onsets between P and PP in
	iSS	C		39	36	our long-period registrations.
	eSKKP	A		40	53.5	
	iSSS	C		43	10	
	ePKPPKP	A		45	49	
	iSSSS	C		46	44	
	i(Sa)	C		47	20	

Day	Phase		h m s	Remarks
cont. 3.	LmH	B	22 02.0	
	LmV	B	02.3	
4.	ePKP2	A	01 02 44	Kermadec Islands
				H = 00 42.2 (UPP)
				D = ca. 159°
4.	eP	A	03 28 59	Central Mid-Atlantic Ridge
	epP	A	29 09	1.32°S 23.91°W
	e	A	30 09	H = 03 18 52.2 h = normal MAG=4.7
	eS	C	37 16	D = 60.0° Az = 25.3° (USCGS); h = 40 km
	LQ	C	44.0	AN USSR gives: Southeast of Northatlantic
	eLR	C	47.3	Ridge 1.6°S 19.1°W
	LmH	C	(54)	H = 03 19 04 h = normal
	LmV	C	(54)	D = 58.3°
4.	-ePKIKP	A	04 11 29	Kermadec Islands 31.44°S 179.45°W
	ePKHKP	A	11 43	H = 03 51 58.9 h = 231 km MAG=5.5
	+iPKP2	A	12 07.5	D = 159.1° Az = 340.0° (USCGS); h = 230 km
	eipPKP2	A	13 05	PV1:1.8s 46.0nm PV2:1.5s 20.1nm
	e(PP)	A	15 41	PV3:1.4s 104nm PV4:2.4s 222nm
	e	C	16 20	AN USSR gives: Kermadec Islands Region
	e	A	16 48	31.4°S 178.5°W
	e	A	17 28	H = 03 51 34 h = normal
				D = 159.4°
4.	e	A	11 58 19.5	Explosion?
	eSg	A	58 20.5	
4.	e(Sg)	A	12 14 31.5	Explosion?
	e	A	14 38	
4.	eSg	A	12 55 03.5	Probably explosion.
	e	A	55 06.5	
4.	eiPKIKP	A	13 19 03.5	New Ireland Region 4.72°S 153.20°E
				H = 13 00 11.3 h = 70 km MAG=4.6
				D = 124.1° Az = 331.5° (USCGS)

Day	Phase		h m s	Remarks
4.	eiP	A	19 41 16	Near East Coast of Kamchatka 54.79°N 159.08°E H = 19 30 13.7 h = 182 km MAG=4.6 D = 71.5° Az = 338.8° (USCGS) PV:0.9s 11.8nm MPV=4.6
5.	eSg	A	02 48 47.5	
	e	A	48 49	
5.	eP	A	08 35 21	Turkey 36.73°N 29.33°E
	e	A	35 30	H = 08 31 01.1 h = 24 km MAG=4.4
	e	A	35 40	D = 18.8° Az = 323.1° (USCGS)
	e	A	35 50	PV:(2.0s 66nm)
	LmH	C	41.3	
	LmV	C	41.3	
5.	ePn	A	11 38 21.5	Northern Italy 45.60°N 13.92°E
	ei	A	38 31	H = 11 37 03.0 h = 38 km MAG=4.0
	iPg	A	38 44.0	D = 5.3° Az = 343.9° (USCGS)
	eiSn	A	39 19.5	PgV:0.9s 28.3nm SnH:0.6s 254nm
	iSg	A	39 50.5	SgH:1.0s 205nm
	iLg2	A	39 59.0	Exceptionally clear Sn-onset in the NS-
	ei	A	40 02.6	component and Sg-onset in the EW-component. The same in the following two aftershocks. BCIS gives: Yugoslavia 45.7°N 14.2°E h = 11 37 04
5.	eiSn	A	15 20 32.5	Yugoslavia 45.7°N 14.2°E
	iSg	A	21 03.2	H = 15 18 16 (BCIS)
	iLg2	A	21 11.6	D = 5.3° SnH:0.7s 45.9nm SgH:1.0s 49.4nm Lg2H:0.9s 82.5nm
5.	ePn	A	15 22 19	Yugoslavia 45.7°N 14.2°E
	iPg	A	22 41.0	H = 15 21 01 (BCIS)
	i	A	22 56.9	D = 5.3°
	iSn	A	23 16.3	PnV:0.8s 21.2nm PgV:0.9s 28.3nm
	iSg	A	23 47.9	SnH:0.6s 215nm SgH:1.0s 214nm
	iLg2	A	23 56.5	Lg2H:0.9s 308nm

Day	Phase		h m s	Remarks
5.	eSg	A	17 57 58.5	Aftershock to the preceding Yugoslavia earthquakes.
				TRIESTE gives: D = 35 km
6.	eP	A	01 54 16	India-East Pakistan Border Region
	i	A	54 26.0	24.1°N 91.7°E
	ei	A	54 32	H = 01 43 31.8 h = 18 km MAG=5.0 (USCGS)
				D = 65.6°
				AN USSR gives: Assam 25.8°N 92.3°E
				H = 01 43 48 h = normal MAG=4 1/2
				D = 65.3°
				P is followed after 10 s by a much larger onset. A double shock in the same focus?
6.	eP	A	03 31 07	Kurile Islands 46.72°N 153.99°E
	e	A	31 17.5	H = 03 19 12.4 h = normal MAG=4.8
				D = 77.7° Az = 336.6° (USCGS); h = 40 km
				PV:1.0s 21.3nm
				MPV=5.2
6.	eP	A	05 03 29	Crete 35.0°N 23.0°E
	e	A	03 38	H = 04 59 24.7 h = normal MAG=4.8 (USCGS)
	e	A	03 47.5	D = 17.7°
	eS	C	06 50	LmH:13s 2.4 µm LmV:14s 2.1 µm
	LQ	C	(07.2)	MLH=4.6 MLV=4.7
	LmH	B	10.7	e 03 44.5 i 03 55.0 i 04 03.0 e 06 55
	LmV	B	11.7	
6.	-eiP	A	07 41 43	Andaman Islands 14.67°N 93.58°E
	+ipP	A	41 54.5	H = 07 30 10.8 h = normal MAG=5.6
	ei(PcP)	A	42 00	D = 73.7° Az = 318.9° (USCGS); h = 44 km
	i(pPcP)	A	42 12.5	LmH:18s 1.8 µm LmV:20s 1.7 µm
	eS	C	51 12	MLH=5.5 MLV=5.4
	eSS	C	55 56	e 42 28 e 44 40 e 51 30
	LmH	B	08 19.7	
	LmV	B	19.7	
6.	e(P)	A	07 59 29	

Day	Phase		h m s	Remarks
6.	eP	A	09 37 26.5	Fox Islands/Aleutian Is. 52.43°N 168.63°W
				H = 09 25 34.2 h = 42 km MAG=4.2
				D = 77.3° Az = 359.8° (USCGS)
				PV:1.0s 14.2nm
				MPV=5.1
6.	e(Pg)	A	11 34 00	Probably explosion.
	e(Sg)	A	34 12	
	e	A	34 18	
6.	+eiP	A	17 36 32.5	Fox Islands/Aleutian Is. 52.57°N 168.54°W
	i(pP)	A	36 44.2	H = 17 24 40.1 h = normal MAG=4.8
	LmH	C	18 18.9	D = 77.2° Az = 359.9° (USCGS);
	LmV	C	19.4	(h = 43 km)
				PV:1.1s 33.7nm
				LmH:18s 0.3 μm LmV:16s 0.2 μm
				MPV=5.4 MLH=4.6 MLV=4.6
7.	eP	A	00 35 06.5	Albania 40.6°N 19.4°E
	e	A	35 26	H = 00 32 22 h = normal MAG=4.4 (USCGS)
	LmH	C	40.9	D = 11.5°
	LmV	C	40.9	LmH:15s 0.4 μm LmV:16s 0.3 μm
				MLH=3.4
7.	eP	A	07 26 00	Celebes Sea 2.69°N 124.30°E
	epP	B	27 04	H = 07 12 36.6 h = 274 km MAG=5.8
	e(pP)	A	27 14	D = 102.1° Az = 323.1° (USCGS); h = 266 km
	ePP	B	30 16	LmH:17.5s 1.5 μm LmV:15s 1.2 μm
	epPP	B	31 13	AN USSR gives: Indonesia 2.6°N 124.3°E
	esPP	B	31 44	H = 07 12 29 h = 202 km
	iSKS	B	36 12	D = 102.1°
	eSKKS	B	36 49	ei 30 21 ei 31 25 e 36 12 e 44 16
	eiS	B	37 17	
	eiSP	B	38 45	
	esSP	B	40 23	
	e(sPPS)	B	41 14	
	iPKKP	A	41 59.2	
	eSS	B	44 32	
	esSS	B	46 25	

Day	Phase		h m s	Remarks
cont.				
7.	eSa	B	07 54 53	
	LmH	B	08 12.2	
	LmV	B	19	
7.	eP	A	08 08 33	
7.	ePKP2	A	09 54 43	Kermadec Islands Region 30.5°S 177.6°W
	e	A	54 49.5	H = 09 34 12.1 h = 27 km MAG=4.7 (USCGS)
	e	A	54 56.5	D = 158.7°
7.	ePKIKP	A	11 27 19	Kermadec Islands 31.28°S 179.63°E
	ePKHKP	A	27 33	H = 11 08 18.2 h = 430 km MAG=5.1
	+iPKP2	A	27 59.0	D = 158.7° Az = 338.7° (USCGS); h = 436 km
	epPKP2	A	29 44	i 28 09.0 ei 28 28.5 e 30 35 e 31 37
	eSKP	A	30 20	e 31 46 e 47 24
	ePP	A	31 39	
7.	iPg	A	13 04 05.7	Explosion
	iSg	A	04 19.3	D = 1.0°
7.	eP	A	13 57 18	
7.	eiP	A	14 12 05.5	Sicily 37.94°N 15.30°E
	ePP	A	12 16	H = 14 09 02.8 h = 53 km MAG=4.3
	ei	A	12 31.5	D = 13.0° Az = 349.5° (USCGS)
	e(SS)	C	14.9	PV:1.0s 19.0nm
	LmH	B	18	LmH:15s 1.3 µm LmV:16s 1.9 µm
	LmV	B	18	MLH=4.1
7.	eiPg	A	14 25 40.5	Explosion/GDR
	iSg	A	25 49.8	D = ca. 0.7°
	ei	A	25 51.5	LmH:1.4s 120nm LmV:1.4s 109nm
	eL	A	25 53	
	LmH	A	26 03	
	LmV	A	26 03	

Day	Phase		h m s	Remarks
8.	iP1	A	02 07 34.5	Greece-Albania Border Region
	iP2	A	07 35.8	40.74°N 20.17°E
	ei	A	07 46.5	H = 02 04 49.1 h = 30 km MAG=4.7
	eiX	A	08 27	D = 11.6° Az = 331.8° (USCGS)
	iS	A	09 42.4	PV2:1.0s 33.2nm
	eiY	A	10 33	LmV:15s 3.0 μm
	LmH	B	11.8	X and Y are very clear unidentified pha-
	LmV	B	13.7	ses in the short-period records. P and S
				of a second shock in the same focus?
8.	e(P)	A	05 15 55	Probably Dodecanese Islands (UPP)
	LmH	C	22	
	LmV	B	24	
8.	eP	A	09 54(52)	Greece 39.17°N 21.66°E
	e	A	55 00	H = 09 51 43.1 h = 45 km MAG=4.4
	LmV	B	10 01.2	D = 13.5° Az = 331.6° (USCGS)
	LmH	B	01.3	LmH:13s 0.6 μm LmV:12s 0.5 μm
				MLH=3.9
				The second onset is much greater than the
				first one.
8.	iPg	A	10 40 26.2	Explosion
	iSg	A	40 42.0	D = ca. 1.2°
8.	eSg	A	12 43 55	Explosion/ČSSR
				BRATISLAVA gives: D = 75 km
8.	eP	A	12 53(03)	East of Severnaja Zemlja 78.20°N 126.34°E
	e	A	53 12.5	H = 12 44 44.8 h = normal MAG=4.5
	e	A	53 44	D = 45.6° Az = 305.9° (USCGS)
				The first onset is much smaller than the
				second one.
8.	e(Sg)	A	17 22 34.5	Probably explosion.
	e	A	22 38.5	
	eL	A	22 46	
	e	A	22 51	

Day	Phase		h m s	Remarks
8.	eP	C	22 51 36	West Caroline Islands 12.21°N 140.78°E
	eP	A	51 38.5	H = 22 37 39.5 h = 27 km MAG=5.3
	ePP	C	55 51	D = 103.4° Az = 329.5° (USCGS)
	ePP	A	55 55.5	LmH:16s 3.6 µm LmV:17s 4.4 µm
	eS	C	23 02 13	MLH=6.0 MLV=6.1
	eS	B	02 15.5	
	ePS	C	04 57	
	iPS	C	05 07	
	iPPS	B	05 48	
	ePPS	C	05 50	
	eSS	C	10 40	
	eSSS	C	14 40	
	eLQ	C	23 28	
	eLR	C	27.0	
	LmH	B	41.9	
	LmV	B	43.2	
9.	ePKP	A	10 18 09.5	Fiji Islands 21.12°S 176.66°W
				H = 09 58 34.0 h = 118 km MAG=4.1
				D = 149.8° Az = 349.5° (USCGS)
				PV:1.0s 14.2nm
9.	−eiP	A	10 19 37.5	Santiago del Estero Prov./Argentina
	−eiP	B	19 38	27.70°S 63.14°W
	−eiP	C	19 39	H = 10 06 44.1 h = 578 km MAG=5.8
	epP	B	21 46	D = 102.0° Az = 38.9° (USCGS)
	esP	BC	22 44	PV(A):1.3s 55.6nm PV(B):10s 1.9 µm
	+iPP	B	23 52.5	PH(B):10s 0.6 µm sPV(B):12s 1.4 µm
	+iPP	AC	23 54.5	PPV(B):8s 3.1 µm sSPV(B):16s 9.9 µm
	ei(pPP)	BC	25 36	MPV(A)=6.0 MPV(B)=6.7 MPH(B)=6.5
	epPP	C	25 44	MPP(B)=6.7
	isPP	B	26 43.5	sSP is the greatest onset in our long-
	eSKS	B	29 16	period records. Sa is very clear.
	eSKS	C	29 22	
	e(S)	B	30 26	
	eiS	C	30 36	
	eiSP	BC	32 00	
	ePS	C	33 00	
	isSKS	B	33 24	

Day	Phase		h m s	Remarks
cont.				
9.	esS	B	10 34 20	
	isS	C	34 24	
	i(sSP)	C	35 39	
	esSP	BC	35 56	
	ePKKP	A	36 12	
	epPKKP	A	38 08	
	iSa	C	48.0	
9.	e(Sg)	A	12 48 44	Explosion/GDR
	iL	A	48 47.5	LmH:1.2s 212nm LmV:1.2s 153nm
	LmH	A	48 49	
	LmV	A	48 49	
9.	iL	A	12 49 22.0	Explosion/GDR
9.	iSg	A	12 50 07.0	Explosion/GDR
	eiL	A	50 09.5	LmH:1.2s 194nm LmV:1.2s 128nm
	LmH	A	50 12	
	LmV	A	50 12	
9.	eP	B	14 57 56	West Caroline Islands 12.3°N 140.7°E
	ePP	B	15 02 14	H = 14 43 57.7 h = normal MAG=5.4 (USCGS)
	ePS	C	11 20	D = 103.3°
	eiPPS	B	12 08	PPV:10s 0.5 μm PPSV:12s 0.7 μm
	ePPS	C	12 10	LmH:17s 2.2 μm LmV:18s 3.2 μm
	eSS	C	16 56	MPPV=6.0 MLH=5.8 MLV=5.9
	eSSS	C	21.0	
	LQ	C	30	
	LR	C	33	
	LmH	B	48.9	
	LmV	B	49.4	
9.	ePKP	A	17 12 02	South Pacific Cordillera 54.78°S 136.03°W
	e(pPKP)	A	12 10	H = 16 52 01.3 h = normal MAG=5.4
	ePKP2	A	12 45	D = 160.1° Az = 88.6° (USCGS)
	epPKP2	A	12 57	LmH:19s 4.2 μm LmV:19s 5.1 μm
	ei	A	13 20	MLH=6.2 MLV=6.4
	iPP	B	16 26	(pPKP) is much stronger than PKP. X is a

Day	Phase		h m s	Remarks
cont.				
9.	eX	B	17 17 00	clear unidentified phase following PP.
	ePP2	B	19 47	PP2 is the PP traveled along the longer
	eSKKS	C	23(15)	path to the station (D2 = 360° - D).
	eSKSP	C	26 50	AN USSR gives: South Pacific Cordillera
	ePPS	C	29 40	55.4°S 139.1°W
	ePPPS	C	31 48	H = 16 52 04 h = normal
	eiSS	C	36 32	D = 161.8°
	ePSS	C	37 28	
	eSSS	C	42 48	
	e(Sa)	C	57.1	
	LmH	B	18 18.9	
	LmV	B	18.9	
9.	LmV	C	23(25)	LmH:19s 0.3/um LmV:20s 0.2/um
	LmH	C	(30)	
10.	LmH	C	03 38	
	LmV	C	38	
10.	e(P)	A	04 54 59	Near earthquake (Yugoslavia?)
	e(S)	A	57 30	TRIESTE gives: D = 400 km
	e	A	57 36	
	e	A	57 46.5	
10.	(LmH)	C	18 39	<u>Luzon/Philippine Islands</u> (USCGS)
11.	ePKP	A	01 42 28	<u>New Hebrides Islands</u> 21.35°S 173.82°E
	e	A	42 37.5	H = 01 22 43.7 h = 32 km MAG=4.8
	LmH	C	02 46.5	D = 147.6° Az = 338.7° (USCGS)
	LmV	C	46.5	LmH:22s 0.2/um LmV:24s 0.2/um
				MLH=4.7 MLV=4.8
11.	+eiPKP	A	04 56 56.5	<u>Loyalty Islands</u> 21.36°S 169.67°E
	e	A	57 04.5	H = 04 37 16.4 h = 11 km MAG=5.0
	e	A	57 11	D = 146.1° Az = 334.8° (USCGS)
	LmH	C	05 57.5	PV:1.8s 61.2nm
	LmV	C	57.5	LmH:25s 0.3/um LmV:26s 0.3/um
				MLH=4.9 MLV=4.9

Day	Phase		h m s	Remarks
11.	eP	A	06 20 34	West Pakistan 27.48°N 66.43°E
	epP	A	20 43	H = 06 12 00.5 h = 36 km MAG=4.6
	eX	A	20 54.5	D = 47.2° Az = 314.8° (USCGS); h = 40 km
	ePP	A	22 26	XV:1.1s 17.8nm
	LmH	B	44.2	LmH:16s 0.6 µm LmV:14s 0.3 µm
	LmV	B	46.8	MLH=4.6 MLV=4.5
				X is the greatest onset in the short-period vertical component.
11.	eiP	A	07 04 07	Algeria 36.45°N 2.82°E
	e	A	04 13.5	H = 07 00 28.7 h = normal MAG=4.6
	e	A	04 16	D = 15.5° Az = 21.3° (USCGS)
	e	A	04 24	LmH:14s 0.5 µm LmV:10s 0.4 µm
	LmH	B	10.0	MLH=3.8 MLV=4.1
	LmV	B	11.9	
11.	ePKP	A	07 11(56)	New Hebrides Islands 21.37°S 174.01°E
	e	A	12 11	H = 06 52 11.5 h = 15 km MAG=4.8
	e	A	12 48	D = 147.7° Az = 338.9° (USCGS)
	eSSS	C	40.0	LmH:22s 0.9 µm LmV:22s 0.9 µm
	LmH	B	08 16.0	MLH=5.4 MLV=5.5
	LmV	B	16	
11.	iPKP2	A	10 14 34.0	South of Kermadec Islands 32.8°S 178.5°W
				H = 09 53 54.2 h = 35 km MAG=4.4 (USCGS)
				D = 160.8°
11.	ePKP	A	10 34 14	New Hebrides Islands 21.31°S 173.67°E
	e	A	34 25	H = 10 14 30.4 h = 34 km MAG=4.8
	e	A	35 10	D = 147.5° Az = 338.6° (USCGS)
	LmH	C	11 38	
	LmV	C	38	
11.	ePKP	A	11 33 27.5	New Hebrides Islands 18.67°S 169.25°E
				H = 11 14 23.7 h = 245 km MAG=5.0
				D = 143.5° Az = 336.0° (USCGS)
11.	eP1	A	13 03 10	Mongolia 45.00°N 99.30°E
	eP2	A	03 14.5	H = 12 53 34.6 h = normal MAG=4.8

Day	Phase		h m s	Remarks
cont.				
11.	ePcP	A	13 04 09	D = 55.8° Az = 309.7° (USCGS)
	ePP	A	05 12	LmH:13s 0.7 μm LmV:13s 0.6 μm
	LmH	B	27.6	MLH=4.9 MLV=4.9
	LmV	B	28.0	e 04 15 ei 06 27.5
				Multiple P. P1 is much smaller than P2.
11.	eSg	A	14 58 48	Probably explosion.
11.	e	A	15 49 36.5	Explosion?
	ei	A	49 39	
	ei	A	49 43	
	ei	A	49 52	
11.	e(Sg)	A	17 33 16.5	Near earthquake?
	e	A	33 26.5	
	e	A	33 37	
11.	eP	A	20 05 31.5	<u>Red Sea</u> 20.33°N 38.67°E
				H = 19 58 23.2 h = normal
				D = 37.0° Az = 331.3° (USCGS)
11.	ePKP	A	21 40(45)	<u>Tonga Islands</u> 17.55°S 173.06°W
	e	A	41 16	H = 21 21 02.8 h = normal MAG=4.2 (USCGS)
				D = 146.7°
11.	eP	A	23 47 15	<u>Jan Mayen Islands</u> 71.18°N 6.14°W
	LmV	B	58.5	H = 23 42 21.6 h = normal MAG=4.4
				D = 22.2° Az = 149.0° (USCGS)
12.	eP	A	00 35 11	<u>South Atlantic Ridge</u> 22.78°S 10.48°W
	e	A	35 24	H = 00 23 27.7 h = normal MAG=4.9
	e	A	35 29	D = 75.7° Az = 14.3° (USCGS)
	e	A	35 41	LmH:18s 0.8 μm LmV:18s 1.2 μm
	ei	A	35 57	MLH=5.1 MLV=5.3
	LmV	B	01 08.2	
	LmH	B	08.5	

Day	Phase		h m s	Remarks
12.	+iP	A	02 55 33.2	Kurile Islands 44.64°N 149.80°E
	e	A	55 38.5	H = 02 43 33.1 h = 25 km MAG=5.1
	iPcP	A	55 41.3	D = 78.4° Az = 334.3° (USCGS)
	ei	A	55 45.5	PV:1.2s 71.4nm
	ei	A	55 55	LmH:19.5s 0.6 μm LmV:18s 0.4 μm
	LmH	C	03 29.9	MPV=5.7 MLH=4.9 MLV=4.8
	LmV	C	35.0	
12.	e(P)	A	08 29 11	
12.	e(Pb)	A	09 32 34.5	Taunus/GFR 50.1°N 7.9°E
	ePg	A	32 39	H = 09 31 49 (BCIS)
	iSn	A	32 58.0	D = 2.5°
	e	A	33 04	
	eiSg	A	33 07.5	
12.	e(Sg)	A	10 34 29.5	
	LmV	A	34 55	
12.	-eiP	A	11 21 28.5	Ascension Islands 4.98°S 11.46°W
	epP	A	21 35.5	H = 11 11 31.3 h = normal MAG=4.9
	LmH	C	44	D = 58.9° Az = 17.0° (USCGS); h = 28 km
	LmV	C	45	PV:1.1s 19.2nm
				MPV=5.1
12.	e(P)	A	13 00 36	
12.	iP1	A	14 49 52	Greece 39.29°N 21.22°E
	iP2	A	50 00.0	H = 14 46 41.3 h = 14 km MAG=4.5
	e	A	52 08	D = 13.2° Az = 332.3° (USCGS)
	eS	A	52 17	Multiple P, P1 much smaller than P2. It
	e(SS)	A	52 35	is possible that e(SS) is the Sg of a
	e	A	53 47	small explosion.
	LmH	B	54.6	
	LmV	B	56.3	
12.	e(P)	A	16 18 38	(P)V:0.9s 14.2nm
12.	iSg	A	16 45 18	
	LmV	A	45 35	

Day	Phase		h m s	Remarks
12.	ePKIKP	A	22 08 42	New Britain Region 5.49°S 151.70°E
	epPKIKP	A	08 52	H = 21 49 47.6 h = 50 km MAG=5.2
	eiPP	C	10 31	D = 124.0° Az = 330.5° (USCGS); h = 36 km
	ePPP	C	13 08	PSH:24s 1.5 /um
	eSKKS	C	17.5	LmH:22s 3.1 /um LmV:19s 3.6 /um
	ePS	C	20 24	MLH=5.9 MLV=6.1
	eSPP	C	21 52	
	ePKKS	C	22 46	
	eSS	C	27 30	
	ePSS	C	27 52	
	eSSS	C	32 20	
	LmH	B	23 04	
	LmV	B	06.7	
13.	iSg	A	12 47 23.5	Explosion?
	LmH	A	47 45	
	LmV	A	48 00	
13.	iSg	A	12 51 43.5	Explosion/GDR
	eiL	A	51 46.3	
13.	eL	A	12 51 58.5	Explosion/GDR
13.	iPg	A	13 55 47.8	Explosion
	iSg	A	55 56.5	D = 0.7°
	eiL	A	56 05	SgV:0.8s 47.2nm
	LmH	A	56 11	LmV:1.4s 135nm
	LmV	A	56 11	
13.	e	A	15 28 07	
13.	iPg	A	16 59 40.0	Explosion
	eiSg	A	59 59	D = 1.4°
13.	+eP	A	18 52 59	Near Islands/Aleutian Is. 52.70°N 172.45°E
	ei	A	53 09	H = 18 41 15.4 h = 34 km MAG=5.7
	ePS	C	19 03 20	D = 75.8° Az = 347.5° (USCGS)
	eSS	C	07 44	PV:1.4s 90.5nm
	LmH	B	24.0	LmH:22s 1.6 /um LmV:24s 1.4 /um
	LmV	B	24.0	MPV=5.7 MLH=5.3 MLV=5.2

Day	Phase		h m s	Remarks
13.	eSP	B	20 26 09	South Sandwich Islands 56.0°S 27.4°W
	eSS	C	32.3	H = 19 57 47.9 h = 148 km MAG=5.3 (USCGS)
				D = 111.1°
13.	eP	A	20 27 09.5	PV:1.4s 15.4nm
13.	e(P)	A	23 16 48	
14.	ePKIKP	A	01 01 31	South of Kermadec Islands 32.9°S 178.4°W
	eiPKP2	A	02 19	H = 00 41 40.7 h = 40 km (USCGS)
	ei	A	02 27	D = 160.9°
14.	e(P)	A	11 02 13	
14.	LmH	B	11 24	LmH:18s 1.1 µm LmV:16s 0.5 µm
	LmV	B	30.2	
14.	e(Sg)	A	13 03 00	Explosion
	LmH	A	03 21	
14.	e(P)	A	13 20 47	
14.	e(P)	A	13 23 01	
14.	eP	A	14 36 15.5	Southern Greece 36.1°N 21.9°E
	e	A	36 24.5	H = 14 32 31.0 h = 102 km MAG=4.5 (USCGS)
	LmH	B	43.7	D = 16.4°
	LmV	B	43.9	PV:1.1s 9.6nm
				LmH:17s 0.9 µm LmV:17s 0.9 µm
				MLH=4.0 MLV=4.2
14.	eSKS	C	14 39 32	Off Coast of Ecuador 1.6°N 84.9°W
	eS	C	40.2	H = 14 16 06.0 h = 40 km MAG=4.8 (USCGS)
	ePS	C	41 35	D = 92.9°
	e	C	45 55	LmH:24s 1.9 µm LmV:24s 2.2 µm
	eSS	C	46 45	MLH=5.5 MLV=5.6
	e(SSS)	C	50.4	
	LmH	B	15 04	
	LmV	B	04	

Day	Phase		h m s	Remarks
14.	+eP	A	14 57 20.5	Southern Iran 28.44°N 57.12°E
				H = 14 49 41.9 h = normal MAG=4.7
				D = 40.7° Az = 315.8° (USCGS)
				PV:1.1s 12nm
				MPV=4.7
14.	ePKIKP	A	15 54 29	New Hebrides Islands 15.36°S 167.48°E
				H = 15 35 17.3 h = 142 km MAG=4.9
				D = 139.8° Az = 336.2° (USCGS)
14.	e(P)	A	19 02 21.5	Southern Greece 36 1/4°N 21 1/2°E
	e	A	02 44	H = 18 58 24 (BCIS)
	LmH	B	09.8	D = 16.1°
	LmV	B	09.8	(P)V:1.3s 11.1nm
				LmH:16s 0.6 μm LmV:16s 0.8 μm
				MLH=3.9 MLV=4.2
				The first onset of P must be about 12 s earlier than e(P).
14.	ePg	A	20 22 38	Swiss 46.4°N 7.4°E
	iSn	A	23 06.3	H = 20 20 58 (BCIS)
	eSg	A	23 39.5	D = 5.1°
15.	eP	A	00 40 59	Near East Coast of Honshu/Japan
	i	A	41 07	35.63°N 140.43°E
	epP	A	41 11	H = 00 28 39.8 h = 59 km MAG=5.2 (USCGS)
	esP	B	41 18	D = 82.8° h = ca. 50 km
	ePP	B	44 10	PV:2.6s 125.0nm
	esPP	B	44 28	LmH:16s 1.9 μm LmV:17s 1.5 μm
	eS	C	51 14	MPV=5.6 MLH=5.6 MLV=5.4
	eScS	C	51 30	
	ePPS	C	52 30	
	eSS	C	56.9	
	eSSS	C	01 00 20	
	LmH	B	20.5	
	LmV	B	21	
15.	e	A	01 40 30	
	e	A	40 37	
	e	A	40 43.5	

Day	Phase		h m s	Remarks
15.	e	A	08 32 51	
15.	e	A	09 09 54.5	
15.	e(P)	A	09 30 55.5	(P)V:0.8s 9.4nm
15.	e(P)	A	10 26 26	(P)V:0.8s 9.4nm
15.	+eiP	A	10 43 13.0	Bhutan 27.40°N 91.80°E
	e	A	43 16	H = 10 32 48.7 h = 57 km MAG=5.8 (USCGS)
	ipP	A	43 19.0	D = 64.3° h = 23 km
	isP	A	43 21.3	PV:1.3s 137nm pPV:1.0s 54.5nm
	ePcP	A	43 49	sPV:2.0s 187nm
	ePP	A	45 30	LmH:13.5s 1.7 μm LmV:14s 1.7 μm
	iS	C	51 45	MPV=6.0 MLH=5.5 MLV=5.4
	eScS	C	53.1	pP has a distinctly shorter period than P.
	eSS	C	55(44)	If the phase interpreted above as sP would
	eSSS	C	58(40)	be really pP than h = 33 km.
	LmV	B	11 13	
	LmH	B	14.5	
15.	e(P)	A	11 12 18.5	
15.	eL	A	13 53 06	Explosion
	e	A	53 08	
15.	ei	A	14 53 29	
15.	ei	A	18 26 18	
15.	i(P)	A	22 10 03.5	
16.	eP	A	00 09 51	Taiwan 24.08°N 120.70°E
	epP	A	10 00.5	H = 23 57 30.1 h = 50 km MAG=5.0
	LmH	C	51.7	D = 83.0° Az = 322.7° (USCGS); h = 35 km
	LmV	C	51.7	PV:1.4s 16.7nm pPV:1.9s 36.7nm
				LmH:17s 0.6 μm LmV:16s 0.7 μm
				MPV=5.0 MLH=5.1 MLV=5.2

Day	Phase		h m s	Remarks
16.	ePP	A	03 59 46.5	Ceram Sea 2.0°S 128.9°E
	ePS	C	09 00	H = 03 40 55.3 h = 50 km MAG=5.4 (USCGS)
	eSSS	C	19.2	D = 108.6°
	LmH	C	04 41.6	LmH:27s 0.6 /um LmV:18s 0.5 /um
	LmV	C	50.8	MLH=5.1 MLV=5.1
16.	+iP	A	04 11 44.0	Eastern Kazakh SSR 50.01°N 77.82°E
	e	A	12 19.5	H = 04 03 58.0 h = 0 km MAG=5.3
				D = 41.0° Az = 297.4° (USCGS)
				PV:0.65s 33.2nm
				MPV=5.4
				Underground explosion.
16.	iPg	A	06 54 37.4	Rock burst/Peißenberg, GFR 47.8°N 11.1°E
	eiSg	A	55 15	H = 06 53 42 (BCIS)
				D = 2.9°
16.	eipP	A	08 44 04	Andreanof Is./Aleutian Is. 52.0°N 176.4°W
	e	A	44 31	H = 08 31 58.4 h = 65 km MAG=5.4 (USCGS)
	e	A	44 48.5	D = 77.3°
	LmV	C	09 12.5	LmH:30s 0.5 /um LmV:30s 0.6 /um
	LmH	C	12.8	MLH=4.7 MLV=4.8
16.	ei	A	10 45 41	Explosion?
	e	A	45 57	
16.	LmH	C	10 55	Traces. LmH:24s 0.2 /um LmV:23s 0.3 /um
	LmV	C	55	
16.	ei(P)	A	11 33 07	
16.	eL	A	13 36 11	Explosion/GDR
16.	LmH	A	13 36 20	Explosion/GDR
	LmV	A	36 20	LmH:1.1s 220nm LmV:1.1s 202nm
16.	eiL	A	13 36 38	Explosion/GDR
16.	eiL	A	13 36 47	Explosion/GDR

Day	Phase		h m s	Remarks
16.	eS	C	18 48 12	South of Panama 5.3°N 82.4°W
	ePS	C	49.4	H = 18 24 47.9 h = normal MAG=4.6 (USCGS)
	eSS	C	54.3	D = 88.5°
	LmV	C	19 09.5	LmH:25s 0.4 µm LmV:24s 0.3 µm
	LmH	C	10	MLH=4.7 MLV=4.7
16.	e(Sn)	A	20 21 34	Little Carpathians. Provisional epicentre
	e(Sb)	A	21 48	48.6°N 17.3°E
	eSg	A	22 05.5	H = 20 19 43 (PRUHONICE)
	e	A	22 13	D = 4.2°
16.	ePKP2	A	23 51 18	Kermadec Islands 31.5°S 179.8°E
				H = 23 31 27.3 h = 377 km MAG=4.0 (USCGS)
				D = 158.9°
17.	ePKP	A	01 41 13.5	Tonga Islands 18.56°S 174.98°W
	e	A	41 24	H = 01 21 52.4 h = 200 km MAG=4.1
				D = 147.6° Az = 352.2° (USCGS)
17.	LmH	C	02 08.5	Bonin Islands 27.5°N 142.4°E
	LmV	C	08.7	H = 01 09 08.4 h = 33 km MAG=4.8 (USCGS)
				D = 90.8°
				LmH:16s 0.2 µm LmV:14s 0.2 µm
				MLH=4.7 MLV=4.7
17.	eiP	A	08 09 02	Chiapas/Mexico 17.24°N 94.08°W
	eipP	A	09 17	H = 07 56 22.7 h = 45 km MAG=5.2 (USCGS)
	epPP	A	12 40	D = 86.2° h = 56 km
	eS	C	19 26	PV:1.9s 60nm pPV:2.4s 91nm
	eSS	C	25.2	LmH:20s 0.5 µm LmV:20.5s 0.6 µm
	eLQ	C	32.0	MPV=5.5 MLH=4.9 MLV=5.0
	LmH	B	48.1	i 10 10
	LmV	B	48.1	
17.	LmH	C	17 38.3	Probably Gulf of California (USCGS)
	LmV	C	38.5	
17.	LmH	C	22 21	Traces. Probably Mid-Indian Rise (USCGS)

Day	Phase		h m s	Remarks
18.	eP	A	02 10 43	Ethiopia 15.69°N 39.03°E
				H = 02 02 59.8 h = normal MAG=4.8
				D = 41.3° Az = 333.6° (USCGS)
18.	LmH	C	07 47	Traces. Gulf of California (USCGS)
	LmV	C	52.3	
18.	iPg	A	09 35 30.8	Explosion/GDR
	iSg	A	35 32.7	D = 16 km
				SgV:0.7s 394nm
18.	eiPg	A	15 02 46	Explosion/Tabarz, GDR
	eiSg	A	02 56	Yield: 6.2 t
	ei	A	02 59.5	D = 0.8°
	LmH	A	03 07	
	LmV	A	03 09	
18.	ePKIKP	A	15 51 57	East New Guinea Region 5.9°S 146.6°E
	ePP	A	53 29	H = 15 33 06.5 h = 39 km MAG=5.5 (USCGS)
	ePP	C	53 31	D = 121.8°
	ePS	C	16 03 25	LmH:44s 2.8 µm LmV:22s 2.0 µm
	eSPP	C	04 44	MLH=5.3 MLV=5.5
	eSS	C	10 12	e 53 48 e 54 43 e 11 00 e 16 28
	eSSP	C	10 32	e 19 30
	eSSS	C	15 00	AN USSR gives: New Guinea 6.0°S 146.8°E
	eLQ	C	23.5	H = 15 33 20 h = 176 km
	LmH	C	30.5	D = 122.1°
	eLR	C	32.5	
	LmV	C	46.5	
18.	eSg	A	17 03 31	
	LmH	A	03 44	
	LmV	A	03 44	
18.	e(P)	A	17 41 46	
18.	ePKHKP	A	19 32 37.5	Fiji Islands 20.72°S 178.37°W
	e	A	32 49	H = 19 13 52.5 h = 562 km MAG=4.0
				D = 149.1° Az = 347.6° (USCGS)
				PV:1.2s 15.3nm

Day	Phase		h m s	Remarks
18.	eSg	A	22 52 24.5	Rock burst/Upper Silesia (KRAKOW)
	e	A	52 27	
	e	A	52 37	
19.	ePKHKP	A	01 05 12	South of Fiji Islands 24.65°S 177.34°W
	e	A	05 41	H = 00 45 31.1 h = 139 km MAG=4.7
				D = 153.1° Az = 347.3° (USCGS)
				PV:1.2s 12.8nm
19.	eiP	A	03 41 13.4	Near East Coast of Honshu/Japan
	ipP	A	41 23.8	37.27°N 141.69°E
	LmH	C	04 21.5	H = 03 28 57.4 h = 53 km MAG=4.9
	LmV	C	21.5	D = 82.1° Az = 330.5° (USCGS); h = 39 km
				PV:1.1s 9.6nm
				MPV=4.8
19.	e(P)	A	08 11 25	
19.	ei(Pg)	A	09 02 26	Explosion?
	eiSg	A	03 05.5	(D = 3.0°)
	ei	A	03 09.5	PRUHONICE gives: iPg 09 01 55 iSg 02 14
				D = 1.5°
19.	+iP	B	11 08 01	Hokkaido/Japan 42.97°N 145.17°E
	eiP1	A	08 01	H = 10 56 08.6 h = 84 km MAG=5.9
	+iP2	A	08 04.1	D = 78.4° Az = 331.9° (USCGS);
	+ipP	B	08 21.6	h = ca. 85 km
	eipP	A	08 23	PV:8s 2.2 µm PV2:1.5s 476nm
	ei(sP)	A	08 35	SH(B):11s 4.8 µm
	ePP	AC	10 57.5	LmH:16.5s 9.0 µm LmV:17s 6.4 µm
	eS	B	17 43	MPV=6.1 MPV2=6.2 MSH=6.5 MLH=6.2
	iS	ABC	17 48	MLV=6.1
	iScS	C	18 08	Multiple P in the short-period records,
	i(sS)	BC	18 36	P2 much greater than P1.
	eiSS	BC	22 52	
	eSSS	C	26 14	
	iSa	B	27 36	
	eLQ	C	29.0	
	eLR	C	33.0	

Day	Phase		h m s	Remarks
cont.				
19.	LmH	B	11 42.7	
	LmV	B	47.0	
19.	ePKIKP	A	13 03 50	South Sandwich Islands Region
	ePP	A	04 47	57.8°S 23.4°W
	LmH	C	49.4	H = 12 45 35.3 h = normal MAG=5.7 (USCGS)
	LmV	C	49.5	D = 111.7°
				LmH:18s 0.7 μm LmV:18s 1.0 μm
				MLH=5.3 MLV=5.5
19.	eiP	A	19 14 42	Southern Sumatra 1.56°S 100.51°E
	e	A	14 55.5	H = 19 01 47.5 h = 83 km MAG=5.0
	e(pP)	A	15 03	D = 90.5° Az = 320.5° (USCGS); (h = 79 km)
	e	A	15 11	PV:1.7s 74.5nm
	eS	C	25.6	MPV=5.6
	eLQ	C	38.5	If we interpret e 14 55.5 as pP then the
	LmH	C	20 00	focal depth would be h = 50 km.
19.	eSS	C	19 59 12	Atlantic-Indian Rise 36.3°S 52.2°E
	LmH	C	20 21.8	H = 19 28 45.2 h = normal MAG=5.4 (USCGS)
				D = 93.7°
				LmH:23s 0.4 μm
				MLH=4.8
20.	iP	A	00 44 57.7	Honshu/Japan 35.99°N 139.86°E
	e(pP)	A	45 09.5	H = 00 32 44.3 h = 94 km MAG=4.9
	ei	A	45 16	D = 82.4° Az = 329.7° (USCGS); (h = 44 km)
	e	A	45 31.5	PV:1.5s 23.5nm pPV:1.7s 21.9nm
				MPV=4.8
				AN USSR gives: Japan 35.6°N 140.2°E
				H = 00 32 35 h = normal
				D = 82.8°
20.	iPn	A	06 11 04.3	Yugoslavia 44.27°N 17.70°E
	eiSn	A	12 26	H = 06 09 10.7 h = 11 km MAG=4.2
	e(Sb)	A	12 50	D = 7.6° Az = 329.3° (USCGS)
	e	A	12 57	LmH:12s 0.7 μm
	e(Sg)	B	13 16	MLH=3.5
	LmH	B	13.5	

Day	Phase		h m s	Remarks
20.	ePKIKP	B	09 59 11.5	<u>Auckland Islands</u> 49.78°S 163.43°E
	ePKIKP	A	59 12	H = 09 39 15.2 h = 30 km MAG=6.1
	i	B	59 26	D = 162.0° Az = 283.6° (USCGS)
	e(PKP2)	A	10 00 02	PV(A):2.7s 286nm
	i	B	00 14	LmH:18s 9.4 µm LmV:20s 9.7 µm
	ePP	C	03 50	MLH=6.6 MLV=6.6
	ePPP	C	07 25	e 59 22 i 59 53 e 00 09 e 03 55
	eiSS	B	24 14	PKIKP has unusually long periods (T =
	iPSS	C	25 08	2 - 4 s) in our short-period records
	eSS2	C	29 42	(Z-component). PKP2 is not clear.
	iSSS	B	30 24	AN USSR gives: 49.2°S 166.2°E
	LmH	B	11 22.4	H = 09 39 13 h = normal MAG=6.5
	LmV	B	22.4	D = 163.5°
20.	ePKIKP	A	10 50 51	<u>Auckland Islands</u> 49.78°S 163.44°E
	ePKP2	A	51 41.5	H = 10 30 53.4 h = 19 km MAG=5.8
	e	A	51 55	D = 162.0° Az = 283.6° (USCGS)
	e	A	52 16.5	PV:2.5s 203nm
	e	A	52 54	
20.	eiPKP	A	10 56 43.5	<u>New Hebrides Islands</u> 20.81°S 169.83°E
	epPKP	A	57 18	H = 10 37 20.3 h = 129 km MAG=5.9
	eisPKP	A	57 29.5	D = 145.6° Az = 335.3° (USCGS); h = 125 km
	iPP	A	11 00 11.5	PV1:1.6s 98.4nm PV3:1.5s 134nm
	e(pPP)	A	00 43	pPKP is much smaller than sPKP. If we interpret ei 57 29.5 as pPKP then the focal depth will be h = 176 km.
20.	ePKP2	A	12 27 45.5	<u>Auckland Islands Region</u> 49.8°S 163.8°E
	e	A	27 48	H = 12 06 52.7 h = normal MAG=5.2 (USCGS)
	e	A	27 55	D = 162.0°
				PV:1.2s 13.8nm
20.	iSg	A	12 54 32.0	<u>Explosion/GDR</u>
	eiL	A	54 34.5	
20.	iL	A	12 54 45.0	<u>Explosion/GDR</u>
20.	iL	A	12 54 55.0	<u>Explosion/GDR</u>

Day	Phase		h m s	Remarks
20.	e(Sg)	A	16 35 55.5	
20.	eP	A	17 24 07	North Atlantic Ocean 23.7°N 44.2°W
				H = 17 14 30.4 h = normal MAG=4.5 (USCGS)
				D = 36.8°
20.	ePn	A	22 45 21.5	Little Carpathians/ČSSR 48.4°N 17.1°E
	e	A	45 26.5	H = 22 44 14 (BCIS)
	e	A	46 17	D = 4.2°
	eSb	A	46 24	PnV:0.8s 10.4nm
	iSg	A	46 34.0	
21.	e(P)	A	05 52 27	
	e	A	52 34	
21.	e(Pg)	A	11 12 56	Explosion
	e(Sg)	A	13 11	(D = ca. 1.1°)
22.	+iP	A	05 11 44	Eastern Kazakh SSR 50.03°N 77.61°E
	ePn	A	13 17	H = 05 03 57.9 h = 0 km MAG=5.3
				D = 40.9° Az = 297.2° (USCGS)
				PV:0.6s 28.5nm
				MPV=5.3
				Underground explosion.
22.	iP	A	08 17 54.5	Central Mid-Atlantic Ridge
	epP	A	18 01.5	0.68°S 20.05°W
				H = 08 08 04.3 h = normal MAG=5.3
				D = 57.8° Az = 23.3° (USCGS); h = 28 km
				PV:1.4s 61.3nm
				MPV=5.5
22.	e	A	08 24 16	Traces. Probably near earthquake.
	e	A	24 41	TRIESTE gives: D = 600 km
	e	A	24 55	
	e	A	25 13	
22.	e(P)	A	10 29 43	Explosion?
				(P)V:0.9s 14.3nm

Day	Phase		h m s	Remarks
22.	+iP	AB	10 29 56.3	Kurile Islands 44.51°N 149.41°E
	eiPcP	A	30 06.5	H = 10 17 59.9 h = 60 km MAG=5.6
	i	A	30 21.5	D = 78.4° Az = 334.1° (USCGS)
	eS	B	39 44	PV(A):1.3s 144.5nm PV(B):6s 1.3 µm
	eSS	C	44 56	LmH:17.5s 8.7 µm LmV:18s 8.4 µm
	LQ	C	51	MPV(A)=5.9 MPV(B)=6.2 MLH=6.2 MLV=6.1
	LmH	B	11 08.5	e 40 08 e 40 20 e 49(00)
	LmV	B	08.5	
22.	iP	A	11 31 21.5	Kurile Islands 44.3°N 149.4°E
	e	A	31 34	H = 11 19 21.4 h = 50 km MAG=4.4 (USCGS)
	LmH	B		D = 78.4° h = 47 km
22.	+eP	A	12 46 49.5	Kurile Islands 44.42°N 149.36°E
	e	A	46 55.5	H = 12 34 51.6 h = 51 km MAG=4.8
	epP	A	47 03	D = 78.4° Az = 334.1° (USCGS); h = 51 km
	e	A	47 06	PV:1.3s 38.9nm
	LmH	B	13 20.5	MPV=5.4
22.	e(P)	A	16 56 26	
22.	LmH	C	20 43	Tibet 31.9°N 94.6°E
				H = 20 09 13 h = normal (USCGS)
				D = 61.8°
				LmH:27s 0.2 µm
				MLH=4.2
22.	eiP	A	22 19 48	Hindu Kush 36.2°N 71.4°E
	e	A	19 58	H = 22 11 48.3 h = 127 km MAG=4.7 (USCGS)
				D = 44.6°
23.	LmH	C	00(56.5)	Traces. Northern Easter Is. Cordillera (USCGS)
23.	ePKP	A	03 41 36	Fiji Islands 17.70°S 178.68°W
				H = 03 22 59.7 h = 567 km MAG=5.0
				D = 146.1° Az = 348.2° (USCGS)

Day	Phase		h m s	Remarks
23.	ePKIKP	A	07 15 21.5	Fiji Islands 21.78°S 179.67°W
	eiPKHKP	A	15 27.5	H = 06 56 43.6 h = 595 km MAG=5.4
	ePKP2	A	15 36	D = 149.9° Az = 345.6° (USCGS);
	epPKP	A	17 42	h = ca. 610 km
	e	A	17 47	AN USSR gives: Fiji Islands Region
	e	A	17 52	23.3°S 178.8°W
	e	A	17 59	H = 06 55 38 h = normal MAG=5.4
				D = 151.5°
				PV2:1.2s 102nm PV3:1.2s 61.2nm
23.	e(PKP)	A	07 21 52	Auckland Islands 49.70°S 164.03°E
	ePKIKP	B	22 01	H = 07 02 03.3 h = 15 km MAG=5.7
	ePKP2	B	22 50	D = 162.3° Az = 283.6° (USCGS)
	ePP	C	26 36	LmH:30s 0.6 µm LmV:22s 0.7 µm
	ePPP	C	30 16	MLH=5.1 MLV=5.4
	eSKKS	C	33 20	
	eSKSP	C	37.0	
	ePPS	C	40.2	
	eSS	C	46 52	
	ePSS	C	47 52	
	eSSS	C	53.0	
	e	C	53.5	
	LmH	C	08 31	
	LmV	C	38	
23.	ePKHKP	A	07 58 32.5	South of Fiji Islands 22.08°S 179.67°W
	ePKP2	A	58 41	H = 07 39 47.8 h = 600 km MAG=4.6
				D = 150.2° Az = 345.6° (USCGS)
				PV1:1.3s 33.3nm PV2:1.2s 28.0nm
23.	eiPg	A	09 01 07.5	Explosion/ČSSR 50°29'N 13°57.2'E
	iSg	A	01 27.2	Yield: 6 t (PRUHONICE)
				D = 1.5°
23.	eiL	A	13 32 59	Explosion/GDR
23.	iSg	A	13 33 19.0	Explosion/GDR
	iL	A	33 21.6	
23.	eiL	A	13 33 43.5	Explosion/GDR

Day	Phase		h m s	Remarks
23.	iL	A	13 33 57.0	Explosion/GDR
	LmH	A	33 59.5	LmH:1.2s 190nm LmV:1.3s 128nm
	LmV	A	33 59.5	
23.	iPg	A	14 10 24.0	Explosion/GDR
	iSg	A	10 29.0	D = 0.4°
	iL	A	10 32	
	LmH	A	10 33.5	
	LmV	A	10 33.5	
23.	e(P)	A	15 46 32	
23.	ePKP	A	23 03 05	Kermadec Islands
				h = 380 km (UPP)
24.	LmH	C	01 56	North of Halmahera (USCGS)
24.	e	A	13 26(11)	Probably near earthquake.
	e	A	26 15	TRIESTE gives: iPg 13 24 42.8 iSg 24 50.8
	e(Sg)	A	27 01	D = 60 km
24.	eP	A	17 13 39	Morocco 32.47°N 5.72°W
	e	A	14 05	H = 17 08 41.8 h = normal MAG=4.3
				D = 22.2° Az = 30.1° (USCGS)
24.	e(P)	A	18 49 08	Near earthquake.
	eiSg	A	49 51	TRIESTE gives: D = 190 km
24.	LmH	C	21 16.0	Bonin Islands 27.6°N 141.5°E
	LmV	C	16	H = 20 17 49.4 h = normal MAG=4.5 (USCGS)
				D = 90.4°
				LmH:16s 0.2 μm
				MLH=4.7
24.	eP	A	21 52 38	PV:1.5s 20.1nm
24.	eP	A	22 13 57	Albania 40.80°N 19.70°E
	e	C	17 49	H = 22 11 17.0 h = 16 km MAG=4.4
	LmV	B	19.9	D = 11.3° Az = 332.9° (USCGS)

Day	Phase		h m s	Remarks
cont.				
24.	LmH	B	22 20	LmH:15s 1.1 /um LmV:18s 1.1 /um
				MLH=3.9
24.	ePn	A	22 29 05	Alps near Bergamo 46.0°N 9.4°E
	iPg	A	29 20.5	H = 22 27 48 (BCIS)
	iSn	A	29 58.0	D = 4.9°
	eLg1	A	30 23.5	LmH:8s 0.8 /um LmV:8s 1.0 /um
	e(Sg)	A	30 26	MLH=3.4
	eLg2	A	30 30	ei 29 28 ei 29 57 e 30 29
	LmH	B	31 10	Very clear Sn.
	LmV	B	31 10	
25.	ePKP	A	04 57 56	Tonga Islands 15.10°S 173.35°W
	epPKP	A	58 12	H = 04 38 26.2 h = 63 km MAG=5.0
				D = 144.3° Az = 354.6° (USCGS); h = 57 km
25.	ePKP	A	06 37 20	Fiji Islands 20.29°S 177.72°W
				H = 06 18 26.1 h = 462 km MAG=4.2
				D = 148.8° Az = 348.5° (USCGS)
25.	LmH	B	08 45	Leeward Islands (USCGS)
	LmV	B	45	
25.	e	A	09 02 46	Leeward Islands 17.67°N 61.59°W
	eS	C	11 18	H = 08 51 49.4 h = 48 km MAG=4.8
	eScS	C	12 28	D = 65.9° Az = 41.9° (USCGS)
	eLQ	C	18 40	LmH:20s 0.6 /um LmV:20s 0.8 /um
	LmH	B	21.4	MLH=4.8 MLV=4.9
	LmV	B	26.5	
25.	e	A	09 29 15	
25.	eiL	A	11 01 07	Explosion
25.	e(P)	A	13 16 46	Talaud Islands 3.60°N 126.61°E
				H = 13 03 06.9 h = 78 km MAG=5.3
				D = 102.7° Az = 323.7° (USCGS)

Day	Phase		h m s	Remarks
25.	iPg	A	14 27 07.5	Explosion
	eiSg	A	27 23	D = ca. 1.2°
25.	LmH	C	17 59	Traces. East New Guinea Region (USCGS)
25.	eP	A	19 56 36.5	Kenai Peninsula, Alaska 60.30°N 151.38°W
				H = 19 45 40.6 h = 70 km MAG=4.0
				D = 68.6° Az = 11.5° (USCGS)
				PV:1.3s 16.7nm
				MPV=5.1
25.	e	A	22 07 16	Traces
	e	A	07 37.5	
	e	A	07 45.5	
25.	eiPg	A	20 48 08	Explosion
	iSg	A	48 15	D = 0.5°
26.	eP	A	05 08 15	Yugoslavia 41.70°N 21.18°E
	e	A	08 24.5	H = 05 05 36.2 h = 39 km MAG=4.4
	e	B	11 40	D = 11.1° Az = 326.8° (USCGS)
	eLg2	A	11 48	LmH:11.5s 1.2 µm LmV:12s 1.2 µm
	e	B	12 27	MLH=3.1
	LmH	C	13.0	AN USSR gives: Yugoslavia 43.3°N 22.2°E
	LmV	C	13.0	H = 05 05 54 h = normal MAG=ca.4
				D = 10.5°
26.	eP	A	06 58 49	Kurile Islands 46.93°N 150.56°E
	e	A	58 53	H = 06 47 11.6 h = 136 km MAG=4.7
	e	A	58 59	D = 76.6° Az = 334.6° (USCGS)
	e	A	59 06.5	PV2:1.0s 14.2nm
	e	A	59 32	MPV2=5.1
				The first onset of P is much smaller than the second one.
26.	e	A	07 18 22.5	
26.	eP	A	11 11 32.5	

Day	Phase		h m s	Remarks
26.	e(pPP)	A	11 30 45	Chile-Argentina Border Region
	eS	C	38 00	33.6°S 70.5°W
	eSS	C	45 35	H = 11 11 23.7 h = 84 km MAG=5.8 (USCGS)
	e	C	49 05	D = 110.6°
	eSSS	C	49(50)	LmH:20s 0.3 μm LmV:20s 0.4 μm
	eLQ	C	56	MLH=4.8 MLV=5.0
	LmV	B	12 16.4	
	LmH	B	17.5	
26.	e(P)	A	11 58 38.5	
26.	eP	C	16 25 40	Near Coast of Central Chile
	ePKIKP	A	29 49	30.04°S 71.53°W
	eiPP	B	30 10	H = 16 11 23.9 h = 55 km MAG=5.7
	eiPP	A	30 11	D = 108.6° Az = 41.8° (USCGS)
	epPP	A	30 24.5	PPV(B):5s 0.8 μm pPPV:2.1s 76nm
	esPP	A	30 30	sPPV:2.2s 175nm
	eSKS	C	36 20	LmH:17.5s 5.5 μm LmH:20s 7.5 μm
	esSKS	C	36 44	MPPV=6.7 MLH=6.2 MLV=6.3
	eS	C	37 38	e 26 12 e 30 06 i 38 16 i 40 00
	eSP	B	39 32	e 49 00 e 53(00)
	eSPP	C	40(40)	
	ePKKP	A	41 12	
	eSS	C	45 27	
	eSSS	C	49 40	
	eiSa	C	56.0	
	LmH	B	17 16.7	
	LmV	B	17.1	
26.	+ePKIKP	A	17 24 50	Solomon Islands 7.10°S 155.80°E
				H = 17 05 55.0 h = 94 km MAG=5.7
				D = 127.4° Az = 332.0° (USCGS)
				PV:1.2s 25.5nm
26.	iSg	A	17 45 00	Probably explosion.
27.	eP	A	07 28 58	Crete 34.36°N 26.58°E
	e	A	29 05.5	H = 07 24 29.9 h = 20 km MAG=4.7
	e	A	29 08.5	D = 19.6° Az = 330.7° (USCGS)

Day	Phase		h m s	Remarks
cont.				
27.	e	A	07 29 13	LmH:14s 0.4 μm LmV:15s 0.3 μm
	LmV	C	38.4	MLH=3.8 MLV=3.8
	LmH	C	38.5	
27.	e(P)	A	10 36 58.5	(P)V:0.7s 9.5nm
27.	iPg	A	12 49 32.0	<u>Explosion/GDR</u>
	iSg	A	49 37.0	D = 0.4°
	LmH	A	49 41.5	LmH:1.1s 198nm LmV:1.1s 188nm
	LmV	A	49 41.5	
27.	e	A	15 48 40	Traces
	e	A	48 51	
27.	e	A	15 54 00	Traces
	e	A	54 47	
	e	A	56 15	
	e	A	56 34	
27.	eP	A	16 00 35.5	PV:1.2s 10.2nm
27.	+eP	A	17 12 18	<u>Nevada</u>
	e	A	13 28	H = 17 00 00
	ePP	A	15 22	D = ca. 81°
	LmH	C	49.7	PV:1.4s 67.5nm
	LmV	C	49.7	MPV= 5.6
				Underground explosion.
28.	-eP	A	03 02 11.5	<u>Alma-Ata Region</u> 41.97°N 79.51°E
	e	A	02 30	H = 02 53 48.4 h = normal MAG=4.8
	e	A	02 35	D = 46.2° Az = 305.2° (USCGS)
	e	A	04 10	PV:1.4s 24.5nm
	eSS	C	12 28	LmH:15s 2.0 μm LmV:13s 1.4 μm
	LQ	C	13.8	MPV=5.1 MLH=5.2 MLV=5.2
	LmH	B	19.1	
	LmV	B	22.2	

Day	Phase		h m s	Remarks
28.	eP	A	03 12 22	<u>Fox Islands/Aleutian Is.</u> 52.25°N 171.02°W
				H = 03 00 30.5 h = 48 km MAG=5.1
				D = 77.5° Az = 358.3° (USCGS)
				PV:0.9s 37.8nm
				MPV=5.5
28.	eP	C	05 12.6	<u>New Britain</u> 6.59°S 153.40°E
	ePKIKP	A	15 55	H = 04 56 56.3 h = 44 km MAG=5.9
	ePP	A	17 48	D = 125.8° Az = 330.9° (USCGS)
	+iPP	B	17 49	PV2:1.3s 22.2nm
	iPP	C	17 51	LmH:24s 3.6 µm LmV:23s 4.1 µm
	ePPP	C	20 30	MLH=6.0 MLV=6.0
	eSKS	C	22 55	e 28 30 e 28 53 e 29 27
	eSKKS	C	24 44	
	eSP	C	27 40	
	eSPP	C	29 12	
	eSKKP	A	29 20	
	eSS	C	34 48	
	eSSS	C	39 30	
	LmH	B	06 02	
	LmV	B	12.2	
28.	e	A	05 25 37	
28.	iPg	A	12 00 40.5	Explosion
	iSg	A	00 56.5	D = ca. 1.2°
28.	eP	A	12 04 05	
	e	A	05 40	
28.	+eiP	A	15 55 59	<u>Gulf of Alaska</u> 59.48°N 147.11°W
	+iP	C	56 00	H = 15 44 55.7 h = 28 km MAG=5.6
	ePP	C	58 28	D = 68.9° Az = 14.3° (USCGS)
	iPP	B	58 34	PV:1.8s 131nm
	iS	B	16 05 04	LmH:17s 3.4 µm LmV:18s 3.8 µm
	iS	C	05 05.5	MPV=5.8 MLH=5.7 MLV=5.7
	iScS	BC	05 53	e 06 05 e 09 08 e 13 12
	e(SS)	C	09 36	Complicated SS- and SSS-groups.
	e(SSS)	C	12 40	

Day	Phase		h m s	Remarks
cont.				
28.	ePKPPKP	A	16 24 12	
	e	A	24 17	
	LmV	B	29.7	
	LmH	B	29.8	
28.	eP	A	22 26 53	LmH:10s 0.4 /um LmV:10s 0.4 /um
	LmH	C	38.5	
	LmV	C	38.5	
28.	eP	A	23 59 40	LmV:12s 0.3 /um
	LmH	C	00 10.5	
	LmV	C	11.1	
29.	+eP	A	05 31 41	<u>Off Coast of Central America</u>
	eS	C	42 18	12.34°N 91.21°W
	LmH	B	06 13	H = 05 18 49.6 h = normal MAG=5.2
	LmV	B	13	D = 88.5° Az = 38.4° (USCGS)
				PV:1.5s 47.0nm
				LmV:18s 0.3 /um
				MPV=5.5 MLV=4.7
29.	eP	A	05 43 34.5	
29.	ePKP	A	07 39 22	<u>Tonga Islands</u> 19.79°S 173.99°W
				H = 07 19 34.8 h = normal MAG=4.6 (USCGS)
				D = 148.1°
29.	eiSg	A	10 03 47	<u>Explosion/ČSSR</u> 49°15.7'N 14°55.4'E
				Yield: 12.5 t (PRUHONICE)
				D = 2.6°
29.	LmH	B	11 14.1	
	LmV	B	14.3	
29.	LmH	C	11(29)	Traces. Probably Mexico (USCGS)
29.	eP	A	12 07 14	

Day	Phase		h m s	Remarks
29.	e	A	15 15 33	Probably to Central Mid-Atlantic Ridge
	e	A	15 43	4.6°N 32.5°W
				H = 15 05 18.9 h = normal MAG=4.9 (USCGS)
				D = 59.0°
				The first onset of P must be about 14 s
				earlier than the onset e 15 33.
29.	LmV	C	24 01.5	Auckland Island Region (USCGS)
	LmH	C	(03)	
30.	eP	A	02 39 35	Iceland 63.60°N 22.82°W
	eS	C	43 44	H = 02 34 38.7 h = normal MAG=4.3
	LmH	B	49.4	D = 22.4° Az = 109.1° (USCGS)
	LmV	B	51.2	LmH:12s 1.0 μm LmV:13s 0.9 μm
				MLH=4.5 MLV=4.5
30.	eP	A	04 24 41	Iceland 63.73°N 22.90°W
	eS	C	28 48	H = 04 19 43.2 h = normal MAG=4.4
	LmH	B	34.5	D = 22.5° Az = 109.3° (USCGS)
	LmV	B	36.2	LmH:13.5s 1.1 μm LmV:14s 1.0 μm
				MLH=4.5 MLV=4.6
30.	eP	A	04 35(02)	Iceland 63.84°N 22.65°W
	e	A	35 05	H = 04 30 07.5 h = normal MAG=4.4 (USCGS)
	e	A	35 12	D = 22.4°
	LmH	B	46.8	LmH:11s 0.6 μm LmV:11s 0.6 μm
	LmV	B	46.8	MLH=4.2 MLV=4.4
30.	+eiP	A	08 09 47.2	Ryukyu Islands 28.91°N 129.92°E
	e(pP)	A	09 55.5	H = 07 57 19.9 h = 32 km MAG=5.5
	e(sP)	A	10 00	D = 83.9° Az = 325.7° (USCGS)
	eS	C	20 12	PV:1.9s 126.8nm (sP)V:2.0s 148nm
	ePS	C	21 08	LmH:14s 6.5 μm LmV:18s 6.7 μm
	eSS	C	25 40	MPV=5.7 MLH=6.2 MLV=6.1
	eSSS	C	29(40)	e 09 48 e 30 12
	LmV	B	51.1	If we interpret (sP) as pP then the focal
	LmH	B	52.1	depth will be h = 47 km.

Day	Phase		h m s	Remarks
30.	e	A	12 07 07	
	e	A	07 10	
30.	e	A	12 35 03	Explosion/GDR
	eiL	A	35 11	
30.	eiL	A	12 35 40	Explosion/GDR
30.	iL	A	12 35 59.3	Explosion/GDR

Day	Phase		h m s	Remarks
1.	e(P)	A	07 06 22.5	
1.	e(P)	A	08 53 00	
	i	A	53 22.0	
1.	e(Pg)	A	11 09 31.5	Probably explosion
	iSg	A	09 45.5	(D = 1.1°)
1.	ePKP	A	12 15 28	Tonga Islands 15.14°S 174.08°W
				H = 11 56 02.8 h = 90 km MAG=4.6 (USCGS)
				D = 144.3°
1.	i(P)	A	17 29 45.2	
	e	A	30 45	
	ei	A	30 58.5	
1.	ePn	A	22 47 16	Northern Italy 44.30°N 11.12°E
	eSn	A	48 28	H = 22 45 43.6 h = normal MAG=4.2
	eSb	A	48 47	D = 6.4° Az = 2.9° (USCGS)
	eiSg	A	49 12	
	eiLg2	A	49 19	
	e(Rg)	A	49 30.5	
2.	ePKIKP	A	00 31 30.5	Fiji Islands 20.96°S 178.84°W
	-iPKHKP	A	31 35.5	H = 00 12 52.8 h = 604 km MAG=5.2
	+iPKP2	A	31 42.5	D = 149.2° Az = 346.9° (USCGS);
	epPKP	A	33 57	h = ca. 620 km
	esPKP	B	34 52	PKIKPV:1.1s 26.4nm PKHKPV:1.3s 150nm
	ePP	A	35 08	PKP2V:1.1s 67.3nm pPKHKPV:1.6s 37.9nm
	eiPP	B	35 12	i 31 48.0 e 32 06 e 32 13 e 41.1
	esPP	C	38 15	e 14.6
	eSKKS	C	41.1	
	ePSKS	C	45 13	
	ePPS	C	48 30	
	eSS	C	53 30	
	esSS	C	57 20	
	eSSS	C	59.0	
2.	e(Sg)	A	11 17 50	Probably explosion

Day	Phase		h m s	Remarks
2.	e(Pg)	A	12 52 10	Explosion
	e(Sg)	A	52 18	(D = 0.6°)
	eL	A	52 27	
2.	ePP	A	15 15 06.5	New Britain Region 6.7°S 153.4°E
	e(pPP)	A	15 23.5	H = 14 54 08.4 h = 27 km MAG=5.3 (USCGS)
	LmH	C	16 04	D = 126.0° (h = 60 km)
				LmH:30s 0.2 µm
				MLH=4.6
2.	eP	A	15 52 38	East China Sea 27.97°N 127.92°E
				H = 15 40 20.6 h = 109 km MAG=4.9 (USCGS)
				D = 83.7°
2.	ePn	A	20 15 16.5	Southern Italy 41.63°N 14.00°E
	ePb	A	15 51	H = 20 13 02.2 h = normal MAG=4.5
	ePg	A	16 04	D = 9.2° Az = 350.5° (USCGS)
	e(Sn)	A	17 06.5	PnV:1.0s 9.5nm
	eSx	A	17 10	ei 15 55 e 16 15 e 17 15 e 17 29.5
	eSb	A	17 45.5	e 18 11.5 e 18 25
	eLg2	A	18 09	
3.	e(P)	A	03 41 45	(P)V:1.5s 20.2nm
3.	LmV	C	03 50.6	LmH:16s 0.5 µm LmV:16s 0.2 µm
	LmH	C	51	
3.	e(Sg)	A	10 19 07.8	Explosion?
	LmH	A	19 33	e 19 08
3.	i(Pg)	A	12 58 06.0	Explosion
	i(Sg)	A	58 14.4	(D = 0.6°)
	LmH	A	58 28	
3.	e	A	13 17 20	
3.	iPg	A	14 22 15.8	Explosion
	eiSg	A	22 31	D = ca. 1.1°

Day	Phase		h m s	Remarks
3.	ePKHKP	A	15 05 34.5	Fiji Islands 21.90°S 179.53°W
	ePKP2	A	05 42	H = 14 46 46.7 h = 553 km MAG=4.7
	ei	A	06 33	D = 150.0° Az = 345.7° (USCGS)
				PKHKPV:1.0s 9.5 nm
3.	-eP	A	18 28 44.5	Costa Rica 10.91°N 85.91°W
	eX	A	29 08	H = 18 16 03.2 h = 21 km MAG=5.8
	eY	A	29 16	D = 86.4° Az = 39.2° (USCGS)
	eZ	A	29 29	PV(A):2.0s 106nm PV(C):15s 2.1 /um
	ePP	A	32(00)	PH(C):15s 1.0/um XV:2.2s 132nm
	ePP	C	32 06	YV:2.0s 139nm ZV:1.7s 70.2nm
	i(S)	B	39 19	PPV(C):16s 0.9/um (S)H(C):22s 4.9/um
	i(S)	C	39 21	PSH:29s 4.5/um SSH:38s 8.3/um
	eiPS	C	40 23	MPV(A)=5.7 MPV(C)=6.1 MPH(C)=6.3
	eSS	C	45.0	MPPV=5.9 (MSH=6.4)
	eSSS	C	48.6	e 29 58 e 30 04 i 40 36 i 41 16
	eLQ	C	53	ei 41 50
	eiLR	C	56 50	X, Y and Z are three clear unidentified phases after P in our short-period registration. If we interpret X as pP and Y as sP, then the focal depth will be about 85 km. SS is totally polarized in the SV-plane. Therefore we were not certain to interpret the onset at 18 39 21, which was also SV-polarized, as SKS, as is usually done. If we interpret this onset as S, then the J.-B.travel-time residual will be only +1 sec., but +9 sec. if we interpret this onset as SKS.
	LmH	B	19 06.3	
	LmV	C	06.3	
3.	e(Pg)	A	20 03 45.5	Probably near earthquake.
	ei(Sg)	A	04 19	VIENNA gives: i(Pg) 20 03 14.5 i(Sg) 03 30
	LmV	A	04 47	e 03 46 e 04 14.5 e 04 20
3.	e	C	21 30 20	
	e	C	30 29	
4.	e	A	04 58 29	

Day	Phase		h m s	Remarks
4.	eP	A	06 14 57.5	Off Coast of Costa Rica 10.72°N 85.97°W
	LmH	C	58.3	H = 06 02 16.4 h = 33 km MAG=5.3 (USCGS)
				D = 86.6°
				LmH:16s 0.7 /um
				MLH=5.2
4.	iPg	A	09 13 44.0	Explosion/GDR
	iSg	A	13 45.8	SgV:0.5s 376nm SgH:0.6s 374nm
	i	A	13 46.0	
4.	ePg	A	10 30 56	Upper Silesia 50.4°N 18.9°E
	eSb	A	31 41.5	H = 10 29 29 (BCIS)
	eSg	A	31 56	D = 4.6°
				e 31 58.5 e 32 09
4.	e	A	14 17 50	Probably explosion
	eSg	A	17 58.5	e 18 10
	LmH	A	18 14	
	LmV	A	18 14	
4.	eP	C	17 39.9	New Ireland Region 5.72°S 153.89°E
	ePKIKP	A	40 18	H = 17 21 20.7 h = 52 km MAG=5.5
	ePKIKP	C	40 20	D = 125.3° Az = 331.5° (USCGS)
	ePP	A	42 09.5	PV:30s 1.3 /um PKIKPV(C):24s 1.9 /um
	iPP	C	42 12	PPV(C):28s 12.9 /um PPH(C):28s 8.3 /um
	e(PKS)	C	43 35	(PKS)H:26s 8.3 /um LQH:75s 50.7 /um
	e(PPP)	A	44 47	LmH:19s 14.4 /um LmV:23s 25.6 /um
	eSKKS	C	49.0	MPPV=6.9 MPPH=6.9 MLH=6.6 MLV=7.0
	iSP	C	52 07	e 40 25 e 40 57.5 e 41 15 e 41 22
	iPS	C	52 12	e 53 28 i 57 19 e 59 16
	eiSPP	C	53 37	The begin of surface waves has a pronoun-
	eSKKP	A	53 45	ced long-period character. P is the diff-
	iSS	C	59 13	racted P around the core. The onset (PKS)
	iSSS	C	18 03.9	is very clear in the horizontal components
	iLQ	C	14 50	of the long-period seismographes of type C
	LmV	C	35.5	but 20 s earlier than PKS expected from
	LmH	B	19 09.4	the DF branch from the I.-B.-tables.

Day	Phase		h m s	Remarks
4.	eP	A	21 52 33.5	Iceland 63.70°N 18.96°W
	LmH	C	22 01.5	H = 21 47 53.0 h = normal MAG=4.5
	LmV	C	01.5	D = 20.8° Az = 114.5° (USCGS)
				PV:1.8s 40.8nm
				LmH:16s 1.2 μm
				MPV=4.5 MLH=4.4
5.	e	A	11 04 04	
5.	eP1	A	12 04 16.5	Ionian Sea 37.75°N 20.68°E
	iP2	A	04 25.5	H = 12 00 51.2 h = 15 km MAG=5.0
	eS	A	06 55	D = 14.4° Az = 336.3° (USCGS)
	eLi	B	07 50	PV2:1.5s 87.2nm PH2:1.5s 63.2nm
	eLg1	B	08 26	LmH:15s 3.6 μm LmV:17s 2.6 μm
	LmH	B	09.7	MLH=4.6
	LmV	B	11.1	Multiple P, the second onset much stronger than the first one.
5.	e	A	12 31 16	Explosion?
	eSg	A	31 53	
	e	A	31 55	
5.	e(Sg)	A	14 46 45	Explosion?
	e	A	46 47	
5.	+eP	A	16 06 59	Kurile Islands 45.43°N 150.72°E
	e	A	07 02.5	H = 15 55 02.8 h = normal MAG=5.3
	ePcP	A	07 10	D = 78.0° Az = 334.8° (USCGS)
	e	A	07 17.5	PV:1.3s 46.5nm
	LmH	C	44.3	LmH:22s 0.6 μm LmV:18s 0.7 μm
	LmV	B	44.5	MPV=5.5 MLH=4.9 MLV=5.0
5.	ePKHKP	A	18 47 05.5	South of Fiji Islands 21.95°S 176.96°W
	epPKHKP	A	48 08	H = 18 27 38.3 h = 214 km MAG=4.9
				D = 150.6° Az = 348.9° (USCGS); h = 229 km
				PKHKPV:1.3s 22.1nm
6.	e(P)	A	04 11 40	Mid-Indian Rise 10.27°S 66.44°E
	+iP	A	11 44	H = 03 59 51.0 h = normal MAG=5.1
	ei(pP)	A	12 02	D = 77.0° Az = 327.7° (USCGS);

Day	Phase		h m s	Remarks
cont				
6.	e(sP)	A	04 12 13	(h = ca. 75 km)
	eS	C	21.3	PV:1.2s 25.5nm (pP)V:1.4s 18.4nm
	eSS	C	26.1	(sP)V:1.3s 22.2nm SH:25s 1.0 µm
	eLQ	C	30.3	LmH:55s 3.1 µm
	LmH	C	34	MPV=5.2 MSH=5.5 MLH=5.2
				e(P) is a small-amplitude precursor.
6.	eP	A	07 06 23.5	<u>Central Russia</u> 57.69°N 65.27°E
	e	A	06 51	H = 07 00 02.3 h = normal MAG=4.7
	iPn	A	07 18.5	D = 31.4° Az = 280.6° (USCGS)
	ePP	A	07 25.5	PV:1.3s 8.3nm
	i	A	08 02.2	MPV=4.5
				Clear Pn in our short-period VSJ-II record.
6.	LmH	C	08 47	Traces; <u>West Caroline Islands</u> (USCGS)
6.	iPg	A	12 33 20.5	Explosion/<u>GDR</u>
	iSg	A	33 27.5	D = 0.5°
	iL	A	33 28.5	
6.	eiL	A	12 33 42	Explosion/<u>GDR</u>
6.	iPg	A	13 58 03.0	Explosion/<u>GDR</u>
	iSg	A	58 17.0	D = ca. 1.1°
6.	e	A	14 06 41.5	
	e	A	06 47	
	eSg	A	07 05	
6.	iSn	A	15 05 20.7	Explosion/<u>GFR</u> 47.63°N 11.15°E
	iSg	A	05 36.2	H = 15 05 00 Yield: 7 t
				D = 3.0°
6.	e	A	17 50 37	
6.	LmH	C	17 59	LmH:18s 0.14 µm

Day	Phase		h m s	Remarks
7.	-ePP	A	01 32 38	Near Coast of Central Chile
	e(pPP)	A	33 05.5	29.56°S 71.05°W
	ePS	C	42 10	H = 01 14 04.1 h = 42 km MAG=5.3
	LQ	C	02 01	D = 107.9° Az = 41.6° (USCGS);
	LmH	B	18	(h = 66 km)
	LmV	C	19.4	PPV:1.6s 56.7nm
				LmH:20s 0.4 μm LmV:19s 0.4 μm
				MPPV=6.0 MLH=5.0 MLV=5.0
7.	ePKP	A	02 58 22	Samoa Islands 16.73°S 172.90°W
	epPKP	A	58 32	H = 02 38 44.1 h = normal MAG=4.4
	e	A	58 50	D = 146.0° Az = 354.9° (USCGS);
				h = 36 km
7.	+iP	A	08 39 46.9	Kurile Islands 49.20°N 156.27°E
	i	A	39 50	H = 08 28 01.2 h = normal MAG=5.3
	e(pP)	A	39 57.5	D = 76.0° Az = 337.7° (USCGS);
	i(PcP)	A	39 59.8	(h = 48 km)
	eS	C	49.6	PV:1.2s 133nm
	ePS	C	50.1	LmH:16s 1.3 μm LmV:17s 1.4 μm
	eSS	C	54.6	MPV=5.9 MLH=5.3 MLV=5.4
	LmV	B	09 17.5	If we interpret (PcP) as pP then the
	LmH	B	17.6	focal depth would be h = 50 km.
7.	eiP	A	09 18 37.5	Kurile Islands 49.16°N 156.35°E
	e	A	18 41	H = 09 06 52.3 h = normal MAG=4.9
	e(pP)	A	18 47	D = 76.1° Az = 337.7° (USCGS);
	e(PcP)	A	18 49.5	(h = 36 km)
	e	A	18 52.5	PV:1.3s 72.2nm
	LmV	B	58.4	LmH:16s 0.9 μm LmV:16s 0.8 μm
	LmH	B	58.5	MPV=5.4 MLH=5.2 MLV=5.2
				If we interpret (PcP) as pP then the
				focal depth would be h = 45 km.
7.	ePKP	A	10 51 43.5	Fiji Islands 17.28°S 178.94°W
	iPKP	A	51 45.5	H = 10 33 08.2 h = 563 km MAG=4.9
	e	A	52 04	D = 145.7° Az = 348.1° (USCGS)
	ei	A	52 57	PKPV:1.1s 48.1nm

Day	Phase		h m s	Remarks
7.	eP	A	14 48 34.5	Off East Coast of Kamchatka
	e	A	48 37	52.17°N 160.36°E
	e	A	49 15	H = 14 36 59.5 h = normal MAG=4.4
	LmH	C	15 25.6	D = 74.2° Az = 339.9° (USCGS)
	LmV	C	27.7	PV:1.2s 12.8nm
				LmH:16s 0.6 µm LmV:15s 0.4 µm
				MPV=4.9 MLH=4.9 MLV=4.9
7.	LmH	C	20 59.3	Traces; New Ireland (USCGS)
8.	+iP	A	05 41 42.4	Kurile Islands 49.28°N 156.32°E
				H = 05 29 58.4 h = 42 km MAG=4.4
				D = 76.0° Az = 337.7° (USCGS)
				PV:1.2s 20.4nm
				MPV=5.1
8.	eSn	A	09 58 44	Hunsrück/GFR 49.8°N 6.9°E
	e(Sg)	A	58 58.5	H = 09 57 18 (BCIS)
	eL	A	59 05	D = 3.2°
	e	A	59 07	
8.	eSg	A	14 00 11	Switzerland 46.4°N 7.5°E
	e	A	00 22	H = 13 57 29 (BCIS)
				D = 5.0°
8.	ePKIKP	A	17 18 36	East New Guinea Region 9.54°S 148.82°E
	e	A	18 38	H = 16 59 34.4 h = 17 km MAG=5.5
	ePP	A	20 30.5	D = 126.0° Az = 327.7° (USCGS)
	eSS	C	37.7	PV:1.3s 11.1nm PPV:2.2s 87.7nm
	LmH	C	18 08.5	LmH:24s 1.3 µm
				MPPV=5.8 MLH=5.4
8.	e	A	18 27(50)	Solomon Islands 5.60°S 153.96°E
	ePP	A	29 03	H = 18 08 18.1 h = 70 km MAG=5.1 (USCGS)
	ePS	C	38.9	D = 125.2°
	ePPS	C	40.5	PPV:2.3s 96.5nm
	eSS	C	46.2	LmH:20s 1.0 µm LmV:20s 1.2 µm
	eSSS	C	51.0	MPPV=5.9 MLH=5.5 MLV=5.6
	LmH	B	19 26.7	
	LmV	B	26.9	

Day	Phase		h m s	Remarks
8.	LmH	C	19 55.5	Traces
8.	iP	A	21 20 58.0	Kurile Islands 49.22°N 153.30°E
	epP	A	21 09	H = 21 09 13.1 h = normal MAG=4.7
	LmH	C	22 00.3	D = 76.0° Az = 337.7° (USCGS); h = 40 km
				PV:1.1s 16.8nm
				MPV=5.1
8.	eP	A	21 31 32.5	Kurile Islands 49.25°N 156.28°E
	LmH	C	11.2	H = 21 19 46.3 h = 23 km MAG=4.5
				D = 76.0° Az = 337.7° (USCGS)
				PV:1.2s 12.8nm
				MPV=4.9
9.	e	A	02 22 34	
9.	eP	A	08 36 35.5	North Atlantic Ridge 19.37°N 46.15°W
				H = 08 27 05.4 h = normal MAG=4.9
				D = 54.9° Az = 41.2° (USCGS)
				PV:2.0s 26.5nm
				MPV=5.0
9.	ePn	A	10 03 45	Austria 47.89°N 11.12°E
	iPb	A	03 52.5	H = 10 03 02.8 h = 8 km
	iPg	A	03 56.0	D = 2.8° Az = 6.5° (USCGS)
	eiSn	A	04 19.5	Bavaria/GFR 47.8°N 11.1°E (STU)
	eiSg	A	04 33	PbV:0.6s 11.9nm
				e 03 54.5 e 04 02.5 e 04 16.5
9.	eP	A	10 48(39)	Greece 38.3°N 21.2°E
	e	A	50 18	H = 10 45 16 ML=3.7 (ATH)
	LmH	C	54.7	D = 14.1°
	LmV	C	55.0	LmH:18s 0.3 µm LmV:12s 0.4 µm
				MLH=3.5
9.	ePKIKP	A	13 46 56	Solomon Islands 5.75°S 153.96°E
	e	A	47 29	H = 13 27 56.7 h = 41 km MAG=4.9 (USCGS)
	LmH	B	14 44.3	D = 125.3°
	LmV	B	44.5	LmH:21s 1.9 µm LmV:21s 1.7 µm
				MLH=5.6 MLV=5.6

Day	Phase		h m s	Remarks
9.	-eiP	A	14 21 36.5	Kamchatka 54.06°N 155.06°E
	ePcP	A	21 55	H = 14 10 57.4 h = 393 km MAG=5.2
	e	A	22 03.5	D = 71.3° Az = 336.4° (USCGS); h = 400 km
	epP	A	23 05	PV:1.1s 45.6nm
	e	A	23 47	MPV=5.5
	e	A	24 22	
9.	e	A	14 49 51	
	e	A	49 53	
9.	-eiPKP	A	17 40 21	Fiji Islands 21.08°S 179.35°W
	eipPKP	A	42 52	H = 17 21 49.5 h = 654 km MAG=6.3
	esPKP	A	43 56	D = 149.3° Az = 346.3° (USCGS);
	eisPP	A	47 12	h = 656 km
	iSKKS	B	49 55	
	iSKKKS	B	50 28	
	iPSKS	B	54 22	
	eiPa	C	57 15	
	eisSS	C	18 06 20	
	eiSSS	C	08 35	
	eisSSS	C	12 14	
	e(Sa)	C	22 10	
9.	eP	A	18 16 58.5	PV:2.0s 81.5nm
9.	ePKIKP	A	18 51 44	Fiji Islands 21.27°S 179.27°W
	iPKHKP	A	51 49.5	H = 18 33 08.2 h = 619 km MAG=5.1
	eiPKP2	A	51 56	D = 149.5° Az = 346.3° (USCGS);
	epPKHKP	A	54 12.5	h = 630 km
	epPKP2	A	54 20	PKIKPV:1.0s 19.0nm PKHKPV:1.4s 123nm
	e	A	54 31	PKP2V:1.4s 67.5nm
				AN USSR gives: Tonga Islands Region
				21.0°S 175.9°W
				H = 18 32 00 h = normal
				D = 149.8°
10.	LmH	B	01 48	Eastern China 36.88°N 115.07°E
				H = 01 07 49.7 h = normal MAG=4.6 (USCGS)
				D = 69.9°
				LmH:15s 0.4 µm MLH=4.8

Day	Phase		h m s	Remarks
10.	+ePKP	A	06 46 14.5	New Hebrides Islands 18.11°S 171.82°E
	e	A	46 45.5	H = 06 26 46.3 h = 63 km MAG=5.2
				D = 143.9° Az = 338.5° (USCGS)
				PV:1.6s 53.2nm
10.	eP	A	06 59 17	Near East Coast of Honshu 36.87°N 140.99°E
	LmH	C	07 39	H = 06 46 58.5 h = 25 km MAG=5.0 (USCGS)
				D = 82.0°
				PV:1.5s 13.4nm
				LmH:33s 0.3 μm
				MPV=4.9 MLH=4.5
11.	eP	A	07 53 00	Dodecanese Islands 36.05°N 27.20°E
	e	A	53 09.5	H = 07 48 44.8 h = 35 km MAG=4.7
	e	A	53 18	D = 18.4° Az = 327.2° (USCGS)
	eS	C	56 21	PV:1.4s 18.4nm
	e	C	56 28	LmH:10s 0.4 μm LmV:11s 0.3 μm
	LmV	C	08 01.5	MLH=4.0 MLV=4.0
	LmH	C	01.6	
11.	-eP	A	16 05 06.5	South of Honshu/Japan 30.38°N 142.59°E
	e	A	05(40)	H = 15 52 16.8 h = 32 km MAG=5.5
	ePP	A	08 35	D = 88.4° Az = 331.3° (USCGS)
	e	A	08 39	PV:1.6s 34.4nm
	LmH	C	44.5	LmH:17s 0.9 μm LmV:16s 0.7 μm
	LmV	C	50.5	MPV=5.3 MLH=5.3 MLV=5.2
11.	eP	A	18 53 15	Kurile Islands 44.37°N 145.98°E
	e	A	53 15.5	H = 18 41 21.4 h = normal MAG=4.6
	LmH	C	19 27	D = 77.4° Az = 332.2° (USCGS)
				LmH:17s 0.3 μm
				MLH=4.6
11.	eP	A	20 40 24	Peru - Brazil Border Region
				10.29°S 71.16°W
				H = 20 28 10.2 h = 585 km MAG=5.0
				D = 93.3° Az = 39.2° (USCGS)
				PV:2.0s 39.8nm
				MPV=5.4

Day	Phase		h m s	Remarks
12.	ePKIKP	A	06 53 40.5	Fiji Islands 21.13°S 179.19°W
	iPKHKP	A	53 46.1	H = 06 35 06.7 h = 636 km MAG=5.6
	iPKP2	A	53 53.0	D = 149.3° Az = 346.5° (USCGS);
	epPKP	C	56 10	(h = ca. 680 km)
	esPKP	C	57 10	PKIKPV:2.2s 158nm PKHKPV:1.8s 866nm
	e(PP)	A	57 22.5	PKP2V:1.4s 338nm
	epPP	C	07 00 32	i 53 50 e 55 20 e 55 22 e(pPKHKP) 56 15.5
	eSKKS	C	03 15	e(pPKP2) 56 24.5 e(sPKIKP) 57 21
	ePKKS	C	06 10	e(sPKHKP) 57 29.5 e(sPKP2) 57 36
	ePSKS	C	07 40	e 57 51 e 57 58
	ePPS	C	10 55	PKP and the deep-phases are multiple in
	eSS	C	15.7	our short-period records.
	esSS	C	19 30	
	esSSS	C	25 20	
	e	C	25.8	
12.	iPg	A	12 01 03.2	Explosion
	eiSg	A	01 12	D = ca. 0.6°
12.	ePg	A	12 54 49	Explosion 51.30°N 13.18°E
	e(Sg)	A	55 05.5	D = 1.1° Yield: 1.05 t
12.	iP	B	13 04 23.5	Northwest of Kurile Islands
	-iP	A	04 24.5	52.23°N 152.52°E
	-iPcP	A	04 42.5	H = 12 53 46.9 h = 476 km MAG=5.5
	epP	A	06 05	D = 72.3° Az = 335.1° (USCGS); h = 480 km
	eisP	A	06 52.5	PV(A):1.3s 516nm PcPV:1.1s 86.5nm
	eS	C	13 09	LmH:24s 0.5 μm
	esS	C	16.1	MPV=6.0
	eLQ	C	23 08	e 06 20.5 e 06 28 e 07 02 i 13 10
	LmH	C	30.7	ei 22 30
12.	eP	A	18 46 18	Banda Sea 7.12°S 129.77°E
	+ePKIKP	A	50 10	H = 18 31 37.1 h = 45 km MAG=6.2
	ePP	C	51 04	D = 113.1° Az = 322.4° (USCGS)
	ePS	C	19 00 40	PV:1.6s 22.7nm PKIKPV:1.5s 74.0nm
	e(SS)	C	06 35	PPV:12s 0.9 μm
	e	C	07.2	LmH:30s 2.7 μm LmV:30s 1.2 μm
	LmH	C	35	MPV=6.6 MPPV=6.3 MLH=5.7 MLV=5.3
	LmV	C	37.5	e 01 05 e 01 37

Day	Phase		h m s	Remarks
13.	eP	A	03 32 56	Southern Sinkiang Prov./China
	e	A	34 54	39.67°N 74.45°E
	LmH	C	51.5	H = 03 24 47.0 h = normal MAG=5.2
	LmV	C	54.5	D = 44.4° Az = 305.9° (USCGS)
				PV:1.4s 9.2nm
				LmH:20s 0.3 μm LmV:18s 0.3 μm
				MPV=4.5 MLH=4.3 MLV=4.4
13.	e	A	11 33 20.5	Explosion/ČSSR 49.57°N 14.58°E
	e	A	33 22	H = 11 32 00 (PRU)
				D = 2.2°
13.	eP	A	20 01 02	Honshu/Japan 36.51°N 138.22°E
	LmH	B	38.2	H = 19 48 48.1 h = 30 km MAG=4.6
	LmV	B	43	D = 81.3° Az = 328.9° (USCGS)
				LmH:14s 1.2 μm LmV:12s 0.5 μm
				MLH=5.3 MLV=5.1
14.	eP	A	03 41 48	Leeward Islands 17.28°N 60.81°W
	eS	C	50 30	H = 03 31 04.5 h = 29 km MAG=5.3
	eSS	C	54.7	D = 65.7° Az = 41.8° (USCGS)
	LmH	B	04 00.7	LmH:20s 1.0 μm LmV:18s 0.8 μm
	LmV	B	09.7	MLH=5.0 MLV=5.0
				e 41 52.5 e 42 00 e 42 16 e 43 44
14.	LmH	C	17 33	Traces; New Hebrides Islands (USCGS)
	LmV	C	33	D = 140.2°
14.	eP	A	23 39 26.5	Kenya 3.25°S 30.20°E
	LmH	C	24 05.5	H = 23 29 30.0 h = normal MAG=5.2 (USCGS)
	LmV	C	11.8	D = 58.4°
15.	eP	A	03 41 02.5	Off East Coast of Kamchatka
				52.25°N 160.76°E
				H = 03 29 27.9 h = 35 km MAG=4.4 (USCGS)
				D = 74.2°
15.	eP1	A	08 13 11	Near Coast of Nicaragua 11.86°N 86.02°W
	+iP	B	13 12	H = 08 00 50.3 h = 162 km MAG=6.2
	iP2	A	13 13.0	D = 85.7° Az = 39.3° (USCGS); h = 164 km

Day	Phase		h m s	Remarks
cont				
15.	epP	A	08 13 51	PV(B):12s 10.6 μm PPV:4s 10.7 μm
	e(sP)	B	14 54	PPH:3.5s 6.1 μm SKSH:16s 21.6 μm
	-iPP	B	16 40	SPH:15s 36.5 μm (sSP)H:18s 48.0 μm
	ipPP	B	17 16	PKKPV:1.8s 86.7nm pPKKPV:1.5s 26.8nm
	isPP	B	17 35	LmH:17.5s 21.5 μm LmV:18s 22.7 μm
	iSKS	B	23 22	MPV=6.9 MPPV=7.7 MPPH=7.8
	e(S)	B	23(32)	e 14 54 e 16 09.5 e 16 13.5 e 16 28
	eiSP	B	24 22	e 16 41 e 16 48 e 16 53 e 17 08
	e(sSP)	B	25 25	e 23 04 e 23 12 e 23 28 e 31 01
	eiSS	C	29 09	e 39 13
	ePKKP	A	31 15	Multiple P in our short-period records,
	epPKKP	A	31 56	the first onset 2 sec before the stronger
	ePKPPKP	A	39 27	second onset.
	eSKPPKP	A	42 23	
	LmH	B	48.9	
	LmV	B	49.5	
15.	eP	A	17 50 37.5	<u>Fox Islands/Aleutian</u> 52.09°N 169.53°W
				H = 17 38 43.1 h = 32 km MAG=4.5
				D = 77.6° Az = 359.3° (USCGS)
				PV:1.3s 11.1nm
				MPV=4.8
15.	iP	A	21 54 30.5	<u>Northern China</u> 36.83°N 105.02°E
				H = 21 43 55.0 h = normal MAG=5.1
				D = 64.5° Az = 315.3° (USCGS)
				PV:1.3s 16.7nm
				MPV=5.1
15.	ePKP	A	23 22 55	<u>Tonga Islands</u> 17.63°S 173.24°W
	e	A	23 22	H = 23 03 15.0 h = 40 km MAG=4.6
				D = 146.8° Az = 354.4° (USCGS)
				PKPV:1.8s 56.0nm
15.	eP	A	23 55 09	<u>Molucca Sea</u> 1.78°S 126.28°E
	LmH	C	24 36	H = 23 36 49.0 h = 51 km MAG=5.3
				D = 106.8° Az = 322.8° (USCGS)

Day	Phase		h m s	Remarks
16.	e(P)	A	13 39 21	Vancouver Island 49.27°N 129.14°W
	i	A	39 24	H = 13 27 35.6 h = normal MAG=5.2
	e	A	39 28.5	D = 74.9° Az = 24.7° (USCGS)
	eS	C	48.9	LmH:15s 2.2 μm LmV:18s 2.2 μm
	eSS	C	53.6	MLH=5.6 MLV=5.5
	eSSS	C	57.5	The first onset of P must be about 6 sec.
	LmV	B	14 13.3	earlier.
	LmH	B	14.7	
16.	e(sP)	A	17 12 36	Halmahera 1.73°N 127.46°E
	e	A	16 17	H = 16 58 02.0 h = 120 km MAG=5.6 (USCGS)
	ePP	A	16 19.5	D = 104.8°
	LmH	C	55	PPV:1.4s 24.5nm
				MPPV=5.6
16.	eP	A	20 29 24	Mascarene Islands 17.28°S 66.56°E
	e	A	29 29	H = 20 16 56.1 h = 18 km MAG=5.2
	eS	C	39.6	D = 83.0° Az = 328.3° (USCGS)
	eSS	C	45.2	PV:1.5s 20.1nm
	eLQ	C	50.7	MPV=5.0
16.	eP	A	21 41 06	Kurile Islands 49.07°N 156.27°E
	LmH	C	22 21.7	H = 21 29 20.5 h = 39 km MAG=4.2
				D = 76.2° Az = 337.7° (USCGS)
				PV:0.9s 14.1nm
				LmH:20s 0.5 μm
				MPV=5.1 MLH=4.9
16.	ePKP	A	22 35 25	Fiji Islands 17.96°S 178.38°W
				H = 22 16 48.8 h = 593 km MAG=4.2
				D = 146.4° Az = 348.5° (USCGS)
				PKPV:1.8s 30.6nm
16.	eP	A	23 43 02	Kurile Islands 43.95°N 150.11°E
	epP	A	43 14.5	H = 23 31 00.1 h = normal MAG=4.3
	LmH	C	24 16.4	D = 79.1° Az = 334.6° (USCGS)
				PV:1.9s 47.1nm pPV:1.4s 18.4nm
				LmH:20s 0.8 μm
				MPV=5.3 MLH=5.0

Day	Phase		h m s	Remarks
17.	+iP	A	05 11 46.0	Eastern Kazakh SSR 49.82°N 78.10°E
	ePn	A	13 19	H = 05 03 58.0 h = 0 km MAG=5.7
				D = 41.2° Az = 297.1° (USCGS)
				PV:0.8s 108nm PH:1.0s 65.0nm
				MPV=5.9 MPH=5.8
				Underground explosion.
17.	-eiPKHKP	A	14 27 37.5	Fiji Islands 21.15°S 179.12°W
	eiPKP2	A	27 44	H = 14 08 58.4 h = 636 km MAG=4.8
				D = 149.4° Az = 346.5° (USCGS)
				PKHKPV:1.2s 71.4nm PKP2V:1.0s 37.9nm
17. to 18.				Seismological station not operating from October 17 at 15^h00^m to October 18 13^h00^m.
18.	eP	A	14 42 18	Nevada; Underground explosion
				H = 14 30 00 (UPP)
				D = 81.1°
				PV:1.5s 57.1nm
				MPV=5.4
18.	e(Sn)	A	18 59 03	Austria 47.9°N 16.3°E
	e	A	59 21.5	H = 18 57 12 (BCIS)
	eiSg	A	59 26	D = 4.1°
18.	ePKIKP	A	22 26 21	South of Kermadec Islands 33.88°S 179.57°W
	ePKP2	A	27 06	H = 22 06 23.5 h = 26 km MAG=5.4
	e(sPKP2)	A	27 18	D = 161.4° Az = 337.3° (USCGS)
	e	A	27 29	PKIKPV:1.4s 18.4nm PKP2V:0.9s 14.1nm
	LmV	C	23 39.2	(sPKP2)V:1.3s 27.8nm
	LmH	C	40.1	LmH:22s 0.7 μm LmV:22s 0.7 μm
				MLH=5.3 MLV=5.5
18.	ePKIKP	A	23 54 31	New Hebrides Islands 13.87°S 166.52°E
				H = 23 35 11.0 h = 87 km MAG=5.0
				D = 138.0° Az = 336.2° (USCGS)
				PV:1.4s 12.3nm

Day	Phase		h m s	Remarks
19.	eSg	A	07 42 51	Rheinland/GFR 50.6°N 6.8°E
	iLg	A	42 58	H = 07 41 08 (BCIS)
				D = 3.1°
19.	e(Pg)	A	16 07 15.5	Explosion
	e(Sg)	A	07 30	(D = ca. 1.1°)
19.	LmH	C	16 38	Traces; South Sandwich Islands (USCGS)
20.	ePKP	A	00 39 25.5	Loyalty Islands 21.42°S 170.07°E
	e	A	39 29	H = 00 19 57.2 h = 130 km (USCGS)
				D = 146.3°
20.	ePKIKP	A	01 21 21	South Sandwich Islands Region
	ePP	A	22 16.5	58.6°S 25.0°W
	ePS	C	31.9	H = 01 02 43.8 h = 12 km MAG=5.6 (USCGS)
	eSS	C	38.2	D = 113.0°
	eSSS	C	42.5	PPV:1.4s 18.4nm
	LmV	B	02 04.5	LmH:18s 1.0 /um LmV:20s 2.1 /um
	LmH	B	04.8	MPPV=5.6 MLH=5.5 MLV=5.8
20.	eP	A	06 52 36	Turkey 37.93°N 37.74°E
	eS	B	56.8	H = 06 47 38.0 h = normal MAG=4.8
	e	B	58.0	D = 22.5° Az = 312.8° (USCGS)
	LmH	B	07 04.2	PV:2.0s 66.2nm
	LmV	B	04.2	LmH:14s 0.9 /um LmV:14s 1.2 /um
				MPV=4.7 MLH=4.4 MLV=4.6
20.	eP	A	07 50 10	Taurus/Anatolie 38.4°N 36.6°E
	e	A	50 20	H = 07 44 57 (BCIS)
				D = 22.7°
20.	iPg	A	08 29 34.5	Explosion
	eiSg	A	29 50.5	D = ca. 1.2°
	i	A	29 51	
20.	ePKIKP	A	16 15 14	Fiji Islands Region 20.6°S 178.1°W
	eiPKHKP	A	15 20	H = 15 56 33.4 h = 556 km MAG=5.0 (USCGS)
	ePKP2	A	15 26	D = 149.2°

Day	Phase		h m s	Remarks
cont				
20.	epPKP	A	16 17 31	PKIKPV:1.3s 19.4nm PKHKPV:1.3s 106nm
				PKP2V:1.3s 33.4nm
21.	eSg	A	00 50 15	Near earthquake/Probably Upper Silesia
	LmH	A	50 31	(D = ca. 4.9°)
				PRU gives: ePn 00 48 31 ePg 48 39
				Sg 49 19
				VIE gives: iPg 48 34 iSg 49 08 (D = 280 km)
21.	ePKP	A	02 53 43	Near Coast of Northern Chile
	eS	C	03 01 35	27.74°S 71.82°W
	ePS	C	03.5	H = 02 35 12.3 h = 13 km MAG=5.4
	eSS	C	09.0	D = 107.0° Az = 41.4° (USCGS)
	eSSS	C	12.9	PKPV:1.1s 7.2nm
	e(SSSS)	C	16.2	LmH:20s 0.9 μm LmV:20s 1.1 μm
	LmH	C	33.8	MLH=5.3 MLV=5.4
	LmV	C	33.8	
21.	+iP	A	05 06 03.5	Novaya Zemlya 73.37°N 54.81°E
	iPn(8.28)	A	06 30.5	H = 04 59 58.1 h = 0 km MAG=5.9
	ei(PP)	A	06 47.5	D = 29.3° Az = 243.0° (USCGS)
	eSn(4.68)	A	11 34	PV:1.0s 71nm
	eSS	B	12 20	LmH:15s 1.4 μm
	eLi(3.71)	B	14 36	MPV=5.4 MLH=4.7
	eLg1(3.54)	B	15 18	Underground explosion.
	eLg(3.44)	B	15 44	Clear developed higher mode surface
	iLg2(3.32)	B	16 16	waves and teleseismic Pn and Sn.
	LmH	B	17.9	
21.	i	A	12 34 24	Explosion/GDR
	iSg	A	34 28	
	iL	A	34 31	
21.	ePn	A	16 56 58	Yugoslavia 43.26°N 16.79°E
	eSn	A	58 29	H = 16 54 58.3 h = 34 km MAG=4.3
	eSb	A	59 02	D = 8.2° Az = 336.2° (USCGS)
	eSg	A	59 26.5	
21.	LmV	B	18 08	Traces; West Caroline Islands (USCGS)

Day	Phase		h m s	Remarks
21.	ePKHKP	A	18 59 27	South of Fiji Islands 24.77°S 177.31°W
	e	A	59 33	H = 18 39 40.3 h = 107 km MAG=4.8
				D = 153.3° Az = 347.3° (USCGS)
21.	e(P)	A	21 20 05	
22.	eP	A	01 05 25	Yujuy Province/Argentina 22.26°S 65.75°W
	e	A	05 45	H = 00 52 10.9 h = 259 km MAG=5.2
	eSKS	C	15.7	D = 99.3° Az = 39.0° (USCGS)
	eS	C	16.6	PV:1.3s 22.2nm
	esSKS	C	17 40	MPV=5.5
	e(sSP)	C	19.7	
22.	LmV	B	03 53.5	Traces
22.	eP	A	05 41 46	Greece 36.8°N 21.4°E
	e	A	41 54	H = 05 38 05 ML=3.8 (ATH)
				D = 15.8°
22.	eP	A	18 56 48	North Atlantic Ridge 30.98°N 41.44°W
	eS	C	19 03 24	H = 18 48 44.5 h = normal MAG=4.7
	LmH	B	12.5	D = 43.6° Az = 47.5° (USCGS)
	LmV	B	12.5	PV:1.7s 21.9nm
				LmH:18s 0.4 μm LmV:19s 1.3 μm
				MPV=4.8 MLH=4.4 MLV=4.9
22.	e(P)	A	21 24 12	
22.	LmH	C	21 32	Traces
22.	-iPKP	A	22 37 40.5	Tonga Islands 17.96°S 174.78°W
				H = 22 18 14.4 h = 145 km MAG=4.5
				D = 147.0° Az = 352.5° (USCGS)
22.	eP	A	23 16 43.5	Ryukyu Islands 27.38°N 128.25°E
	-iP	A	16 44.5	H = 23 04 14.2 h = 34 km MAG=5.2
	e	A	16 50	D = 84.3° Az = 325.1° (USCGS); h = 40 km
	epP	A	16 54.5	PV:1.3s 27.8nm
	LmH	B	59.5	LmH:15s 0.3 μm LmV:17s 0.4 μm
	LmV	B	59.5	MPV=5.2 MLH=4.8 MLV=4.9

Day	Phase		h m s	Remarks
23.	eP	A	03 05 30	Kurile Islands 43.36°N 146.89°E
	e	A	05 48	H = 02 53 30.7 h = normal MAG=5.0
				D = 78.6° Az = 332.8° (USCGS)
				PV:1.0s 14.2nm
				MPV=5.1
23.	-iP	A	08 39 07.5	Bonin Islands 28.86°N 139.07°E
	epP	A	40 52.5	H = 08 27 06.2 h = 463 km MAG=5.3
	ePP	A	42 41	D = 88.2° Az = 329.6° (USCGS); h = 468 km
	esPP	C	45.0	PV:1.4s 79.7nm PPV:1.9s 94.0nm
	eSKS	C	48 52	MPV=5.3 MPPV=5.6
	eSP	B	50 12	e 44 31.5 e 55 12
	esSKS	C	52.0	
	esSP	C	53.3	
	eSS	C	55 08	
	esSS	C	58.0	
	eSSS	C	58.7	
	eSSSS	C	09 02 10	
	LmV	C	21.5	
	LmH	C	22.5	
23.	ePg	A	10 41 21	Explosion
	iSg	A	41 37	D = ca. 1.2°
23.	ePg	A	12 16 35.5	Explosion
	eiSg	A	16 53.5	D = ca. 1.4°
24.	ePKIKP	A	03 32 53	Kermadec Islands 31.3°S 179.7°W
	ePKHKP	A	33 07	H = 03 13 26.5 h = 250 km MAG=5.4 (USCGS)
	ePKP2	A	33 32	D = 158.9°
				PKIKPV:1.5s 20.1nm PKHKPV:1.2s 10.2nm
				PKP2V:1.4s 30.7nm
24.	eP	A	06 17 59	Greece 38.9°N 22.0°E
	e	A	18 04.5	H = 06 14 43.0 h = 29 km MAG=4.3 (USCGS)
	e	A	18 08	D = 13.9°
	e	A	18 11	LmV:12s 0.7μm
	LmH	C	24	
	LmV	B	24.0	

Day	Phase		h	m	s	Remarks
24.	e	A	07	35	40	
	e	A		35	48	
	e	A		36	40	
24.	+eP	A	11	04	20	Southern Sumatra 3.13°S 101.46°E
	epP	A		04	34	H = 10 51 15.1 h = 63 km MAG=5.5
	e	A		04	38.5	D = 92.3° Az = 320.4° (USCGS); h = 50 km
						PV:1.2s 20.4nm pPV:1.3s 16.7nm
						MPV=5.3
24.	e	A	12	59	50.5	Explosion
	eiSg	A		59	59	
	eL	A	13	00	05	
	LmH	A		00	20	
24.	eSg	A	16	20	42	Yugoslavia 46.1°N 15.9°E
						H = 16 17 47 (BCIS)
						D = 5.4°
25.	+iP	C	01	11	44	Taiwan 24.46°N 122.19°E
	+eiP1	A		11	44.3	H = 00 59 22.6 h = 65 km MAG=6.0
	+iP	B		11	45	D = 83.5° Az = 323.1° (USCGS)
	iP2	A		11	47.0	PV(B):10s 10.7μm P2V:1.8s 316nm
	-iPP	B		14	55	SH:15s 17.3μm
	ePPP	C		16.8		LmH:18s 63.5μm LmV:18s 55.8μm
	e(Pa)	C		18.6		MPV=6.9 MP2V=6.1 MSH=7.0 MLH=7.1
	iS	B		22	00	MLV=7.0
	iPS	B		22	44	Multiple P in our short-period registra-
	iPPS	B		23	20	tion. The first onset, 2.7 sec. earlier,
	iSS	C		27	28	is much smaller than the second one.
	i	C		28	22	Probably multiple shocks in the same fo-
	iSSS	C		31	54	cus region.
	eiPKPPKP	A		38	05.5	
	eSKPPKP	A		41(24)		
	e	A		41	52	
	LmH	B		47.5		
	LmV	B		54.2		
25.	+i(P)	A	01	30	05.8	Aftershock or PKKP of the preceding earthquake? (P)V:1.2s 58.7nm

Day	Phase		h m s	Remarks
25.	ePKP	A	01 51 21	New Hebrides Islands 18.34°S 171.95°E H = 01 31 50.0 h = 41 km MAG=4.9 D = 144.2° Az = 338.5° (USCGS)
25.	eP	A	02 09 45	Taiwan 24.30°N 122.19°E
	e	A	09 50.5	H = 01 57 23.1 h = 67 km MAG=5.3 D = 83.6° Az = 323.1° (USCGS) PV:1.5s 16.8nm MPV=5.0
25.	eP	A	09 33 42	Rat Islands/Aleutian 51.41°N 176.46°E H = 09 21 48.6 h = 33 km MAG=4.8 D = 77.5° Az = 350.2° (USCGS) PV:1.0s 9.5nm MPV=4.9
25.	ePKP2	A	09 36 46	Off East Coast of N. Island, N.Z. 37.06°S 177.47°E H = 09 16 16.0 h = 189 km MAG=5.1 (USCGS) D = 163.0°
25.	e	A	11 10 41	
	e	A	10 46	
25.	e	A	11 26 05	
25.	e	A	11 38 58	
25.	e	A	12 26 53	
26.	eP	A	00 34 44	Taiwan Region 24.48°N 122.24°E
	e	A	35 26	H = 00 22 21.6 h = 63 km MAG=5.6
	ePP	A	37 58	D = 83.5° Az = 323.1° (USCGS)
	LmH	C	01 04	PV:1.8s 61.2nm
	LmV	C	08.7	LmH:40s 2.5/um LmV:28s 1.1/um MPV=5.5 MLH=5.3 MLV=5.1
26.	e(P)	A	00 53 02.5	Aftershock or PKKP of the preceding earthquake? (P)V:1.4s 12.3nm

Day	Phase		h m s	Remarks
26.	+eP1	A	04 59 52	Turkey 37.25°N 29.14°E
	+eP2	A	59 53	H = 04 55 38.3 h = 35 km MAG=5.1
	eP3	A	59 58	D = 18.3° Az = 322.5° (USCGS)
	e(S)	C	05 03.4	P2V:(1.4s) 52.1nm P3V:1.6s 114nm
	LmH	C	07.7	LmH:16s 2.4/um LmV:14s 2.5/um
	LmV	C	07.8	MP2V=4.5 MP3V=4.7 MLH=4.6 MLV=4.8
26.	LmH	C	12 56.4	Leeward Islands 17.7°N 60.9°E
	LmV	C	56.5	H = 12 21 33.0 h = 33 km MAG=5.2 (USCGS)
				D = 65.4°
				LmH:18s 0.7/um LmV:18s 0.8/um
				MLH=4.9 MLV=5.0
26.	iPg	A	14 10 20.5	Explosion
	iSg	A	10 29	D = ca. 0.7°
26.	LmH	C	14 14	Leeward Islands 17.6°N 61.0°W
	LmV	C	20	H = 13 44 45.1 h = 37 km MAG=5.3 (USCGS)
				D = 65.6°
				LmH:24s 0.9/um LmV:20s 1.2/um
				MLH=4.9 MLV=5.1
26.	eP	A	17 36 08	Molucca Sea 0.2°S 125.2°E
	e	A	36 13	H = 17 22 05.3 h = 42 km MAG=5.6 (USCGS)
	e	A	36 17	D = 104.9°
	e	A	39 58	PV:1.4s 24.5nm
	ePP	A	40 30	LmH:30s 1.0/um LmV:20s 0.8/um
	LmH	C	18(19)	MPV=5.9 MLH=5.2 MLV=5.3
	LmV	C	26.5	e 40 38 e 40 41
26.	e(pP)	A	20 29 37	North Atlantic Ridge 17.42°N 46.44°W
				H = 20 19 43.7 h = normal MAG=4.7 (USCGS)
				D = 56.5°
				(pP)V:1.2s 12.7nm
				The first onset of P must be 7 sec. earlier.
27.	ei	A	08 01 57	Probably to Carpathian Mountains
	e	A	05 44	45.9°N 26 3/4° E
				H = 07 59 53 (BCIS) D = 11.1°

Day	Phase		h m s	Remarks
27.	e	A	12 06 56	
27.	e(Pg)	A	14 44 41	Explosion
	eSg	A	45 02	(D = ca. 1.6°)
27.	eP	A	20 53 54.5	Taiwan 24.31°N 122.15°E
				H = 20 41 33.0 h = 74 km MAG=4.9
				D = 83.7° Az = 323.1° (USCGS)
				PV:1.2s 17.9nm
				MPV=5.1
28.	eP	A	18 51 13	North Atlantic Ridge 24.87°N 45.91°W
	e	A	51 18	H = 18 42 13.7 h = normal MAG=5.1
	e	A	51 25	D = 50.7° Az = 43.9° (USCGS)
				PV:1.3s 27.8nm
				MPV=5.2
29.	ePKP	A	08 04 43	South of Fiji Islands 19.26°S 175.75°E
				H = 07 45 04.5 h = 42 km MAG=5.0
				D = 146.2° Az = 341.7° (USCGS)
29.	e(PS)	C	13 06.7	South Sandwich Islands Region
	e(SS)	C	13.2	60.8°S 23.1°W
	LmH	B	42.5	H = 12 37 22.2 h = 33 km MAG=5.3 (USCGS)
	LmV	B	42.5	D = 114.6°
				LmH:18s 0.8 µm LmV:18s 0.7 µm
				MLH=5.3 MLV=5.4
30.	ePKP	A	02 56 24	Loyalty Islands 22.02°S 170.12°E
	e	A	56 25.5	H = 02 36 45.1 h = 32 km MAG=4.4
	ePP	A	03 01 05	D = 146.8° Az = 334.8° (USCGS)
	LmH	C	54	PKPV:1.2s 17.9nm
	LmV	C	55	LmH:28s 0.2 µm LmV:28s 0.2 µm
				MLH=4.6 MLV=4.7
30.	+iP	A	06 11 46	Eastern Kazakhstan 49.8°N 78.1°E
	ePn	A	13 19	H = 06 03 57.9 h = 0 km MAG=5.5 (USCGS)
				D = 41.3°
				PV:0.8s 61.3nm PH:0.9s 27.9nm
				MPV=5.6 MPH=5.5

Day	Phase		h m s	Remarks
31.	LmH	C	00 24	Traces; West New Guinea Region (USCGS)
31.	e(P)	A	04 02 48	
31.	ePKP	A	10 34 24	South of Fiji Islands 19.74°S 177.33°E
	e	A	34 26	H = 10 14 43.8 h = 40 km MAG=5.4
	e	A	34 30	D = 147.1° Az = 343.0° (USCGS)
	e	A	34 47	PKPV:1.5s 40.5nm
31.	eP1	A	21 11 07.5	Sicily 37.85°N 14.55°E
	ePL	C	11 08	H = 21 08 07.2 h = normal MAG=4.8
	iP2	A	11 10.7	D = 13.0° Az = 351.6° (USCGS)
	eS	C	13 20	P1V:1.2s 15.3nm P2V:1.2s 53.6nm
	eSS	C	13 40	LmH:17s 10.6/um LmV:14.5s 8.4/um
	LmH	B	16.1	MLH=5.0
	LmV	B	16.9	Multiple P in our short-period record. Clear PL wave in the long-period record of type C.

Day	Phase		h m s	Remarks
1.	eiPg	A	10 35 38	Explosion
	eiSg	A	35 54	D = ca. 1.2°
1.	e(Pg)	A	12 48 30	Explosion
	e(Sg)	A	48 40	
1.	-iPKHKP	A	15 19 37.5	South of Fiji Islands 23.00°S 176.83°W
	ePKP2	A	19 45	H = 14 59 58.9 h = 140 km MAG=5.3
	e	A	20 07	D = 151.6° Az = 348.7° (USCGS)
				PKHKPV:1.0s 47.3nm PKP2V:1.0s 35.6nm
1.	eP	A	16 21 03	Kurile Islands 48.16°N 154.44°E
	LQ	C	46.0	H = 16 09 16.7 h = 47 km MAG=5.3
	LmH	C	55.0	D = 76.5° Az = 336.7° (USCGS)
	LmV	C	58.0	PV:1.4s 43.0nm
				LmH:21s 0.6 µm LmV:20s 0.8 µm
				MPV=5.4 MLH=4.9 MLV=5.1
1.	-eiP	A	16 42 44	Kurile Islands 48.27°N 154.36°E
	LQ	C	17 07	H = 16 30 57.1 h = 40 km MAG=5.5
	LmH	C	19.7	D = 76.4° Az = 336.6° (USCGS)
				PV:1.0s 42.7nm
				MPV=5.5
1.	ePP	A	19 16 38	West New Guinea Region 4.84°S 135.68°E
	eSS	C	32.9	H = 18 56 54.8 h = 14 km MAG=5.8 (USCGS)
	LmH	C	59.8	D = 114.8°
				LmH:22s 1.2 µm
				MLH=5.6
1.	eP	A	19 29 38.5	Near East Coast of Honshu/Japan
	epP	A	29 51.5	37.07°N 141.28°E
				H = 19 17 24.7 h = 72 km MAG=4.7
				D = 82.1° Az = 330.3° (USCGS); h = 52 km
				PV:1.3s 13.9nm pPV:1.1s 16.9nm
				MPV=4.9

Day	Phase		h m s	Remarks
2.	eiP1	A	03 10 53.5	Greenland Sea 73.22°N 7.66°E
	iP2	A	10 58.0	H = 03 05 51.0 h = normal MAG=4.3
				D = 22.7° Az = 173.5° (USCGS)
				P1V:1.4s 24.5nm P2V:1.5s 33.6nm
3.	-iPKHKP	A	07 51 55	New Hebrides Islands 18.67°S 169.01°E
	-iPKIKP	A	51 56.2	H = 07 32 50.1 h = 230 km MAG=5.3
	+iPKP	B	51 57	D = 143.4° Az = 335.8° (USCGS)
	ePSKS	C	08 05.0	PKIKPV:1.6s 116nm PKPV:5.5s 1.3 μm
	ePPS	C	07.8	LmH:43s 1.7 μm
	eSS	C	13 50	
	e(PSS)	C	14 50	
	e(PSPS)	C	15 16	
	LmH	C	38	
3.	eiSn	A	13 22 57.5	Yugoslavia 45.5°N 14.6°E
	eiSg	A	23 32.5	H = 13 20 33 (BCIS)
				D = 5.6°
3.	ePPS	C	23 06 12	South Sandwich Islands Region
	eSS	C	11.3	56.07°S 24.24°W
	LmH	C	26.5	H = 22 37 49.6 h = 155 km MAG=5.4 (USCGS)
				D = 107.2°
				LmH:48s 0.7 μm
4.	-eP	A	05 19 39	Taiwan 24.29°N 122.20°E
	LmH	C	06 01	H = 05 07 18.0 h = 76 km MAG=5.0
				D = 83.6° Az = 323.2° (USCGS)
				PV:1.2s 25.5nm
				LmH:25s 0.3 μm
				MPV=5.2 MLH=4.5
4.	+eiPKIKP	A	10 35 51	Fiji Islands 17.84°S 178.99°W
	+iPKHKP	A	35 53.3	H = 10 17 14.7 h = 573 km MAG=5.4
	eipPKP	B	38 00	D = 146.2° Az = 347.8° (USCGS)
	esPKP	B	39 00	PKIKPV:1.2s 51.0nm PKHKPV:1.2s 230nm
	ePSKS	C	49 30	
	ePPS	C	52 25	
	eSS	C	57 25	
	LmH	C	11 17.5	

Day	Phase		h m s	Remarks
4.	iPg	A	12 40 17	Explosion/GDR
	iSg	A	40 24	D = ca. 0.5°
4.	+iP	A	13 39 04.0	Near East Coast of Honshu/Japan
	+iP	B	39 04.4	37.35°N 141.65°E
	eipP	A	39 16.5	H = 13 26 47.7 h = 46 km MAG=5.7
	ePP	A	42 06	D = 82.0° Az = 330.5° (USCGS); h = 46 km
	eS	C	49 12	PV(A):1.7s 140nm PV(B):6.5s 1.1 μm
	ePS	C	50 08	SH(B):12s 1.0 μm
	eSS	C	54.5	LmH:18s 4.3 μm LmV:17s 4.0 μm
	LmH	B	14 13.9	MPV(A)=5.8 MPV(B)=6.1 MSH=5.9
	LmV	B	18.7	MLH=5.9 MLV=5.9
4.	+eP1	A	14 42 31.5	Hokkaido/Japan 43.54°N 144.09°E
	iP2	A	42 33.8	H = 14 30 37.5 h = 30 km MAG=5.8
	eipP	A	42 42	D = 77.5° Az = 331.2° (USCGS); h = 40 km
	i(sP)	A	42 49	P1V:1.5s 36.9nm P2V:1.5s 90.5nm
	iX	B	45 00	LmH:14.5s 44.5 μm LmV:19s 21.5 μm
	ePP	A	45 25	MP1V=5.3 MP2V=5.7 MLH=6.9 MLV=6.5
	iY	B	46 40	i 52 30.5 i 53 04 i 01 38
	ePPP	A	47 10	Multiple P in our short-period records,
	eS	B	52 20	the first onset much smaller than the
	iS	B	52 24	second one. X and Y are distinct phases
	i(SKS)	B	52 38	in the long-period records, about 25 sec.
	iSP	B	52 50	and 30 s earlier than expected for PP and
	eSS	C	57.2	PPP, respectively. LQ begins with periods
	eLQ	C	15 02.6	around 90 - 100 sec. Exceptionally clear
	LmH	B	17.1	normal dispersed surface wave train in
	LmV	B	20.9	the registrations of type C.
4.	ei	A	14 57 50	
4.	eP	A	14 57 57	Hokkaido/Japan 43.45°N 144.01°E
	e	A	58 16.5	H = 14 46 01.9 h = normal MAG=5.4
				D = 77.5° Az = 331.2° (USCGS)
				PV:1.8s 51.0nm
				MPV=5.4

Day	Phase		h m s	Remarks
4.	eP	A	16 13 21	Gulf of Alaska 59.41°N 145.04°W
	e(pP)	A	13 26	H = 16 02 19.4 h = normal MAG=4.7
				D = 68.7° Az = 15.7° (USCGS)
				PV:1.6s 26.5nm (pP)V:1.6s 37.9nm
				MPV=5.2
4.	eP	A	16 39 45.5	Peru - Equador Border Region
	eipP	A	40 11.5	2.76°S 77.68°W
	esP	A	40 22.5	H = 16 26 48.2 h = 99 km MAG=6.0
	iSKS	B	50 12	D = 91.7° Az = 39.6° (USCGS); h = 99 km
	iS	B	50 40	PV:2.2s 175nm pPV:1.6s 90.9nm
	e(SP)	C	52 06	SH:8s 1.7 μm SKSH:8s 1.7 μm
	iPS	B	52 18	PSH:11s 1.4 μm
	eSS	C	56.9	LmH:56s 3.1 μm LmV:50s 2.7 μm
	LmH	C	17 11.4	MPV=5.9 MSH=5.9
	LmV	C	11.4	
5.	eP	A	00 29 29	Greece 38.11°N 20.46°E
				H = 00 26 12.9 h = normal MAG=4.3
				D = 14.0° Az = 336.2° (USCGS)
6.	iP	A	04 16 30	Iceland Region 68.48°N 18.22°W
	e	A	16 38.5	H = 04 11 30.0 h = 36 km MAG=4.2 (USCGS)
	e	A	16 44	D = 23.0°
	LmH	B	26.5	
6.	-iP	A	05 54 57	Iceland Region 67.96°N 18.76°W
	e	A	55 03	H = 05 49 49.0 h = normal MAG=4.5 (USCGS)
	LmV	B	06 04.0	D = 22.9°
	LmH	B	04.6	PV:1.2s 20.4nm
				LmH:16s 0.6 μm LmV:(16s) 0.2 μm
				MPV=4.5 MLH=4.1 MLV=3.8
6.	eP	A	10 36 06	Greece - Albania Border Region
	e	A	36 16	39.25°N 20.70°E
	e	A	36 52.5	H = 10 33 01.2 h = 39 km MAG=4.4
	LmH	B	40.8	D = 13.1° Az = 333.6° (USCGS)
	LmV	B	41.6	LmH:13.5s 1.2 μm LmV:12s 0.6 μm
				MLH=4.1
				i 36 28 e 37 23 e 40 05

Day	Phase		h m s	Remarks
6.	ePKP	A	21 51 36	Fiji Islands 17.72°S 178.74°W
				H = 21 32 56.5 h = 549 km MAG=4.5
				D = 146.1° Az = 348.2° (USCGS)
7.	+ePKP	A	04 08 48.5	Samoa Islands 14.89°S 172.98°W
	epPKP	A	09 02.5	H = 03 49 17.4 h = 43 km MAG=5.6
	eSS	C	31 40	D = 144.1° Az = 355.0° (USCGS); h = 50 km
	LmV	B	05 16.3	PKPV:1.5s 80.5nm pPKPV:1.5s 60.4nm
	LmH	B	16.5	LmH:18s 0.8 μm LmV:18s 1.1 μm
				MLH=5.5 MLV=5.7
7.	ePb	A	05 47 17	Etruscian Appenin 44.1°N 11.9°E
	ePg	A	47 39	H = 05 45 24 (BCIS)
	eLg2(3.28)	A	49 08.5	D = 6.6°
	e	A	49 13	
	e	A	49 28	
7.	eiPg	A	13 08 21.5	Explosion
	eiSg	A	08 35.5	D = ca. 1.1°
7.	e(Sg)	A	13 10 39	Near earthquake?
	e	A	10 49.5	
7.	epP	A	20 05 53	Afghanistan - USSR Border Region
	ei	A	05 56.5	36.98°N 71.69°E
	esP	A	06 10	H = 19 57 26.1 h = 136 km MAG=5.3
				D = 44.3° Az = 307.8° (USCGS); h = 140 km
8.	eP	A	02 52 42.5	Kenai Peninsula/Alaska 58.96°N 150.19°W
				H = 02 41 34.1 h = 34 km MAG=4.4
				D = 69.8° Az = 12.2° (USCGS)
8.	eP	A	03 23 10	Caribbean Sea 16.83°N 85.89°W
	e	A	23 25	H = 03 10 53.3 h = 28 km MAG=5.4
	e	A	23(33)	D = 81.8° Az = 39.6° (USCGS)
	eSS	C	38.7	PV:1.6s 30.3nm
	LmH	C	58.5	LmH:19s 0.5 μm LmV:20s 0.5 μm
	LmV	C	58.5	MPV=5.2 MLH=4.9 MLV=4.9

Day	Phase		h m s	Remarks
8.	ePP	A	06 27 03	Aroe Islands Region 5.26°S 133.97°E
	e	A	27 27	H = 06 07 21.4 h = normal MAG=5.9 (USCGS)
	e	A	27 35	D = 114.2°
	ePPS	C	37.6	LmH:22s 1.0 µm LmV:24s 0.7 µm
	eSS	C	43.0	MLH=5.4 MLV=5.2
	LmH	C	07 10.0	
	LmV	C	20.0	
8.	eSS	C	07 51	Southern Pacific Ocean 36.16°S 100.73°W
	eSSS	C	56.0	H = 07 11 59.3 h = normal MAG=4.6 (USCGS)
	LmH	C	08 16.5	D = 130.6°
	LmV	C	16.5	LmH:25s 0.4 µm LmV:25s 0.4 µm
				MLH=5.0 MLV=5.0
8.	eiPg	A	09 01 44.5	Explosion
	eiSg	A	01 58	D = ca. 1.0°
8.	eP	A	15 12 18	Nevada; Underground explosion
				H = 15 00 00 (UPP)
				D = 81.1°
				Obscured by short-period microseisms.
8.	eP	A	17 21 24.5	Rat Islands/Aleutian 51.12°N 178.52°E
	e	A	21 58	H = 17 09 27.1 h = 29 km MAG=5.3
	eS	C	31 18	D = 78.0° Az = 351.5° (USCGS)
	e	C	31 25	PV:1.0s 16.6nm
	eSS	C	36.5	LmH:17s 1.8 µm LmV:17s 1.6 µm
	LmH	B	18 03.1	MPV=5.1 MLH=5.5 MLV=5.4
	LmV	B	03.8	
8.	-eP	A	17 34 32.5	Rat Islands/Aleutian 51.07°N 178.42°E
	eS	C	44.5	H = 17 22 32.1 h = 10 km MAG=5.2
				D = 78.0° Az = 351.5° (USCGS)
				PV:(1.0s) 14.2nm
				MPV=5.1
8.	eiPg	A	17 43 46	North Italy 45.8°N 10.8°E
	eSn	A	44 21	H = 17 42 14 (BCIS)
	eSg	A	44 46.5	D = 4.9°
	eiLg2	A	44 56	

Day	Phase		h m s	Remarks
9.	eiPKIKP	A	02 36 14	Banda Sea 7.18°S 123.63°E
	eiPP	A	37 00	H = 02 18 45.5 h = 560 km MAG=5.8
	eSP	C	45.2	D = 109.4° Az = 321.3° (USCGS)
	eSPP	C	46.4	PKIKPV:1.2s 15.3nm PKKPV:1.0s 9.5nm
	ePKKP	A	47 20	e 39 18 e 40 19
	LmV	C	03 23.5	
	LmH	C	(25)	
9.	ePKP	A	02 40 20.5	Fiji Islands 15.78°S 178.41°E
	e	A	40 27	H = 02 20 51.0 h = 84 km MAG=4.8
				D = 143.6° Az = 345.8° (USCGS)
9.	eP	A	07 58 53	Alaska Peninsula 54.80°N 162.18°W
	e(pP)	A	59 07	H = 07 47 16.0 h = 40 km MAG=4.7
	ei	A	59 35	D = 74.8° Az = 4.1° (USCGS)
				PV:1.4s 15.3nm
				MPV=4.9
9.	eP	A	14 53 06	Dodecanese Islands 35.51°N 27.76°E
	e	A	53 17.5	H = 14 48 44.2 h = 47 km MAG=5.7
	e	A	53 31	D = 19.1° Az = 327.3° (USCGS)
	e	A	53 42	PV:1.2s 12.8nm
	LmH	B	15 03	
	LmV	B	03	
9.	+eiP	A	18 31 53.5	Near East Coast of Honshu/Japan
	ePcP	A	31 57.5	35.47°N 140.09°E
	e(pP)	A	31 07	H = 18 19 35.0 h = 68 km MAG=5.3
	e	A	31 13	D = 83.0° Az = 329.9° (USCGS);
	e	A	31 24	(h = 50 km)
				PV:1.2s 17.9nm
				MPV=5.1
9.	ePKP	A	20 31 02	Tonga Islands 15.08°S 173.19°W
				H = 20 11 32.1 h = 65 km MAG=4.9
				D = 144.3° Az = 354.8° (USCGS)
10.	LmH	C	03 14.5	Iran 35 3/4°N 53 3/4°E
				H = 02 50 52 (BCIS)
				D = ca. 34°

Day	Phase		h m s	Remarks
10.	LmH	B	04 02	LmH:17s 0.3 µm LmV:17s 0.4 µm
	LmV	B	02	
10.	LmH	B	04 17.7	LmH:16s 0.6 µm LmV:18s 0.6 µm
	LmV	B	17.7	
10.	eP	A	04 26 17	Probably North Atlantic Ridge
	e	A	26 25	LmH:18s 0.7 µm LmV:17s 0.7 µm
	LmH	B	36	(MLH=4.3 MLV=4.4)
	LmV	B	36	
10.	eP	A	04 45 55	<u>North Atlantic Ridge</u> 45.07°N 28.08°W
	eS	C	50.6	H = 04 40 15.0 h = normal MAG=4.8
	eLR	C	52.8	D = 27.0° Az = 63.7° (USCGS)
	LmH	B	56.1	PV:1.8s 30.6nm
	LmV	B	56.5	LmH:16.5s 2.3 µm LmV:16s 2.6 µm
				MPV=4.7 MLH=4.8 MLV=5.0
10.	eP	A	05 17 42.5	<u>North Atlantic Ridge</u> 45.04°N 28.10°W
	LmH	B	28	H = 05 12 01.4 h = normal MAG=4.4 (USCGS)
	LmV	B	28	D = 27.0°
				LmH:16s 0.7 µm LmV:16s 0.7 µm
				MLH=4.4 MLV=4.4
10.	eP	A	05 56 09	<u>North Atlantic Ridge</u> 44.77°N 28.12°W
	e	A	56 18	H = 05 50 27.6 h = normal MAG=4.7
	LmH	C	06 06.3	D = 27.1° Az = 63.2° (USCGS)
	LmV	B	06.5	PV:1.5s 16.8nm
				LmH:18s 0.5 µm LmV:16s 0.5 µm
				MPV=4.6 MLH=4.1 MLV=4.3
10.	eP	A	06 14 43	<u>India - East India Border Region</u>
				25.49°N 91.71°E
				H = 06 04 09.4 h = 59 km MAG=4.4
				D = 64.6° Az = 316.0° (USCGS)
				PV:1.0s 7.1nm
				MPV=4.8

Day	Phase		h m s	Remarks
10.	eiSg	A	10 01 17.5	Explosion/ČSSR 48.73°N 14.50°E
				Yield: 12.6 t (PRU)
				D = 2.7°
10.	ePKIKP	A	13 29 52.5	Fiji Islands 17.99°S 178.47°W
	+iPKHKP	A	29 54.8	H = 13 11 18.1 h = 592 km MAG=5.0
	-ePKP2	A	29 57.5	D = 146.4° Az = 348.4° (USCGS)
	e	A	30 08.5	PKIKPV:0.8s 14.2nm PKHKPV:1.2s 51.0nm
				PKP2V:1.0s 26.1nm
10.	eP	A	18 40 39	Central Alaska 62.33°N 151.43°W
				H = 18 29 57.3 h = 90 km MAG=4.9
				D = 66.6° Az = 11.7° (USCGS)
				PV:0.8s 16.5nm
				MPV=5.0
10.	eP1	A	18 50 24	Chagos Archipelago Region 5.98°S 71.38°E
	iP2	A	50 29.8	H = 18 38 37.6 h = 32 km MAG=5.4
	iP3	A	50 33.5	D = 76.2° Az = 325.5° (USCGS)
	eS	B	19 00 08	P1V:1.2s 20.4nm P2V:1.2s 38.3nm
	eSS	C	04 55	LmH(C):40s 2.4 µm LmH(B):18s 0.4 µm
	eSSS	C	08.5	MP1V=5.1 MP2V=5.4 MLH(C)=5.2 MLH(B)=4.8
	eSa(4.45)	C	10 28	e 00 20 e 05 00
	LmH	C	14.2	Multiple P in our short-period records.
	LmH	B	24.3	If we interpret P2 as pP and P3 as sP
				then the focal depth would be about 24 km.
10.	eP	A	21 00 14.5	Kurile Islands 45.27°N 149.77°E
				H = 20 48 25.7 h = 95 km MAG=4.8
				D = 77.8° Az = 334.2° (USCGS)
				PV:0.9s 14.2nm
				MPV=5.1
11.	eP	A	00 33 33	Bonin Islands 28.41°N 138.57°E
				H = 00 21 37.6 h = 529 km MAG=4.3
				D = 88.4° Az = 329.4° (USCGS)

Day	Phase		h m s	Remarks
11.	eP	A	02 37 47	Uganda 2.05°N 31.53°E
	LmH	C	59	H = 02 28 45.6 h = normal MAG=5.1
	LmV	C	03 02	D = 51.3° Az = 343.9° (USCGS)
				PV:1.4s 30.7nm
				MPV=5.2
11.	ePg	A	11 19 59	Explosion/ČSSR 50.35°N 14.03°E
	eSg	A	20 19.5	Yield: 30.8 t (PRU)
				D = 1.6°
11.	-eP	A	12 07 41	Chagos Archipelago Region 6.0°S 71.3°E
	epP	A	07 47.5	H = 11 55 55.6 h = 37 km MAG=5.6
	esP	A	07 51	D = 76.2° Az = 325.5° (USCGS); h = 26 km
	eS	B	17 22	PV:2.0s 66.3nm pPV:1.6s 60.5nm
	eSS	B	22 11	SH:11s 0.7 μm
	eSa	C	27 50	LmH:44s 1.9 μm
	LmH	C	31.5	MPV=5.4 MSH=5.7 MLH=5.1
11.	-eP	A	12 26 43.5	Chagos Archipelago Region 6.01°S 71.29°E
	epP	A	26 51	H = 12 14 57.3 h = 34 km MAG=5.7
	ePP	C	29 33	D = 76.1° Az = 325.5° (USCGS); h = 28 km
	eS	B	36 28	PV:1.5s 73.8nm SH:12s 2.0 μm
	eSS	C	41 10	LmH:16s 1.4 μm LmV:16s 1.9 μm
	eSa	C	46 38	MPV=5.6 MSH=6.1 MLH=5.4 MLV=5.6
	LmH	B	13 11.5	e 29 (00)
	LmV	B	11.7	
11.	eP	A	15 16 56.5	Chagos Archipelago Region 6.11°S 71.35°E
	e(sP)	A	17 06	H = 15 05 10.3 h = normal MAG=5.3
				D = 76.3° Az = 325.5° (USCGS);
				(h = 28 km)
11.	+eP	A	18 11 48.5	Chagos Archipelago Region 6.12°S 71.35°E
	eS	B	21 28	H = 18 00 00.7 h = normal MAG=5.7
	eSS	C	26 16	D = 76.3° Az = 325.5° (USCGS)
	eSa	C	31 50	PV:2.4s 136nm SH:14s 1.1 μm
	LmH	C	35.5	LmH:44s 1.3 μm
				MPV=5.7 MSH=5.8

Day	Phase		h m s	Remarks
11.	+eP	A	20 29 59.5	Chagos Archipelago Region 5.97°S 71.27°E
				H = 20 18 11.1 h = 20 km MAG=5.4
				D = 76.1° Az = 325.5° (USCGS)
				PV:2.0s 59.3nm
				MPV=5.4
11.	e	A	23 40 14	
12.	e(P)	A	00 03 02	
12.	eP	A	02 39 13	Kurile Islands 44.80°N 149.78°E
	e	A	40 24	H = 02 27 16.6 h = 41 km MAG=5.5
				D = 78.2° Az = 334.3° (USCGS)
				PV:1.2s 40.8nm
				MPV=5.4
12.	ePKIKP	A	10 56 29	Tonga Islands 17.20°S 171.99°W
	+iPKP	B	56 30	H = 10 36 52.0 h = 34 km MAG=5.6
	+iPKHKP	A	56 31.5	D = 146.5° Az = 355.8° (USCGS); h = 39 km
	iPKP2	A	56 34.0	PKHKPV:1.4s 255nm PKPV:7s 2.8 μm
	-eipPKP	B	56 41	pPKPV:6s 5.2 μm
	-ipPKP	A	56 41.3	LmH:20s 2.2 μm LmV:19s 2.2 μm
	ePP	A	11 00 08	MLH=5.9 MLV=5.9
	eSKKS	C	06 40	i 56 48.8 e 59 52 e 04 (00)
	ePSKS	C	09.9	
	ePPS	C	12.6	
	eSS	C	19.0	
	eSSS	C	25.0	
	LmH	B	12 02.0	
	LmV	B	05.3	
12.	ePKIKP	A	17 44 14.5	Loyalty Islands 22.82°S 170.74°E
	ePKHKP	A	44 18.5	H = 17 24 31.9 h = 26 km MAG=5.1
	LmH	C	18 43	D = 147.8° Az = 334.8° (USCGS)
12.	e(PKHKP)	A	22 19 22	South of Fiji Islands 25.24°S 177.20°W
	e	A	19 24	H = 21 59 50.7 h = 225 km MAG=4.9
	e(PKP2)	A	19 34	D = 153.7° Az = 347.3° (USCGS)

Day	Phase		h m s	Remarks
13.	eP	A	06 54 40	Turkey 37.87°N 29.13°E
	e	A	55 39	H = 06 50 33.8 h = 46 km MAG=4.6
	LmH	B	07 00.7	D = 17.8° Az = 321.3° (USCGS)
	LmV	B	02.3	PV:1.8s 35.7nm
				LmH:15.5s 2.1 /um LmV:11s 1.5 /um
				MLH=4.5 MLV=4.7
14.	e(P)	A	00 15 10	India - East Pakistan Border Region
	e	A	15 29.5	23.96°N 91.47°E
	LmH	C	40.3	H = 00 04 17.8 h = normal MAG=5.1 (USCGS)
	LmV	C	46.5	D = 65.7°
				LmH:32s 0.5 /um LmV:24s 0.4 /um
				MLH=4.5 MLV=4.6
14.	ePKIKP	A	05 47 07.5	East New Guinea 5.43°S 147.07°E
	ePP	A	48 37.5	H = 05 28 36.9 h = 201 km MAG=5.8
	LmH	C	06 28	D = 121.6° Az = 328.4° (USCGS)
	LmV	C	34.3	PKIKPV:1.4s 30.7nm
				LmH:39s 1.8 /um LmH:25s 0.6 /um
14.	e(Sg)	A	10 34 25	Explosion?
	e	A	34 28	(Sg)V:1.0s 14.2nm
14.	eiPKP	A	20 04 00	Tonga Islands 18.02°S 175.24°W
	epPKP	A	05 02	H = 19 44 45.6 h = 255 km MAG=4.4
				D = 147.0° Az = 352.0° (USCGS)
				PKPV:1.4s 28.6nm
15.	LmH	C	18 15.0	Traces
15.	eP	A	21 46 11	Near Coast of Central Chile
	ePKIKP	A	50 19	28.75°S 71.16°W
	ePP	B	50 43	H = 21 31 51.5 h = 15 km MAG=6.2
	eiSKS	C	56 51	D = 107.4° Az = 41.4° (USCGS)
	eSP	C	59 51	PKIKPV:1.7s 39.5nm PPV:7s 1.4 /um
	eiPS	C	59 58	PKKPV:1.6s 22.8nm
	eSPP	C	22 00 54	LmH:18.5s 7.4 /um LmV:19.5s 10.7 /um
	ePKKP	A	01 49.5	MPPV=6.7 MLH=6.3 MLV=6.4
	eSS	C	05 30	e 50 15 e 50 18.5 e 50 46.5 e 01 53
	eLQ	C	15.7	e 02 07.5

Day	Phase		h m s	Remarks
cont,				
15.	LmV	B	22 35.7	
	LmH	B	36.1	
16.	eP	A	02 29 54	Ethiopia 15.09°N 39.82°E H = 02 22 03.1 h = normal MAG=5.1 D = 42.1° Az = 333.4° (USCGS)
16.	eiPg	A	10 38 07	Explosion
	eiSg	A	38 22.5	D = ca. 1.2°
16.	e(P)	A	12 04 49.5	(P)V:1.3s 38.9nm
16.	e	A	12 10 35	Near earthquake?
	e(Sg)	A	10 40.5	KHC gives: eiPg 09 56.2 eiSg 10 25.6
	e	A	10 41.5	D = 2.3°
16.	eiPKHKP	A	16 49 46	Fiji Islands 20.72°S 178.75°W H = 16 31 03.6 h = 591 km MAG=4.5 D = 149.0° Az = 347.1° (USCGS) PKHKPV:1.1s 12.0nm
16.	eiSg	A	19 52 44.5	Explosion? KHC gives: eiSg 52 30.5 D = 1.2°
16.	eiSg	A	20 05 42.5	Explosion? KHC gives: eiPg 05 10.5 eiSg 05 27.5 D = 1.3°
17.	eP1	A	05 07 25.5	North Atlantic Ridge 28.45°N 43.83°W
	eP2	A	07 30.5	H = 04 58 56.8 h = normal MAG=5.2
	eiS	B	14 17	D = 46.9° Az = 45.9° (USCGS)
	iSS	C	17 56	P1V:1.8s 30.6nm P2V:1.8s 51.0nm
	iLQ	C	18.5	SH:16.5s 3.2 /um
	eLR	C	20.5	LmH:16.5s 5.5 /um LmV:17s 7.6 /um
	LmH	B	24.3	MP1V=5.1 MP2V=5.3 MSH=6.1 MLH=5.6
	LmV	B	24.4	MLV=5.8 e 07 34.5 e 08 34 e 14 24

Day	Phase		h m s	Remarks
17.	ePKIKP	A	09 38 16	Solomon Islands 6.31°S 154.78°E
	e	A	38 33	H = 09 19 24.2 h = 91 km MAG=5.0
	LmH	C	10 25	D = 126.2° Az = 331.8° (USCGS)
				LmH:23s 0.6 μm
				MLH=5.2
17.	ePKP	A	10 29 05.5	New Hebrides Islands 13.72°S 167.28°E
				H = 10 10 04.7 h = 215 km MAG=4.5
				D = 138.2° Az = 336.8° (USCGS)
17.	e	A	12 47 37.5	Explosion?
17.	LmH	C	14 52.6	Taiwan Region 24.00°N 122.26°E
				H = 13 58 35.3 h = 36 km MAG=5.1 (USCGS)
				D = 83.8°
				LmH:16s 1.0 μm
				MLH=5.3
18.	eP1	A	02 35 39	Crete 35.20°N 23.11°E
	eP2	A	35 49	H = 02 31 35.4 h = 43 km MAG=4.5
	e	A	36 06.5	D = 17.5° Az = 335.1° (USCGS)
	e	A	37 04	LmH:27s 4.9 μm
	eLQ	C	39.0	MLH=4.6
	LmH	C	41.5	Successive P-onsets with increasing amplitudes.
18.	e	A	03 20 54.5	
18.	ei(P2)	A	03 37 43.5	Crete 35.31°N 23.02°E
	e	A	37 59	H = 03 33 34.3 h = normal MAG=4.3
				D = 17.4° Az = 335.1° (USCGS)
				The first onset of P must be about 7.5 s earlier than (P2).
18.	e	A	09 36 06	Explosion/ČSSR
	e(Sg)	A	36 22	KHC gives: D = 129 km Yield: 9.3 t
18.	e	A	11 04 40	(Sg)V:0.8s 16.5nm
	e(Sg)	A	04 45.5	

Day	Phase		h m s	Remarks
18.	e(P)	A	12 00 18	(P)V:1.0s 14.2nm
18.	eP	A	12 29 31.5	El Salvador 13.37°N 89.07°W
	LmH	C	13 09.5	H = 12 16 55.4 h = 78 km MAG=5.1
				D = 86.4° Az = 38.8° (USCGS)
18.	e(P)	A	16 56 49.5	
18.	ePKHKP	A	22 00 47	South of Fiji Islands 22.10°S 179.55°W
	e	A	00 55.5	H = 21 41 58.3 h = 553 km MAG=4.3
				D = 150.2° Az = 345.6° (USCGS)
				PKHKPV:0.6s 16.7nm
19.	e(S)	A	01 35 23	Yugoslavia 41 1/4°N 20 3/4°E
	e	A	36 23	H = 01 30.5 (BCIS); M=3.7 (SKOPJE)
	e	A	36 45.5	D = 11.3°
19.	e(Sg)	A	10 10 53.5	Probably near earthquake.
	e	A	11 04	KHC gives: ePg 09 30.5 eiSg 10 14.5
	e	A	11 15	D = 3.4°
19.	+iP	A	12 19 19.4	Near East Coast of Honshu/Japan
	epP	A	19 31.5	36.40°N 141.10°E
	esP	A	19 37	H = 12 06 59.5 h = 41 km MAG=5.5
	ePP	B	22 30	D = 82.6° Az = 330.3° (USCGS); h = 47 km
	eS	C	29 32	PV:1.3s 122nm pPV:1.4s 116nm
	eSS	C	34(40)	sPV:1.2s 56.1nm PPV:5.5s 0.6 /um
	LmH	B	58.8	LmH:16s 2.1 /um LmV:16s 2.6 /um
	LmV	B	13 01.3	MPV=5.9 MPPV=6.2 MLH=5.6 MLV=5.7
				ei 22 12 e 35 (24)
19.	e	A	14 34 33	Etruscan Appenin 44.3°N 10.4°E
	e(Sg)	A	35 07	H = 14 31 31 (BCIS)
				D = 7.4°
19.	ePKHKP	A	15 59 54	Fiji Islands 21.5°S 179.3°W
				H = 15 41 09.6 h = 588 km MAG=4.4
				D = 149.6° Az = 346.3° (USCGS)
				PKHKPV:0.9s 9.4nm

Day	Phase	h m s	Remarks
19.	ePKP B	17 48 58	Loyalty Islands 22.60°S 170.94°E
	ePKP A	49 03	H = 17 29 20.9 h = normal MAG=5.2
	-iPKP B	49 04	D = 147.7° Az = 335.1° (USCGS)
	ePP C	52 30	PKPV(B):12s 8.8 μm
	eSKKS C	59.4	LmH:21.5s 6.5 μm LmV:22s 7.2 μm
	iPSKS C	18 02 46	MLH=6.3 MLV=6.4
	eiPPS C	05.0	e 49 10
	iX C	07 46	
	eSS C	11.6	
	eiSSS C	17 20	
	LmH B	54.5	
	LmV B	55.4	
19.	ei A	18 01 40.	
19.	ePKP A	18 09 02	Loyalty Islands 22.59°S 170.83°E
	ei A	09 05.5	H = 17 49 20.3 h = 38 km MAG=4.5
			D = 147.6° Az = 335.0° (USCGS)
19.	ePKP A	18 52 04	Loyalty Islands 22.60°S 171.17°E
			H = 18 32 18.2 h = 21 km MAG=4.7
			D = 147.8° Az = 335.4° (USCGS)
19.	ePKP A	19 25 40	Loyalty Islands 22.74°S 170.98°E
			H = 19 05 57.1 h = normal MAG=4.4
			D = 147.8° Az = 335.1° (USCGS)
			PKPV:1.1s 7.2nm
19.	ePKIKP A	20 39(18)	Loyalty Islands 22.64°S 170.72°E
	ePKHKP A	39 21.5	H = 20 19 35.0 h = normal MAG=4.3
	ePKP2 A	39 26	D = 147.6° Az = 334.9° (USCGS)
19.	ePKIKP A	22 02 48	Loyalty Islands 22.71°S 170.78°E
	e A	02 49.5	H = 21 43 07.0 h = 31 km MAG=4.6
			D = 147.7° Az = 334.9° (USCGS)
			PKIKPV:0.9s 9.4nm
19.	ePKP A	23 12 42	Loyalty Islands 22.84°S 170.17°E
			H = 22 53 07.3 h = 32 km
			D = 147.6° Az = 334.3° (USCGS)

Day	Phase		h m s	Remarks
20.	eSg	A	01 04 00	GFR 49.5°N 8.5°E H = 01 02 46 (BCIS) D = 2.4°
20.	ePKP	A	02 31 06	Tonga Islands 15.29°S 174.33°W
	LmH	C	03 31	H = 02 11 25.3 h = normal MAG=4.8 D = 144.4° Az = 353.5° (USCGS)
20.	iP	A	10 26 28.5	Sea of Okhotsk 51.16°N 151.27°E H = 10 15 43.0 h = 426 km MAG=4.7 D = 72.9° Az = 334.5° (USCGS) PV:1.0s 30.8nm MPV=4.9
20.	+eiP	A	11 01 08.5	South of Honshu/Japan 32.03°N 140.87°E
	eipP	A	01 26	H = 10 48 31.8 h = 65 km MAG=5.0
	ePP	A	04 29.5	D = 86.3° Az = 330.4° (USCGS); h = 68 km PV:1.4s 24.5nm pPV:1.1s 24.1nm MPV=5.2
20.	ePKP	A	13 25 23	Loyalty Islands 20.54°S 168.79°E H = 13 05 47.1 h = normal MAG=4.4 D = 145.0° Az = 334.5° (USCGS) PKPV:1.3s 13.9nm
21.	iP	A	09 02 59.0	North of Ascension Islands 0.10°N 17.09°W
	e	A	03 18	H = 08 53 22.4 h = normal MAG=4.8 D = 56.0° Az = 21.7° (USCGS) PV:1.2s 20.4nm MPV=5.1
21.	iPg	A	10 37 38.7	Explosion
	iSg	A	37 54.5	D = ca. 1.2°
21.	e(P)	A	14 37 50	(P)V:1.1s 9.6nm
21.	eSg	A	16 55 48	Probably explosion. KHC gives: eiPg 55 16 iSg 55 34.2 D = 1.4° SgV:0.8s 16.5nm

Day	Phase		h m s	Remarks
21.	eP	A	17 07 20	Jan Mayen Region 72.8°N 8.5°E
	iPL	C	07 23	H = 17 02 20 (BCIS)
	iS	C	11 26.5	D = 22.3°
	eLR	C	12.8	PV(A):1.8s 97.0nm PV(B):(10s) 2.0/um
	LmV	B	16.5	SH(B):(13s) 3.6/um
	LmH	B	16.6	LmH:17s 5.2/um LmV:18s 6.3/um
				MPV(A)=5.0 MPV(B)=5.6 MSH=5.5 MLH=5.1
				MLV=5.4
				Multiple P.
21.	eP	A	21 55 51.5	North Atlantic Ridge 48.15°N 27.83°W
	e	A	55 55	H = 21 50 24.3 h = normal MAG=5.0
	e	A	56 10.5	D = 25.6° Az = 69.5° (USCGS)
	LmH	B	22 04.9	LmH:14s 0.7/um LmV:9s 0.6/um
	LmV	B	06.9	MLH=4.4 MLV=4.6
22.	eP	A	06 02 59.5	Kurile Islands (UPP)
				PV:1.6s 31.6nm
				MPV=5.2
22.	e(Sg)	A	10 46 39	Explosion?
				KHC gives: ePg 46 22.5 eiSg 46 41.2
				D = 1.4°
22.	ePKP	A	15 39 07	Loyalty Islands 22.71°S 170.91°E
	+iPKP	B	39 09	H = 15 19 26.8 h = 42 km MAG=5.2
	ei	A	39 13	D = 147.8° Az = 335.0° (USCGS);
	e(pPKP)	B	39 42	(h = 121 km)
	LmH	C	16 32	PKPV(B):7s 3.6/um (pPKP)V:6.5s 2.8/um
	LmV	C	34.5	LmH:40s 1.9/um LmV:35s 1.4/um
				MLH=5.5 MLV=5.5
22.	e(P)	A	15 53 30	(P)V:0.9s 9.4nm (pP)V:1.2s 12.8nm
	e(pP)	A	55 22	(h = ca. 450 km)
22.	ePKP	A	17 00 17	Loyalty Islands 23.01°S 170.98°E
				H = 16 40 33.0 h = 22 km MAG=4.7
				D = 148.1° Az = 334.9° (USCGS)
22.	e	A	21 57 24	

Day	Phase		h m s	Remarks
23.	eP	A	08 06 37.5	
23.	-eiP1	A	08 44 37	Eastern Gulf of Aden 14.53°N 52.07°E
	-iP2	A	44 40.5	H = 08 35 49.5 h = 3 km MAG=5.8
	-iP	B	44 44.3	D = 48.6° Az = 326.6° (USCGS)
	eiP3	A	44 46.8	P2V:2.0s 132nm P3V:1.8s 428nm
	iP4	A	44 57.8	PV(B):7s 5.5 /um SH(B):10.5s 18.8 /um
	iPP	B	46 32	LmH:18.5s 44.2 /um LmV:16s 37.3 /um
	iX	B	46 41	MP2V=5.7 MP3V=6.2 MPV(B)=6.8
	iS	C	51 44	MSH(B)=7.1 MLH=6.5 MLV=6.5
	eiSS	B	55 15	Exceptionally clear multiple P with
	LmH	B	09 07	successively increasing amplitudes.
	LmV	B	11.5	
23.	eP	A	10 14 42.5	PV:1.2s 10.2s
23.	eSg	A	12 02 13	Explosion?
23.	iPg	A	13 07 59.3	Explosion
	eiSg	A	08 13	D = ca. 1.0°
23.	ePn	A	13 19 51.5	South Bavaria/GFR 47.9°N 11.0°E
	e	A	19 53.5	H = 13 19 04 (BCIS)
	ePb	A	19 56.5	D = 2.8°
	eSn	A	20 21	
	eSg	A	20 35	
23.	eSg	A	13 27 04.5	Near earthquake?
				KHC gives: eiSg 26 27.5 D = 1.9°
23.	-eiP1	A	13 48 11.5	North of Svalbard 80.19°N 0.99°W
	-iP2	A	48 19.0	H = 13 42 01.6 h = 10 km MAG=5.8
	iS	C	53 00	D = 30.0° Az = 163.8° (USCGS)
	eiS	B	53 08	P1V:1.4s 172nm P2V:1.8s 449nm
	iLR	C	56 06	SH(B):14.5s 8.7 /um
	LmH	B	14 00.6	LmH:16.5s 15.5 /um LmV:17s 21.7nm
	LmV	B	00.6	MP1V=5.7 MP2V=6.0 MSH=6.3 MLH=5.7
				MLV=6.0
				Multiple P.

Day	Phase		h m s	Remarks
23.	iPg	A	14 20 33.8	Explosion
	eSg	A	20 49	D = ca. 1.1°
	iSg	A	20 49.8	
23.	e(Sg)	A	14 35 07.5	Explosion?
	LmH	A	35 20	
23.	ei(Sg)	A	14 38 30	Explosion?
	e	A	38 39	
	LmH	A	38 45	
23.	e	A	14 48 26	
	e(Sg)	A	48 29	
	e	A	48 30.5	
24.	e	A	00 21 55	
	e	A	22 03.5	
24.	-eiPKP	A	06 01 02.5	Fiji Islands 16.39°S 177.95°W
	i(PKHKP)	A	01 04.0	H = 05 42 14.0 h = 428 km MAG=5.4
	epPKP	A	02 48	D = 145.0° Az = 349.4° (USCGS); h = 440 km
				PKPV:0.8s 23.9nm (PKHKP)V:1.3s 116nm
24.	iPg	A	07 59 27	Explosion
	eiSg	A	59 40.5	D = ca. 1.0°
24.	e(P)	A	14 33 46	(P)V:1.1s 12.0nm
24.	iPg	A	15 05 20.5	Explosion
	iSg	A	05 36.0	D = ca. 1.3°
24.	LmH	C	20 16	LmH:17s 0.2 μm
24.	ePKP	A	22 25 10	Tonga Islands (UPP)
25.	e	A	11 54 57.5	Explosion?
	e(Sg)	A	55 00.5	KHC gives: eiPg 53 36 eiSg 53 58
				D = 1.6°

Day	Phase		h m s	Remarks
25.	eSg	A	12 03 04.5	Explosion?
	e	A	03 30.5	
25.	ePKP	A	13 18 27	Loyalty Islands 22.64°S 170.68°E
				H = 12 58 48.1 h = 64 km MAG=4.8
				D = 147.6° Az = 334.9° (USCGS)
				PKPV:1.2s 12.8nm
26.	+iP	B	00 20 40	Ryukyu Islands 28.62°N 129.99°E
	eS	C	31 02	H = 00 08 09.8 h = normal MAG=5.7
	eSS	C	36.2	D = 84.2° Az = 325.7° (USCGS)
	LmV	B	01 02.6	PV(B):5s 0.9 µm PV(A):1.6s 90.9nm
	LmH	B	02.7	LmH:14.5s 8.2 µm LmV:15.5s 10.8 µm
				MPV(A)=5.7 MPV(B)=6.2 MLH=6.2 MLV=6.4
26.	eP	A	03 07 51	Java 8.09°S 112.94°E
	LmH	C	50	H = 02 53 57.8 h = 80 km MAG=5.7
				D = 103.4° Az = 320.1° (USCGS)
				PV:0.8s 9.4nm
				MPV=6.3
26.	eP	A	03 27 59.5	Greece - Albania Border Region
	e	A	28 06	39.42°N 20.44°E
	eLg1	A	31 36	H = 03 24 56.7 h = 44 km MAG=4.6
	eLg2	A	32 03	D = 12.8° Az = 333.9° (USCGS)
	eRg	A	32 33.5	LmH:9.0s 0.7 µm LmV:11s 0.7 µm
	LmV	B	34.2	MLH=4.1
	LmH	B	34.3	e 31 53 e 32 30.5
26.	eP1	A	08 22 31.5	Kodiak Islands 56.63°N 152.18°W
	eP2	A	22 33	H = 08 11 06.3 h = 28 km MAG=4.9
	ipP	A	22 39.5	D = 72.3° Az = 10.8° (USCGS); h = 30 km
	eX	A	23 14	P2V:1.0s 52.2nm
	eS	C	31 58	LmH:18s 0.7 µm LmV:16s 0.8 µm
	eSS	C	36.8	MPV=5.6 MLH=5.0 MLV=5.1
	LmH	B	58.5	
	LmV	B	09 01.9	

Day	Phase		h m s	Remarks
26.	LmH	C	11 58.8	Halmahera 1.93°S 127.76°E
	LmV	C	12 10.7	H = 10 53 21.9 h = 8 km MAG=5.5 (USCGS)
				D = 107.8°
				LmH:22s 0.8 μm LmV:18s 0.6 μm
				MLH=5.3 MLV=5.2
26.	ePKP	A	12 15 34	Loyalty Islands 22.89°S 171.41°E
	e	A	15 47	H = 11 55 53.3 h = 58 km MAG=4.7
				D = 148.1° Az = 335.4° (USCGS)
				PKPV:1.2s 20.4nm
26.	e(P)	A	15 02 11	
27.	eP	A	04 37 56.5	Southeastern Alaska 60.31°N 140.80°W
	e	A	38 03	H = 04 27 02.4 h = 16 km MAG=4.6
	eiPcP	A	38 25	D = 67.2° Az = 18.7° (USCGS)
	e	A	40 40	PV:1.2s 25.5nm
				MPV=5.3
27.	eP	A	05 21 02	Colorado 39.95°N 104.71°W
	LmH	B	06 17	H = 05 09 22.7 h = 5 km MAG=5.2
	LmV	B	17	D = 74.0° Az = 36.4° (USCGS)
				PV:1.0s 7.1nm
				LmH:20s 0.5 μm LmV:20s 0.8 μm
				MPV=4.8 MLH=4.8 MLV=5.0
27.	e(PKHKP)	A	08 38 32	Tonga Islands 21.34°S 174.27°W
	e	A	38 34.5	H = 08 18 42.4 h = normal MAG=5.4
	e	A	38 42	D = 150.4° Az = 352.4° (USCGS)
	LmH	B	09 46	(PKHKP)V:2.3s 129nm
	LmV	B	53.3	LmH:20s 0.4 μm LmV:18s 0.4 μm
				MLH=5.1 MLV=5.2
27.	ePKP	A	11 27 52	Loyalty Islands 22.75°S 170.80°E
	e	A	27 56.5	H = 11 08 08.4 h = normal MAG=4.6
				D = 147.8° Az = 334.9° (USCGS)
				PKPV:1.2s 12.8nm
27.	ei(P)	A	11 57 47.5	

Day	Phase		h m s	Remarks
27.	eP	A	13 13 07	Eastern Gulf of Aden 14.10°N 52.01°E
	ePP	A	15 03.5	H = 13 04 16.0 h = normal MAG=5.1
				D = 48.9° Az = 326.8° (USCGS)
				PV:1.8s 35.7nm PPV:1.6s 22.7nm
				MPV=5.2 MPPV=4.8
27.	eiPg	A	14 09 43	Explosion
	eiSg	A	09 58.5	D = ca. 1.2°
27.	e(P)	A	18 14 18	(P)V:1.1s 9.6nm
27.	e(Sg)	A	21 47 20	Near earthquake?
	e	A	47 37	
27.	LmH	B	21 53.4	Mid-Indian Rise 12.69°S 66.41°E
				H = 21 03 20.4 h = normal MAG=4.8 (USCGS)
				D = 79.2°
				LmH:16s 0.6 µm
				MLH=5.1
27.	eP	A	21 58 34	Ryukyu Islands 28.54°N 129.58°E
	LmH	C	22 40	H = 21 46 02.9 h = 17 km MAG=5.0
				D = 84.0° Az = 325.6° (USCGS)
				PV:(1.5s) 23.4nm
				MPV=5.1
28.	eP	A	02 33(43)	Chagos Archipelago Region 6.07°S 71.40°E
				H = 02 21 55.8 h = normal MAG=5.1 (USCGS)
				D = 75.7°
28.	+iP	A	02 48 59.5	Kyushu/Japan 32.09°N 130.76°E
	ePcP	A	49 07	H = 02 36 54.1 h = 125 km MAG=5.6
	eX	A	49 20	D = 81.7° Az = 325.8° (USCGS); h = 132 km
	eipP	A	49 32	PV:2.0s 170nm XV:1.6s 37.8nm
	e(pPP)	A	52 39	pPV:1.4s 90.5nm (pPP)V:2.3s 115nm
	e(sPP)	A	52 54	(sPP)V:1.9s 86.6nm
	eS	C	58.9	LmH:15s 1.4 µm
	esS	C	59 50	MPV=5.5 MLH=5.5
	eSS	C	03 04.2	i 49 36 ei 05 15 e 08 32

Day	Phase		h m s	Remarks
cont.				
28.	e(sSS)	C	03 05.1	If we interpret X as pP and pP as sP,
	eSSSS	C	10.3	then the focal depth would be h = 85 km,
	e(sSSSS)	C	11 08	but the interpretation given above seems
	LmH	B	29.6	to be more likely.
28.	e	A	03 18 30	Near earthquake?
	e	A	18 37	
	e	A	18 40	
28.	iPg	A	04 30 38.5	Explosion
	ei	A	30 48.5	D = ca. 1.1^{0}
	iSg	A	30 52.5	PgV:0.4s 14.9nm
	LmH	A	31 19.5	
28.	e(Sg)	A	12 50 02.5	Explosion?
				(Sg)V:0.5s 12.0nm
28.	ei(P)	A	15 52 35	
28.	LmH	B	19 01.3	Northeastern China (USCGS)
28.	eP	A	20 31 03	Kodiak Islands 56.61^{0}N 152.09^{0}W
	e	A	31 10	H = 20 19 36.8 h = 14 km MAG=4.5
	e	A	31 19.5	D = 72.3^{0} Az = 10.8^{0} (USCGS)
				PV:1.3s 16.7nm
				MPV=5.0
29.	eP	A	01 34 21	Leeward Islands 18.41^{0}N 62.41^{0}W
				H = 01 23 34.5 h = 58 km MAG=5.1
				D = 65.9^{0} Az = 42.1^{0} (USCGS)
29.	iPg	A	13 06 05.0	Explosion
	iSg	A	06 18.9	D = ca. 1.0^{0}
29.	iPg	A	13 13 26.0	Explosion
	i	A	13 39.8	D = ca. 1.0^{0}
	iSg	A	13 40.3	KHC gives: ePg 13 45 eiSg 14 12.2
				D = 2.0

Day	Phase		h m s	Remarks
29.	ePKHKP	A	13 43 41	South of Fiji Islands 23.49°S 179.82°W
				H = 13 24 47.0 h = 528 km MAG=4.8
				D = 151.5° Az = 344.7° (USCGS)
29.	e	A	15 20 15	
29.	ePS	C	15 31.1	Bouvet Island Region 53.62°S 6.10°E
	ePPS	C	32.1	H = 15 03 38.0 h = normal MAG=5.4 (USCGS)
	eSS	C	36.8	D = 103.9°
	LmH	C	54	LmH:50s 1.1 µm
				MLH=5.0
29.	ei(Pg)	A	23 43 49	Explosion
	e(Sg)	A	44 06.5	(D = ca. 1.4°)
30.	iPL	C	07 26 28	Albania - Yugoslavia Border Region
	+eiP1	A	26 30	41.47°N 20.54°E
	iP2	A	26 31.4	H = 07 23 51.5 h = 29 km MAG=6.0
	iP3	A	26 33.5	D = 11.1° Az = 329.0° (USCGS)
	iS	B	28 38	P2V:1.2s 97.0nm P3V:1.4s 350nm
	iSS	B	28 56	PV(B):12.5s 10.5 µm
	LmH	B	30.3	LmH:16s 760 µm LmV:16s 417 µm
	LmV	B	31.3	MLH=6.7
				Multiple P with successive increasing amplitude in our short-period records. PL waves have periods of approximately 40 sec.
30.	eiP	A	07 45 31.5	Albania - Yugoslavia Border Region
	ei	A	45 36	41.34°N 20.54°E
	ei	A	45 40	H = 07 42 52.7 h = normal MAG=4.8
	ei	A	45 45	D = 11.2° Az = 329.4° (USCGS)
30.	e	A	07 56 48.5	Probably Albania - Yugoslavia Border Region
30.	e	A	08 06.5	
30.	eP	A	08 14 14	Albania - Yugoslavia Border Region
				H = 08 11 33 (UPP)

Day	Phase		h m s	Remarks
30.	eP	A	09 23 36	Albania - Yugoslavia Border Region
30.	e	A	09 57 40	
	e	A	58 02	
30.	e	A	10 01 17.5	
30.	e(P)	A	11 25 03	
	e	A	25 11	
	e	A	25 18	
	e	A	25 54	
30.	e(S)	C	11 28 32	LmH:44s 5.4/um LmV:16s 2.3/um
	e(SS)	C	33 06	
	eL	C	39 10	
	LmH	C	41.3	
	LmV	B	51.6	
30.	e(P)	A	12 00 21.5	Probably Albania - Yugoslavia Border Region
30.	e(P)	A	12 03 43	(P)V:1.4s 15.3nm
30.	eSg	A	13 30 39	Explosion/ČSSR 49.19°N 13.86°E Yield: 10.1 t (PRU) D = 2.0°
30.	ePKIKP	A	16 06 14	Fiji Islands 17.94°S 178.26°W
	-iPKHKP	A	06 16.8	H = 15 47 44.2 h = 629 km MAG=4.7
	ei	A	06 18.5	D = 146.4° Az = 348.6° (USCGS)
	e	A	06 30	PKHKPV:1.1s 48.1nm
	e	A	07 48	
	e	A	08 59	

Day	Phase		h m s	Remarks
1.	ePg	A	07 55 48.5	Explosion/ČSSR
	eSg	A	56 08	D = ca. 1.4°
	LmH	A	56 25.5	KHC gives: D = 100 km
1.	e(P)	A	09 18 15	Albania - Yugoslavia 40.5°N 19.9°E
	e	A	18(20)	H = 09 15 31 (ATH)
				D = 11.7°
1.	e(P)	A	09 20 40	Traces; Probably near earthquake
	e	A	21 28	
	e	A	21 48	
1.	iPg	A	13 01 39.5	Explosion/GDR 51.29°N 12.73°E
	iSg	A	01 53.5	D = 0.9°
1.	+eiP	A	14 08 31.5	Kurile Islands 49.47°N 154.38°E
	+iP	B	08 32.5	H = 13 57 02.4 h = 136 km MAG=5.9
	epP	B	09 10	D = 75.3° Az = 336.5° (USCGS); h = 160 km
	iX	A	09 47	PV(A):1.6s 1020nm PV(B):6.5s 5.6 μm
	ePP	B	11 22	pPV(A):1.9s 218nm XV:1.5s 131nm
	eY	B	12.1	SH:10s 4.3 μm
	iPPP	B	13 11	LmH:15.5s 4.1 μm LmV:16s 3.3 μm
	eiZ	B	15 00	MPV(A)=6.4 MPV(B)=6.5 MSH=6.3
	iS	B	18 00	
	eSP	B	18 30	
	e(SPP)	B	19(00)	
	i	B	19 19	
	eSS	B	23.1	
	i(Sa)	C	28.0	
	LmV	B	29.3	
	LmH	B	38.0	
1.	eP	A	17 02 49	North Atlantic Ridge 60.2°N 29.6°W
	LmH	C	13.5	H = 16 57 25.2 h = normal MAG=4.5 (USCGS)
				D = 25.0°
1.	ePKP	A	17 11 51	Fiji Islands 17.58°S 178.54°W
				H = 16 53 08.8 h = 504 km MAG=4.2
				D = 146.0° Az = 348.4° (USCGS)

Day	Phase		h m s	Remarks
1.	eP	A	18 33 50.5	Macedonia 41.2°N 20.2°E
	e	A	35 41	H = 18 31 04 (BCIS)
	e	A	36 18	D = 11.2°
1.	eP	A	20 10 33	Albania - Yugoslavia 41.24°N 20.12°E
	e	A	12 45	H = 20 07 50.9 h = 33 km MAG=4.7
				D = 11.1° Az = 330.7° (USCGS)
2.	eP	A	00 26 57	Albania - Yugoslavia 41.34°N 20.57°E
	e	A	27 15	H = 00 24 15.7 h = 29 km MAG=5.1
	LmH	C	30.5	D = 11.2° Az = 329.3° (USCGS)
				LmH:16s 1.1 µm
				MLH=3.8
2.	e(P)	A	00 44 16	Gulf of California 24.13°N 108.63°W
	i	A	44 21.5	H = 00 31 18.9 h = normal MAG=5.1
	LmH	C	01 25.3	D = 88.8° Az = 33.4° (USCGS)
	LmV	B	25.4	LmH:20s 1.3 µm LmV:16s 1.5 µm
				MLH=5.4 MLV=5.5
				The first motion of P must be about 5 s earlier.
2.	e(P)	A	09 29 57	Albania - Yugoslavia 41.37°N 20.32°E
	e	A	30 15.5	H = 09 27 09.9 h = 33 km MAG=4.5
	e	A	31 27.5	D = 11.1° Az = 329.9° (USCGS)
	e(S)	A	32 06.5	The first motion of P must be about 8 s earlier.
	eLg2(3.28)	A	33 24	
	eRg(3.06)	A	33 50	
2.	e(Sg)	A	09 45 19	Explosion/ČSSR
	e	A	45 21	Yield: 8.1 t (PRU)
	e	A	45 22.5	KHC gives: eiPg 44 35.5 eiSg 45 00
				D = 200 km
2.	iP1	A	12 47 22.0	Albania - Yugoslavia 41.33°N 20.33°E
	eP2	A	47 24.5	H = 12 44 42.7 h = 17 km MAG=5.4
	iX	A	47 31.5	D = 11.1° Az = 330.0° (USCGS)
	eS	B	49 31	P1V:1.1s 21.6nm P2V:0.9s 47.2nm
	eLi(3.74)	B	50 12	XV:1.3s 77.8nm
	iLg2(3.32)	B	50 54	LmH:9s 14.0 µm LmH:11s 11.1 µm

Day	Phase		h m s	Remarks
cont				
2.	LmH	B	12 51.4	MLH=5.2
	LmV	B	52.3	i 47 27.5 e 47 36.5 e 49 30 e 50 (00) Multiple P.
2.	eP	A	14 20 44.5	Albania - Yugoslavia 41.36°N 20.09°E
	LmH	B	24.5	H = 14 18 04.5 h = normal MAG=4.4 D = 11.0° Az = 330.5° (USCGS)
2.	eiP	A	14 21 48	Albania - Yugoslavia
	e	A	23 15.5	H = 14 19 07 (UPP)
	e(S)	A	23 53.5	
	ei	B	24.2	
	LmH	B	25.5	
2.	e(P)	A	14 26 43.5	
2.	e(P)	A	16 41 05.5	
2.	eP	A	20 17 02	Northeastern China 37.77°N 115.21°E
	ei	A	17 06.5	H = 20 05 52.4 h = 13 km MAG=5.2
	LmH	B	45.5	D = 69.4° Az = 318.6° (USCGS)
	LmV	B	50.0	LmH:15.5s 6.0/um LmV:13s 3.1/um MLH=5.9 MLV=5.8
3.	e(P)	A	00 06 20.5	
3.	LmV	B	00 17.2	
3.	e(P)	A	00 25 47.5	
3.	LmH	C	00 36.5	
	LmV	B	36.5	
3.	e(P)	A	11 15(11.5)	Traces
	e	A	15 15	
	e	A	15 20.5	
3.	e	A	18 01 37	

Day	Phase		h m s	Remarks
3.	e(P)	A	18 02 14	Probably Albania - Yugoslavia
	e	A	02 44	
	e	A	04 49	
3.	ePx	A	19 51 47	Central Italy 42.25°N 13.62°E
	iPg	A	52 25.2	H = 19 49 38.5 h = 18 km MAG=4.6
	e(Sn)	A	53 11.5	D = 8.5° Az = 351.4° (USCGS)
	iSn	A	53 13.5	e 51 50.5 ei 54 35.5
	eiLi	A	53 48	
	e(Sg)	A	54 16.5	
	eiLg2	A	54 21.5	
3.	ePn	A	21 32 00	Central Italy 42.45°N 13.20°E
	ePx	A	32 05	H = 21 30 00.3 h = normal MAG=4.6
	ePg	A	32 43	D = 8.3° Az = 353.0° (USCGS)
	e(Sn)	A	33 29.5	LmH:18s 2.0 µm
	iSn	A	33 31.0	MLH=3.8
	eiLg2	A	34 42	e 33 25 e 34 08
	LmH	C	35	
3.	ePn	A	22 12(00)	Carpathians/ČSSR 48.7°N 17.5°E
	e(Pb)	A	12 13	H = 22 10 54 (BCIS)
	ei(Sn)	A	12 53	D = 4.3°
	i(Sx)	A	13 02	e 12 06 e 12 15 ei 13 11
	i(Sb)	A	13 10.0	
	i(Sg)	A	13 16.0	
4.	eiSg	A	03 01 09	Switzerland 47.0°N 7.8°E
				H = 02 58 46 (BCIS)
				D = 4.4°
4.	LmH	C	08 11.2	
4.	e(Sg)	A	08 56 57.5	Little Carpathians (KHC)
	e	A	56 59	
4.	e(P)	A	09 00 55.5	Off Coast of Oregon 43.72°N 127.37°W
	e	A	01 00	H = 08 48 45.2 h = normal MAG=4.6
	LmH	C	36.6	D = 79.5° Az = 25.1° (USCGS)
				The first motion of P must be about 6 s earlier.

Day	Phase		h m s	Remarks
4.	e	C	11 26.5	
	e	C	36(14)	
	e	C	39 38	
4.	eP	A	20 29 38	Ryukyu Islands 27.51°N 128.92°E
	e	A	29 43	H = 20 17 05.7 h = normal MAG=4.8
				D = 84.6° Az = 325.4° (USCGS)
4.	eP	A	22 30 36	Andreanof Islands/Aleutians
	e	A	30 42	51.62°N 173.50°W
	epP	A	30 49	H = 22 18 41.0 h = 50 km MAG=4.7
				D = 78.0° Az = 356.7° (USCGS); h = 50 km
				PV:1.2s 23.0nm
				MPV=5.2
5.	eiP1	A	05 24 03	Dodecanese Islands 36.48°N 26.89°E
	iP2	A	24 05.7	H = 05 20 02.9 h = 138 km MAG=4.6
				D = 17.9° Az = 327.0° (USCGS)
				P2V:1.3s 111nm
				Multiple P. The amplitude of P2 is much bigger than that of P1.
5.	iP	A	09 17 09.5	Andreanof Islands/Aleutian
	e	A	17 18.5	51.64°N 173.45°W
	e	A	17 23	H = 09 05 13.1 h = 36 km MAG=5.3
				D = 78.0° Az = 356.7° (USCGS)
				PV:1.0s 40.2nm
				MPV=5.5
5.	iPg	A	13 05 13.3	Explosion
	eiSg	A	05 26	D = ca. 1.0°
5.	eiPg	A	13 36 42.5	Probably explosion
	iSg	A	36 51.5	D = ca. 0.7°
	LmH	A	37 11.5	
5.	ePKHKP	A	17 47 09	Fiji Islands 20.82°S 178.82°W
				H = 17 28 31.5 h = 655 km MAG=4.5
				D = 149.1° Az = 347.0° (USCGS)
				PKHKPV:1.1s 12.0nm

Day	Phase		h m s	Remarks
5.	LmH	C	19 23.5	Gulf of California (USCGS)
6.	epP	A	03 06 02	Near Coast of Nicaragua 12.46°N 87.24°W
	LmH	C	38.4	H = 02 53 06.9 h = 87 km MAG=5.3
				D = 86.0° Az = 39.1° (USCGS)
6.	+ei(P)	A	03 23 50.5	
6.	ePKIKP	A	05 22 22.5	Fiji Islands 21.26°S 178.79°W
	-iPKHKP	A	22 27.8	H = 05 03 40.8 h = 559 km MAG=5.1
	iPKP2	A	22 35.5	D = 149.6° Az = 346.9° (USCGS)
				PKIKPV:2.0s 66.2nm PKHKPV:1.5s 168nm
				PKP2V:1.5s 67.1nm
6.	eSg	A	12 00 34.5	Probably explosion
				SgV:1.2s 20.4nm
6.	eSg	A	14 59 14.5	Explosion
6.	eP	A	15 08 41.5	Off East Coast of Honshu/Japan
				40.93°N 142.99°E
				H = 14 56 37.1 h = 31 km MAG=4.6
				D = 79.4° Az = 330.9° (USCGS)
				PV:1.1s 9.6nm
				MPV=4.8
6.	LmH	C	22 55	Carlsberg Ridge (USCGS)
7.	i	A	04 14 17.3	
7.	eP	A	07 36 19.5	Fox Islands/Aleutian 52.90°N 166.72°W
	epP	A	36 27.5	H = 07 24 28.5 h = 31 km MAG=4.7
				D = 76.8° Az = 1.1° (USCGS); h = 30 km
				PV:1.3s 22.2nm
				MPV=5.1
7.	+iPKP	A	10 00 41.0	Tonga Islands 16.69°S 174.13°W
	e	A	01 04	H = 09 41 14.1 h = 120 km MAG=5.0
	epPKP	A	01 13	D = 145.8° Az = 353.5° (USCGS); h = 116 km
				PKPV:1.0s 47.3nm

Day	Phase		h m s	Remarks
7.	ePKIKP	A	10 08 47.5	New Hebrides Islands 14.62°S 167.26°E H = 09 49 37.0 h = 151 km MAG=5.3 D = 139.0° Az = 336.4° (USCGS)
7.	e(P)	A	10 13 33.5	
7.	iPg	A	08 10 45.0	Explosion
	iSg	A	11 04.0	D = ca. 1.4°
7.	e(Sg)	A	14 01 51.5	Explosion/ČSSR 49.50°N 14.93°E
	e	A	01 55	Yield: 16.2 t (PRU) D = 2.4°
7.	e(Sg)	A	15 40 16	Probably explosion
	i	A	40 18	KHC gives: eiPg 39 45.4 eiSg 40 04.4 D = 1.2°
7.	eP	A	18 06 14	Albania - Yugoslavia 41.28°N 20.21°E
	e	A	06 24	H = 18 03 35.5 h = normal MAG=4.5
	e	A	06 39	D = 11.1° Az = 330.4° (USCGS)
	e(SS)	A	08 46	e 09 23.5 e 09 51.5 e 10 09
8.	e(P)	A	01 42 19	
8.	LmH	C	04 40	Bismark Sea (USCGS)
8.	+iP	A	06 11 46	Eastern Kazakh SSR 49.84°N 78.22°E
	ePn.	A	13 20	H = 06 03 57.1 h = 0 km MAG=5.4 D = 41.3° Az = 297.7° (USCGS) PV:0.8s 78.2nm MPV=5.7 Underground explosion.
8.	iPg	A	09 58 39.2	Explosion/ČSSR
	iSg	A	58 57.5	D = 1.4° KHC gives: eiPg 58 25 D = 36 km Yield: 11.9 t
8.	e(Sg)	A	16 01 51.5	Explosion

Day	Phase		h m s	Remarks
9.	ePn	A	03 12 04	Adriatic Sea 42.01°N 16.54°E
	iPx	A	12 09.0	H = 03 09 52.2 h = normal MAG=4.5
	iSn	A	13 41.5	D = 9.3° Az = 340.2° (USCGS)
	iSx	A	13 50	PxV:0.9s 42.3nm
	ei(Lg1)	A	14 52.5	LmH:11s 1.7 µm LmV:12s 2.2 µm
	iSg	A	14 54.5	MLH=4.1
	LmH	B	16.0	e 12 05.5 i 13 51 e 14 39
	LmV	B	16.1	
9.	ePKIKP	A	05 47 18	South of Fiji Islands 22.23°S 179.40°W
	iPKHKP	A	47 24.5	H = 05 28 38.9 h = 588 km MAG=4.9
	ePKP2	A	47 26.5	D = 150.4° Az = 345.8° (USCGS)
	e	A	47 33.5	PKIKPV:1.5s 20.2nm PKHKPV:1.4s 104nm
	e	A	47 59	PKP2V:1.5s 67.1nm
9.	e	A	08 57 12	Adriatic Sea 45.4°N 14.6°E
	eSn	A	57 27.5	H = 08 55 00 (BCIS)
	e(Sb)	A	57 55.5	D = 5.6°
	eiSg	A	58 06.5	
	ei	A	59 30	
9.	ePKIKP	A	11 10 03.5	Santa Cruz Islands 10.94°S 164.15°E
	eSS	C	30(20)	H = 10 50 46.6 h = normal MAG=5.5
	eSSS	C	35(20)	D = 134.4° Az = 335.7°(USCGS)
	LmH	C	12 11.8	LmH:22s 1.2 µm
				MLH=5.5
9.	eP	A	22 23 00	Unimak Islands 53.78°N 163.25°W
				H = 22 11 12.9 h = 14 km MAG=4.3
				D = 75.9° Az = 3.4° (USCGS)
10.	eP1	A	12 19 09.5	Near Coast of Northern California
	eP2	A	19 15	40.51°N 124.60°W
	eP3	A	19 18	H = 12 06 50.3 h = 5 km MAG=5.8
	ePP1	A	22 16	D = 81.5° Az = 26.5° (USCGS)
	ePP2	A	22 24.5	P3V:1.8s 92.0nm
	eS	C	29 22	LmH:17s 3.5 µm LmV:19s 4.8 µm
	iSP	B	30 12	MP3V=5.6 MLH=5.8 MLV=5.9
	eSPP	B	30 30	e 20 24 e 38 40
	eSS	C	34.5	Successive P-onsets with increasing amplitude.

Day	Phase		h m s	Remarks
cont.				
10.	LmH	B	12 57.3	
	LmV	B	57.4	
10.	LmH	C	15 59.7	Eastern Kazakh SSR 46.32°N 81.91°E
				H = 15 34 53.0 h = 48 km MAG=4.8 (USCGS)
				D = 45.2°
				LmH:21s 0.6 µm
				MLH=4.5
10.	eP	A	18 54 22	Burma 22.51°N 94.80°E
	epP	A	54(59)	H = 18 43 34.4 h = 158 km MAG=5.2 (USCGS)
	e	C	19 04.1	D = 68.7° h = 155 km
	eLQ	C	12.2	LmH:24s 0.4 µm
	LmH	C	23	
10.	-eP	AB	23 01 22	India 17.66°N 73.93°E
	iX	B	03 50	H = 22 51 24.3 h = normal MAG=6.0
	eiPPP	B	05 07	D = 59.0° Az = 318.9° (USCGS)
	eiS	C	09 28	PV(A):1.8s 194nm PV(B):8s 1.6 µm
	iS	B	09 40	LmH:17s 16.7 µm LmV:13s 14.1 µm
	eiSoS	C	11 00	MPV(A)=5.9 MPV(B)=6.2 MLH=6.2 MLV=6.3
	iLQ	C	16 12	i 01 24 i 01 28.5 ei 11 06
	LmH	B	32	
	LmV	B	35.3	
11.	eP	A	00 01 19	India 17.4°N 73.7°E
				H = 23 52 16.7 h = normal MAG=5.0 (USCGS)
				D = 59.2°
11.	iPx	A	02 37 23.1	Switzerland 46.7°N 9.9°E
	iP(b)	A	37 32	H = 02 36 16 (BCIS)
	iSn	A	38 05	D = 4.1°
	iS(b)	A	38 19	
	iSg	A	38 27	
	LmH	A	38 37	
11.	e	A	10 47 10	Explosion?
	e	A	47 16.5	
	e	A	47 23	

Day	Phase		h m s	Remarks
11.	eP	A	19 57 27.5	Eastern Gulf of Aden 13.69°N 51.59°E
	e(pP)	A	57 34.5	H = 19 48 43.3 h = normal MAG=5.2
	ePP	A	59 26	D = 49.0° Az = 327.2° (USCGS);
	LmH	C	20 15	(h = 30 km)
				e 57 48 e 58 08
11.	ePKHKP	A	20 00 41.5	Tonga Islands 20.64°S 174.29°W
	i	A	00 47.2	H = 19 40 53.3 h = normal MAG=5.3
	ePKP2	A	00 51	D = 149.7° Az = 352.5° (USCGS)
	LmH	C	21 05	PKHKPV:1.5s 47.0nm PKP2V:1.6s 37.9nm
11.	+iP	A	22 39 03.5	Eastern Gulf of Aden 13.62°N 51.60°E
	ePP	B	41 02	H = 22 30 18.3 h = normal MAG=5.6
	ePPP	C	41.9	D = 49.1° Az = 327.2° (USCGS)
	eS	B	46 05	PV:1.1s 33.7nm PPV:9s 0.7 µm
	eSS	C	49 50	SH:20s 1.6 µm
	iLR	C	54 25	LmH:16s 2.1 µm LmV:16s 1.8 µm
	LmH	B	23 01.5	MPV=5.3 MPPV=5.5 MSH=5.7 MLH=5.2
	LmV	B	04.6	MLV=5.2
				e 39 07 e 39 42.5 e 50 00
				S has a period of about 40 s in the registrations of type C.
11.	eP	A	23 14 16.5	Jan Mayen 71°N 8°W
				H = 23 09 26 (UPP)
				D = 22.2°
12.	i	A	04 15 43	Near earthquake?
	e(Sg)	A	16 03.5	
12.	iP	A	06 28 37	India 17.59°N 73.91°E
				H = 06 18 37.9 h = 29 km MAG=5.4
				D = 59.1° Az = 318.9° (USCGS)
12.	ePKHKP	A	08 25 59	Loyalty Islands 22.67°S 171.06°E
	ePKP2	A	26 03.5	H = 08 06 16.7 h = 39 km MAG=4.9
	eX	A	27 07	D = 147.8° Az = 335.2° (USCGS)
	eSSS	C	54 20	PKP2V:1.1s 24.0nm XV:1.7s 48.3nm
	LmH	C	09 20.5	LmH:33s 1.1 µm LmV:18s 0.8 µm
	LmV	B	35	MLH=5.4 MLV=5.5 e 26 48 e 49 (00)

Day	Phase		h m s	Remarks
12.	LmH	C	11 05.5	Taiwan (USCGS)
12.	eiSg	A	11 30 56	Explosion/ČSSR KHC gives: D = 34 km
12.	e	A	12 24 25.5	
12.	LmH	C	15 17.5	
12.	eP	A	15 58 55	India 17.43°N 73.92°E H = 15 48 55.5 h = 27 km MAG=5.0 D = 59.2° Az = 319.0° (USCGS) PV:2.0s 53.0nm MPV=5.3
12.	i(Sg)	A	18 48 34	Explosion? KHC gives: ePg 48 02 eiSg 48 18 D = 1.2°
12.	e(P)	A	20 09 06.5	Traces; near earthquake?
	e	A	09 23	
	e	A	10 08	
	e	A	10 27.5	
13.	LmH	B	05 47.6	LmH:16s 0.7/um LmV:16s 0.8/um
	LmV	B	47.6	
13.	eiPg	A	10 44 11	Slovenia 45.9°N 14.8°E
	eSn	A	44 42.5	H = 10 42 28 (BCIS)
	e	A	44 47	D = 5.2°
	eSg	A	45 13	
13.	+iP	A	10 50 01.0	Kurile Islands 47.63°N 152.64°E
	+iPcP	A	50 12.7	H = 10 38 23.4 h = 124 km MAG=5.5
	eiX	A	50 18	D = 76.5° Az = 335.7° (USCGS);
	e	A	50 38	(h = 136 km)
	e(sP)	B	50(50)	PV:1.0s 387nm PcPV:1.0s 104nm
	iY	A	50 56.8	XV:1.1s 96.0nm YV:1.6s 100nm
	e	C	11 09.2	MPV=6.1
	LmH	B	30	

Day	Phase		h m s	Remarks
13.	eP	A	11 09 51	<u>Kurile Islands</u> 49.38°N 154.51°E H = 10 58 21.6 h = 138 km MAG=5.1 D = 75.4° Az = 336.6° (USCGS) PV:1.5s 60.4nm MPV=5.1
13.	e	A	12 58 00	Explosion/<u>GDR</u>
	iSg	A	58 05.3	
13.	eP	A	17 59 00	<u>Sea of Okhotsk</u> 47.09°N 145.72°E H = 17 47 51.2 h = 312 km MAG=4.8 D = 74.9° Az = 331.7° (USCGS) PV:1.8s 46.8nm MPV=4.9
13.	+iPKP	A	19 26 42.1	<u>New Hebrides Islands</u> 19.12°S 168.68°E
	epPKP	A	26 55	H = 19 07 14.4 h = 51 km MAG=5.7
	ePSKS	C	40 10	D = 143.7° Az = 335.2° (USCGS); h = 46 km
	ePPS	C	42 40	PKPV:1.2s 143nm pPKPV:1.6s 98.4nm
	e	C	43.8	LmH:22s 1.0 µm
	eSS	C	48.4	MLH=5.3
	e(PSS)	C	49(10)	
	LmH	B	34.5	
13.	ePKP	A	21 53 50	<u>Fiji Islands</u> 17.72°S 178.11°W H = 21 35 11.4 h = 562 km MAG=4.6 D = 146.2° Az = 348.8° (USCGS) PKPV:1.1s 14.4nm
13.	ePKIKP	A	23 29 18	<u>Loyalty Islands</u> 22.81°S 171.45°E
	ePKHKP	A	29 22	H = 23 09 37.2 h = 43 km MAG=4.3 D = 148.1° Az = 335.5° (USCGS)
14.	eP1	A	02 29 20	<u>Arabian Sea</u> 14.30°N 53.65°E
	eP2	A	29 23.5	H = 02 20 27.9 h = normal MAG=4.9
	ePP	B	31 15	D = 49.6° Az = 326.0° (USCGS)
	eiS	B	36 26	P2V:2.0s 79.5nm SH:8.5s 0.8 µm
	eSS	C	40 18	LmH:16s 0.5 µm LmV:17s 0.7 µm
	LmH	B	56	MP2V=5.5 MSH=5.8 MLH=4.6 MLV=4.8
	LmV	B	58	e 29 42 e 31 29

Day	Phase		h m s	Remarks
14.	eP	A	02 59 20	Crete 34.45°N 26.24°E
	e	A	59 24	H = 02 54 47.6 h = 7 km MAG=4.5
	e	A	59 31.5	D = 19.4° Az = 331.0° (USCGS)
	LmH	B	03 08.5	LmH:13s 0.8 µm LmV:16s 1.0 µm
	LmV	B	08.5	MLH=4.2 MLV=4.4
				Successively increasing amplitudes of P.
14.	eP	A	08 39 38.5	Crete 34.88°N 24.34°E
	eS	B	42 52	H = 08 35 28.7 h = 78 km MAG=4.6
	e	B	46.0	D = 18.3° Az = 333.4° (USCGS)
	LmH	B	47.6	LmH:11s 0.3 µm LmV:11s 0.6 µm
	LmV	B	47.9	MLH=3.9 MLV=4.2
14.	eiSg	A	13 00 43	Probably explosion
				KHC gives: eiPg 00 23.5 eiSg 00 42
				D = 1.4°
14.	eP	A	18 36 38	Near East Coast of Kamchatka
	epP	A	36 43.5	54.57°N 160.43°E
	ePcP	A	36 56	H = 18 25 16.6 h = normal MAG=5.5
	e	A	37 20	D = 72.0° Az = 339.7° (USCGS); h = 21 km
	LmH	C	19 02	PV:0.9s 9.4nm pPV:1.1s 21.6nm
	LmV	C	02	LmH:45s 0.9 µm LmV:45s 0.6 µm
				MPV=4.9 MLH=4.7 MLV=4.6
14.	eP	A	22 10 23	Chagos Archipelago 4.00°S 68.38°E
				H = 21 58 54.5 h = normal MAG=4.9
				D = 72.9° Az = 326.1° (USCGS)
15.	ei	A	10 31 02	Explosion?
	e(Sg)	A	31 25	KHC gives: eiPg 30 19.5 eiSg 30 34
				D = 1.1°
				(Sg)V:1.8s 40.8nm
15.	ePKIKP	A	20 07 03	Kermadec Islands Region 29.08°S 177.57°W
	ePKHKP	A	07(16)	H = 19 47 13.5 h = 61 km MAG=5.3 (USCGS)
	eiPKP2	A	07 34.5	D = 157.3°
				PKP2V:1.1s 19.2nm

Day	Phase		h m s	Remarks
16.	e(Pg)	A	10 37 18	Explosion/<u>ČSSR</u> 51.00°N 14.41°E
	e	A	37 23	Yield: 26.5 t (PRU)
	eSg	A	37 43	D = 1.8°
16.	i(Pg)	A	12 02 36.8	Explosion?
	e(Sg)	A	02 52	(D = ca. 1.1°)
16.	ePg	A	12 54 24.5	Explosion/<u>GDR</u>
16.	ePKP2	A	13 33 48	<u>Kermadec Islands</u> 31.67°S 179.79°E
				H = 13 14 00.0 h = 403 km MAG=4.1 (USCGS)
				D = 159.2°
				PKP2V:1.3s 13.9nm
16.	eSg	A	13 49 21	Probably explosion.
16.	eP	A	21 05 37	<u>Near East Coast of Kamchatka</u>
	e	A	06 28	51.22°N 157.66°E
	eS	C	15 00	H = 20 53 58.3 h = 24 km MAG=5.5
	ePS	C	15 30	D = 74.5° Az = 338.3° (USCGS)
	e(Sa)	C	24.0	PV:1.8s 81.6nm
	LmH	B	43.5	LmH:16s 3.9 µm LmV:16.5s 4.8 µm
	LmV	B	43.6	MPV=5.6 MLH=5.8 MLV=5.9
				i 05 54.5 ei 06 12
17.	eP1	A	00 33 19.5	<u>Afghanistan - USSR Border Region</u>
	iP2	A	33 20.7	36.51°N 71.44°E
	i(pP)	A	33 49.5	H = 00 25 15.1 h = 82 km MAG=5.2
	e(sP)	A	34 03	D = 44.4° Az = 308.1° (USCGS);
	ePP	A	35 06	(h = ca. 130 km)
				P2V:1.8s 122.5nm
				MP2V=5.7
				If we interpret pP as sP, then the focal depth would be h = 88 km.
17.	ePKP	A	11 16 23.5	<u>Loyalty Islands</u> 20.47°S 168.92°E
				H = 10 56 48.2 h = 32 km
				D = 145.0° Az = 334.7° (USCGS)
				PKPV:1.6s 19.0nm

Day	Phase		h m s	Remarks
17.	LmH	C	17 04.5	Samar/Philippine Islands 11.58°N 125.92°E
	LmV	C	04.5	H = 16 03 26.9 h = 3 km MAG=5.3 (USCGS)
				D = 96.0°
				LmH:20s 0.5 µm LmV:18s 0.6 µm
				MLH=5.0 MLV=5.1
17.	ePKP	A	21 12 35	New Hebrides 20.74°S 169.00°E
	e	A	12 41	H = 20 53 01.0 h = normal MAG=4.2
	e	A	12 45.5	D = 145.2° Az = 334.6° (USCGS)
				PKPV:1.6s 41.7nm
18.	ePKP	A	06 44 02	Loyalty Islands 22.35°S 170.79°E
	e	A	44 03.5	H = 06 24 19.7 h = 34 km MAG=4.8
				D = 147.4° Az = 335.2° (USCGS)
18.	eP	A	11 01 10	Nepal 29.15°N 81.92°E
	LmH	B	27.5	H = 10 51 34.8 h = 42 km MAG=5.2
	LmV	B	27.5	D = 55.8° Az = 313.6° (USCGS)
				PV:1.1s 14.4nm
				LmV:18s 0.5 µm
				MPV=5.0 MLV=4.7
18.	LmH	C	11 20.1	Probably to Mariana Islands
				(H = 10 21 54.7 (USCGS))
18.	LmH	C	11 28.5	Probably to Mariana Islands
				(H = 10 30 41.2 (USCGS))
18.	e(Sg)	A	12 01 53.5	Explosion?
18.	eSg	A	12 02 56	Explosion?
	LmH	A	03 07	
18.	eP	A	14 18 41.5	Eastern China 36.15°N 111.67°E
	LmH	B	49.7	H = 14 07 39.5 h = normal MAG=4.9 (USCGS)
	LmV	B	51.2	D = 68.7°
				PV:1.1s 9.7nm
				LmH:14s 0.8 µm LmV:15s 1.1 µm
				MPV=4.9 MLH=5.1 MLV=5.3

Day	Phase		h m s	Remarks
18.	LmH	B	15 10.2	South of Mariana Islands 12.06°N 143.65°E
	LmV	B	10.3	H = 14 04 19.5 h = 12 km MAG=5.5
				D = 104.9° Az = 330.7° (USCGS)
				LmH:21s 1.1 µm LmV:18s 1.4 µm
				MLH=5.4 MLV=5.6
18.	eP	A	17 37 02	Central California 37.04°N 121.77°W
	LmH	B	18 12.3	H = 17 24 31.9 h = 11 km MAG=5.0
				D = 83.5° Az = 27.8° (USCGS)
				PV:2.0s 33.0nm
				LmH:18.5s 1.0 µm
				MPV=5.1 MLH=5.2
19.	+eP	A	03 31 51	Tadzhik SSR 37.49°N 72.00°E
				H = 03 23 49.6 h = 89 km MAG=5.5
				D = 44.1° Az = 307.4° (USCGS)
				PV:1.2s 28.1nm
				MPV=5.1
19.	ePn	A	08 35 10	Albania - Yugoslavia 41.46°N 20.36°E
	e	A	35 26	H = 08 32 30.9 h = 19 km MAG=4.8
	e	A	36 26.5	D = 11.0° Az = 329.5° (USCGS)
	eS	A	37 13	PnV:0.8s 16.5nm
	e(S)	C	37 24	LmH:16s 2.9 µm
	i(SS)	A	37 30.8	MLH=4.3
	e	A	37 46	e 35 16 e 35 18 e 37 28 e 37 41
	eLg1(3.43)	A	38 34	e 38 25
	LmH	C	38.8	
20.	eSg	A	04 33 14	Switzerland 46.9°N 9.8°E
	e	A	33 17.5	H = 04 31 04 (BCIS)
				D = 3.9°
20.	eP	A	05 57 30	Kurile Islands 43.23°N 146.25°E
				H = 05 45 30.1 h = normal MAG=4.7
				D = 78.5° Az = 332.5° (USCGS)
				PV:2.2s 52.6nm
				MPV=5.3

Day	Phase		h m s	Remarks
20.	iPg	A	08 01 21.5	Explosion
	eiSg	A	01 38	D = ca. 1.2°
20.	+i(P)	A	10 32 03.7	(P)V:0.8s 47.8nm
	e	A	32 22	
	e	A	33 34	
20.	eP1	A	11 46 05	Andaman Islands 11.81°N 93.03°E
	eP2	A	46 13	H = 11 34 25.9 h = 61 km MAG=5.4
	i(pP)	A	46 23.0	D = 75.5° Az = 319.4° (USCGS);
	e(sP)	A	46 29	(h = 70 km)
	ePP	B	48(56)	P1V:1.6s 31.6nm P2V:1.3s 58.4nm
	e(SP)	B	56 11	(pP)V:1.6s 110nm (sP)V:2.0s 185nm
	eSS	C	12 01.1	LmH:27s 1.0/um
	eSSS	C	04.4	MP1V=5.2 MP2V=5.6 MLH=5.0
	LmH	C	22.8	e 47 11 e 48 47.5 e 04 47
				Multiple P.
20.	iSg	A	12 57 27.5	Explosion/GDR
20.	ePKP	A	17 26 57	New Hebrides Islands 15.13°S 167.44°E
				H = 17 07 49.1 h = 135 km MAG=5.1
				D = 139.6° Az = 336.3° (USCGS)
20.	LmH	C	23 26.5	
21.	+iPn	A	00 12 13	Yugoslavia 42.15°N 20.70°E
	e	A	12 19	H = 00 09 39.0 h = 19 km MAG=4.7
	e	A	12 52	D = 10.6° Az = 326.7° (USCGS)
	eLg1	B	15 18	PnV:1.1s 14.4nm
	iLg2	B	15 35	LmH:12s 1.4/um LmV:10.5s 2.2/um
	LmH	B	15.7	MLH=4.1
	LmV	B	16.7	e 13 48 e 15 16.5 e 15 32 i 15 43
				e 16 30
21.	e	A	00 27 31.5	
21.	+eiP	BC	02 39 12	Near Coast of Northern Chile
	eP	A	39 15	21.85°S 70.01°W
	-iPP	C	43 20	H = 02 25 21.6 h = normal MAG=6.3

Day	Phase		h m s	Remarks
cont.				
21.	+iPP	B	02 43 23	D = 101.5° Az = 40.0° (USCGS)
	eiPP	A	43 24.5	PV(A):1.6s 49.2nm PV(B):10s 5.8 /um
	iX	A	43 42.7	PH(B):9s 1.9 /um PV(C):29s 11.8 /um
	eiPPP	B	45 42	PH(C):32s 7.4 /um XV:1.9s 587nm
	iSKS	B	49 57	PPV(A):2.7s 1143nm PPV(B):9s 11.1 /um
	iSKKS	B	50 28	SH:11.5s 5.3 /um
	iS	B	50 56	LmH:22s 186 /um LmV:22s 218 /um
	iPS	B	52 35	MPV(A)=5.9 MPV(B)=7.2 MPH(B)=7.0
	ePKKP	A	54 57	MPV(C)=7.0 MPH(C)=7.1 MPPV(A)=6.9
	eY	A	55 55	MPPV(B)=7.3 MSH=7.5 MLH=7.6 MLV=7.6
	eiSS	B	57 51	e 39 19 e 39 22 e 40 04 e 42 28
	LmH	B	03 20.6	e 43 45 e(SS) 57 42 i 03 (00)
	LmV	B	20.7	
21.	eP	A	11 49 33	Northern Colombia 6.98°N 72.12°W
	epP	A	49 39.5	H = 11 37 22.5 h = normal MAG=5.4
	LmH	C	12 23.6	D = 80.7° Az = 39.9° (USCGS); h = 24 km
				PV:1.4s 18.4nm pPV:1.5s 26.8nm
				LmH:20s 0.5 /um
				MPV=5.0 MLH=4.9
21.	iPKHKP	A	12 01 06.5	Fiji Islands 21.08°S 179.23°W
				H = 11 42 25.3 h = 606 km MAG=4.5
				D = 149.3° Az = 346.4° (USCGS)
				PKHKPV:0.8s 11.8nm
21.	e(Sg)	A	12 33 43	Probably explosion.
	e	A	33 48	
	e	A	33 56	
21.	iPg	A	13 14 41.0	Explosion/GDR
	iSg	A	14 43.2	
21.	e	A	14 10 41.5	

Day	Phase		h m s	Remarks
21.	eP	A	16 15 04.5	Kurile Islands 49.18°N 156.17°E
	LmH	C	45.0	H = 16 03 20.5 h = 44 km MAG=4.5
				D = 76.0° Az = 337.6° (USCGS)
				PV:1.2s 20.4nm
				LmH:28s 0.8 μm
				MPV=5.1 MLH=4.9
21.	iP	A	16 24 14.0	Kurile Islands 49.13°N 156.20°E
	epP	A	24 27	H = 16 12 30.7 h = 53 km MAG=4.8
	LmH	C	53.5	D = 76.1° Az = 337.7° (USCGS); h = 48 km
				PV:1.2s 35.7nm
				LmH:24s 0.9 μm
				MPV=5.4 MLH=5.0
21.	eP	A	16 28 24	Kurile Islands 49.18°N 156.17°E
	epP	A	28 42	H = 16 16 41.0 h = 49 km MAG=4.5
	LmH	C	17 02.8	D = 76.0° Az = 337.6° (USCGS); h = 70 km
				PV:1.0s 23.7nm
				LmH:20s 0.7 μm
				MPV=5.3 MLH=5.0
21.	ePKP2	A	18 06 29.5	Kermadec Islands 31.68°S 179.15°W
	eX	A	06 35	H = 17 45 54.4 h = 23 km MAG=5.1 (USCGS)
	eY	A	06 43	D = 159.5°
				XV:2.0s 66.2nm YV:1.7s 30.7nm
21.	eP1	A	23 54 54.5	Andaman Islands 11.83°N 93.14°E
	eP2	A	55 02.5	H = 23 43 11.4 h = 33 km MAG=5.0
	eiP3	A	55 11	D = 75.6° Az = 319.4° (USCGS)
	LmH	C	24 30	P2V:1.3s 11.1nm P3V:1.8s 51.0nm
				MP2V=4.8 MP3V=5.4
				Multiple P with successively increasing amplitudes. If we interpret P2 as pP, then the focal depth would be h = 30 km.
22.	eP	A	07 24 51.5	Albania - Yugoslavia 40 3/4°N 21°E
	e	A	26 07	H = 07 22 03 (ATH)
	e	A	26 51.5	D = 11.9°
	e	A	27 33.5	

Day	Phase		h m s	Remarks
cont				
22.	e	A	07 28 08	
	e(Lg1)	A	28 23	
22.	e(Sb)	A	10 49 30	Switzerland 46.8°N 9.8°E
	eSg	A	49 40	H = 10 47 29 (BCIS)
				D = 4.0°
22.	ePg	A	13 06 02	Explosion/GFR 11.15°E 47.63°N
	eSn	A	06 22	H = 13 05 00 Yield: 15 t
	eiSg	A	06 37.5	D = 3.0°
	i	A	06 40.0	
22.	iPKIKP	A	23 28 58.5	Kermadec Islands 29.9°S 177.4°W
	ePKHKP	A	29 10	H = 23 08 58.0 h = 22 km MAG=5.4 (USCGS)
	eiPKP2	A	29 28	D = 158.1°
	e	A	29 36	PKP2V:1.6s 64.4nm
	e	A	29 40.5	
23.	LmH	C	00 53.5	Probably to Kermadec Islands
	LmV	C	01 03.5	(H = 23 08 58.0)
23.	ePKP	A	01 16 28.5	Tonga Islands 16.79°S 173.29°W
				H = 00 56 53.4 h = 38 km MAG=4.3
				D = 146.0° Az = 354.4° (USCGS)
23.	ePKIKP	A	13 42 08	New Britain 5.23°S 151.84°E
	e	C	49.9	H = 13 23 15.0 h = 61 km MAG=5.5
	eSS	C	14 01.5	D = 123.9° Az = 330.6° (USCGS)
	LmV	B	37.2	PKIKPV:1.2s 12.8nm
	LmH	B	37.5	LmH:20s 1.2 μm LmV:22s 1.8 μm
				MLH=5.6 MLV=5.7
23.	eP	A	16 16 31.5	Kurile Islands 48.25°N 157.28°E
	epP	A	16 36	H = 16 04 37.9 h = 26 km MAG=5.1
				D = 77.2° Az = 338.4° (USCGS); h = 17 km
				PV:1.3s 19.5nm
				MPV=5.1

Day	Phase		h m s	Remarks
24.	ePKIKP	A	02 43 53.5	Fiji Islands 21.03°S 177.98°W
	-iPKHKP	A	43 59	H = 02 24 58.4 h = 428 km MAG=5.0
	ePKP2	A	44 05	D = 149.5° Az = 347.9° (USCGS)
				PKHKPV:1.4s 46.0nm PKP2V:1.7s 48.2nm
24.	eP	A	04 26 55.5	Jan Mayen Island 71.89°N 0.94°W
	ePL	C	27 00	H = 04 22 01.2 h = normal MAG=5.0
	e	A	27 35	D = 22.1° Az = 158.4° (USCGS)
	eS	C	30 56	PV:1.6s 136nm SH:15s 1.7 μm
	LmH	B	36.8	LmH:15s 0.9 μm LmV:15s 1.1 μm
	LmV	B	36.8	MPV=5.2 MSH=5.2 MLH=4.3 MLV=4.6
24.	eP	A	08 45 07	Sakhalin Island 54.54°N 142.54°E
	e	A	45 14	H = 08 34 13.5 h = normal MAG=5.0
	LmH	B	09 13.3	D = 67.4° Az = 328.6° (USCGS)
	LmV	B	17.9	PV:1.7s 57.0nm
				LmH:17s 2.0 μm LmV:14s 1.8 μm
				MPV=5.5 MLH=5.4 MLV=5.5
24.	e(P)	A	15 23 15	Gulf of Alaska 57.25°N 149.50°W
	eX	A	23 21.5	H = 15 11 53.2 h = normal MAG=4.3 (USCGS)
	e	A	23 26.5	D = 71.2°
				XV:1.2s 15.3nm
				The first motion of P must be about 5 s earlier.
24.	eP1	A	20 13 55.5	Leeward Islands 17.38°N 61.11°W
	iP2	A	13 59	H = 20 03 10.9 h = 24 km MAG=6.4
	iPcP	B	14 25	D = 65.8° Az = 41.8° (USCGS)
	eiS	B	22 42.5	P1V:1.2s 25.5nm P2V:2.2s 927nm
	iS	B	22 46.5	SH:16s 5.7 μm ScSH:15.5s 8.8 μm
	i(PPS)	C	23 16	XV:16s 5.1 μm PKPPKPV:2.1s 60.5nm
	iScS	B	23 42	LmH:19s 21.9 μm LmV:19s 31.6 μm
	iX	B	24 08	MP1V=5.3 MP2V=6.6 MSH=6.5 MLH=6.4
	eiSS	C	27 16	MLV=6.6
	LmH	B	37.9	Probably multiple shock in the same focus area.
	LmV	B	38.2	
	ePKPPKP	A	42 35	

Day	Phase		h m s	Remarks
24.	-iP	A	21 43 17.5	Leeward Islands 17.43°N 61.26°W
	LmH	B	22 06.7	H = 21 32 31.3 h = 20 km MAG=5.9
	LmV	B	07	D = 65.9° Az = 41.8° (USCGS)
				PV:2.1s 242nm
				LmH:19s 5.2 μm LmV:20s 5.2 μm
				MPV=6.0 MLH=5.8 MLV=5.8
24.	eP	A	23 59 53	India 17.54°N 73.88°E
				H = 23 49 53.9 h = normal MAG=5.5
				D = 59.1° Az = 318.9° (USCGS)
				PV:0.6s 11.9nm
				MPV=5.2
25.	ePKIKP	A	01 42 28.5	New Ireland 5.28°S 153.67°E
	+eiPP	B	44 18.5	H = 01 23 33.6 h = 64 km MAG=5.7
	eiSKKS	C	51 11	D = 124.8° Az = 331.5° (USCGS)
	eiS	C	52 06	PKIKPV:1.3s 25.0nm PPV:21s 9 μm
	iPS	C	54 13	PPH:21s 6.4 μm
	i(SPP)	C	55 40	LmH:21s 65.0 μm LmV:21s 76.6 μm
	eiSS	C	02 01 12	MPPV=6.6 MPPH=6.8 MLH=7.3 MLV=7.3
	LmV	B	36.7	ei 42 31 e 42 32.5
	LmH	B	38.4	S is the diffracted S around the core.
25.	e(pP)	A	10 55 34	Near Coast of Northern Chile
	i	A	56 23.5	21.48°S 70.37°W
	ePP	A	59 34	H = 10 41 31.6 h = 53 km MAG=5.8
	eiSKS	B	11 05 55	D = 101.4° Az = 40.0° (USCGS)
	eS	C	06 58	PPV:2.5s 102nm SH:14s 1.4 μm
	eiPS	C	08 28	LmH:21s 3.7 μm LmV:22s 4.7 μm
	eiSS	C	13 52	MPPV=5.9 MSH=6.0 MLH=5.8 MLV=5.9
	LmH	B	38.5	e 13 58 e 19 (18)
	LmV	B	39	P must be about 14 s earlier than (pP).
25.	e	A	11 04 07	
25.	eP	A	21 29 34.5	North of Ascension Island 1.90°S 12.73°W
	LmH	C	54.2	H = 21 19 52.4 h = normal MAG=5.0
	LmV	C	55	D = 56.3° Az = 18.4° (USCGS)
				LmH:20s 0.8 μm LmV:18s 0.8 μm
				MLH=4.8 MLV=4.9

Day	Phase		h m s	Remarks
26.	eP	A	09 41 43	Off Coast of Oregon 44.48°N 129.69°W
	LmH	B	10 19.2	H = 09 29 38.5 h = normal MAG=5.1
	LmV	B	19.2	D = 79.5° Az = 23.9° (USCGS)
				PV:1.4s 15.3nm
				LmH:16s 1.4 µm LmV:16s 1.5 µm
				MPV=4.9 MLH=5.4 MLV=5.5
26.	LmV	C	10 06.5	New Ireland Region 5.15°S 153.71°E
				H = 08 52 42.3 h = 59 km MAG=5.2 (USCGS)
				D = 124.7°
				LmV:20s 1.2 µm
				MLV=4.5
26.	LmH	C	10 41.6	Off Coast of Oregon (USCGS)
26.	eP	A	10 52 46	Off Coast of Oregon 44.50°N 129.70°W
	e	A	52 50	H = 10 40 40.6 h = normal MAG=5.0
	LmH	C	11 30	D = 79.5° Az = 23.9° (USCGS)
	LmV	C	30	
26.	ePKP2	A	14 55 04.5	South of Kermadec Islands 32.04°S 177.99°W
				H = 14 34 27.8 h = normal MAG=4.7 (USCGS)
				D = 160.0°
				PKP2V:1.1s 14.4nm
26.	eP	A	22 35 54	Near Islands/Aleutian 51.68°N 174.54°E
				H = 22 24 03.3 h = normal MAG=4.6
				D = 77.0° Az = 348.9° (USCGS)
				PV:1.3s 8.3nm
				MPV=4.9
27.	LmH	C	03 14	Probably to South Pacific Cordillera
	LmV	C	14.3	54.8°S 132.8°W
				H = 01 48 10.7 h = normal MAG=4.8 (USCGS)
				D = 158.3°
				LmH:18s 0.5 µm LmV:20s 0.5 µm
				(MLH=5.3 MLV=5.4)

Day	Phase		h m s	Remarks
27.	LmH	C	03 24.4	Probably to Solomon Islands 5.2°S 154.0°E
	LmV	C	24.9	H = 02 11 26.5 h = 113 km (USCGS)
				D = 124.9°
				LmH:18s 0.5 /um LmV:18s 0.5 /um
27.	e	A	04 47 40	
27.	e(P)	A	07 21 10	TRI: ePn 20 01.8 D = 700 km
				KHC: eP 20 43
27.	-eP	A	09 31 26	Chile - Bolivia Border Region
	-iP	B	31 27	21.20°S 68.30°W
	+eipP	B	31 58	H = 09 17 55.7 h = 135 km MAG=6.4
	iPP	A	35 27.7	D = 100.0° Az = 39.5° (USCGS);
	epPP	B	36 00	h = 125 km
	iSKS	B	42 00	PV(A):1.7s 105nm PV(B):6s 0.9 /um
	iSKKS	B	42 28	PPV:2.0s 139nm SPH(C):18.5s 9.7 /um
	isSKS	B	42 55	SPV(B):14.5s 6.3 /um (PPS)H:25s 11.1 /um
	eiSP	C	44 15	PKKPV:1.9s 41.2nm (pPKKP)V:2.0s 79.4nm
	iSP	B	44 20	LmH:20s 5.8 /um LmV:19s 7.3 /um
	i(PPS)	C	45 14	MPV(A)=6.1 MPV(B)=6.4 MPPV=6.1
	ePKKP	A	47 45	
	e(pPKKP)	A	48 12	
	LmH	B	10 14.5	
	LmV	B	14.7	
27.	LmH	C	15 45.5	New Guinea 3.5°S 141.3°E
				H = 14 39 08.5 h = 36 km MAG=5.1 (USCGS)
				D = 116.9°
				LmH:22s 0.6 /um
				MLH=5.1
27.	-ePKIKP	A	16 42 34	Tonga Islands 22.34°S 174.78°W
	+iPKHKP	A	42 41.5	H = 16 22 48.5 h = normal MAG=6.1
	+iPKP2	A	42 49.5	D = 151.3° Az = 351.5° (USCGS)
	eiSKKS	C	53 14	PKIKPV:1.8s 76.5nm PKHKPV:1.8s 622nm
	e(PSKS)	C	56 12	PKP2V:1.6s 334nm
	eiPPS	C	59 22	LmH:19s 4.2 /um LmV:19s 5.4 /um
	eSS	C	17 05(44)	MLH=6.2 MLV=6.4
	eSSS	C	11 20	e 42 36.5 i 42 43.0 ei 43 00

Day	Phase		h m s	Remarks
cont.				
27.	LmH	B	17 57	
	LmV	B	18 14.7	
27.	ePKHKP	A	19 45 56	Tonga Islands 21.82°S 174.76°W
	e	A	46 08	H = 19 26 05.9 h = normal MAG=4.6
				D = 150.8° Az = 351.7° (USCGS)
28.	ePKP	A	00 17 57	Tonga Islands 16.38°S 174.01°W
				H = 23 58 34.2 h = 150 km MAG=4.2
				D = 145.5° Az = 353.7° (USCGS)
				PKPV:1.2s 15.3nm
28.	eP	A	06 30 48.5	
28.	eP	A	06 38 23.5	Off Coast of Oregon 44.18°N 128.84°W
	eS	B	48 22	H = 06 26 15.8 h = normal MAG=5.4
	eSS	C	53 30	D = 79.5° Az = 24.3° (USCGS)
	eLQ	C	58.6	PV:2.4s 102nm SH(C):25s 2.1 /um
	LmH	B	07 15.9	LmH:18s 5.7 /um LmV:17s 7.2 /um
	LmV	B	16.1	MPV=5.5 MSH=5.8 MLH=6.0 MLV=6.1
28.	eP	A	07 13 45	Off Coast of Oregon 44.2°N 129.0°W
	e	A	13 49	H = 07 01 36.8 h = normal MAG=4.9 (USCGS)
	LmH	B	51	D = 79.5°
	LmV	B	51.3	LmH:16.5s 1.6 /um LmV:16s 1.6 /um
				MLH=5.5 MLV=5.5
28.	e	A	12 16 13	
28.	e(Sg)	A	12 49 43	Explosion?
				KHC gives: iPg 48 26.5 iSg 48 45.5
				D = 1.5°
28.	eP	A	22 23 42	Off Coast of Oregon 44.24°N 128.90°N
	i	A	23 49.4	H = 22 11 33.9 h = normal MAG=5.0
	LmH	B	23 01.3	D = 79.5° Az = 24.3° (USCGS)
	LmV	B	01.3	LmH:16s 1.3 /um LmV:16s 1.2 /um
				MLH=5.4 MLV=5.4

Day	Phase		h m s	Remarks
29.	iPg	A	10 35 29.8	Explosion
	iSg	A	35 46.0	D = ca. 1.2°
29.	eP	A	12 47 41.5	Near East Coast of Kamchatka 56.33°N 163.93°E H = 12 36 26.4 h = 36 km MAG=4.6 D = 70.9° Az = 341.8° (USCGS)
29.	e	A	12 53 42	
29.	e(P)	A	18 33 56	(P)V:1.2s 15.3nm
29.	eP1	A	19 52 02.5	Albania - Yugoslavia 41.46°N 20.40°E
	iP2	A	52 05.5	H = 19 49 23.6 h = 39 km MAG=4.7
	iP3	A	52 11.5	D = 11.0° Az = 329.5° (USCGS)
	LmH	B	55.7	P2V:1.2s 35.7nm
	LmV	C	56.6	LmH:15s 2.2/um LmV:11s 0.8/um MLH=4.2 Multiple P.
29.	ePKHKP	A	20 49 25	Tonga Islands 22.82°S 175.28°W
	ePSKS	C	21 03.3	H = 20 29 32.2 h = 30 km MAG=5.3
	iSS	C	12 40	D = 151.7° Az = 350.7° (USCGS)
	eSSS	C	18 20	PKHKPV:1.5s 20.2nm
	LmH	B	22 04.7	LmH:17s 1.3/um LmV:20s 2.0/um
	LmV	B	05.4	MLH=5.7 MLV=5.9
29.	ePKHKP	A	22 42 58.5	Tonga Islands 22.71°S 175.23°W
	ePKP2	A	43 09	H = 22 23 06.0 h = normal MAG=5.1
	LmH	C	24 08.5	D = 151.6° Az = 350.8° (USCGS) PKP2V:1.5s 23.5nm
29.	LmH	B	23 01.2	Albania - Yugoslavia H = 22 54.9 (UPP)
30.	ePn	A	04 20 47	Northern Italy 44.74°N 12.17°E
	iPg	B	21 18	H = 04 19 21.2 h = normal MAG=5.3
	iSn	B	21 54	D = 5.9° Az = 356.6° (USCGS)
	iSg	B	22 36	LmH:8.5s 24.7/um LmV:10.5s 24.0/um

Day	Phase		h m s	Remarks
cont.				
30.	LmH	B	04 22.9	MLH=4.9
	LmV	B	23.6	Multiple P in our short-period records.
30.	eP	A	21 30 16	Greece 40.7°N 21.6°E
	e	A	30 22	H = 21 27 21.4 h = 53 km MAG=4.3 (USCGS)
	e	A	30 30	D = 12.2°
	e	A	30 56	PV:1.4s 12.3nm
31.	eP	A	02 41 37	Aleutian 51.9°N 171.8°W
	e	A	41 39	H = 02 29 41.2 h = 40 km MAG=4.9 (USCGS)
	epP	A	41 51	D = 77.8°
31.	LmH	C	07 08.5	Probably to South of Greece (UPP).
31.	e	A	11 03 53.5	
31.	ePKIKP	A	15 24 38	Solomon Islands 7.14°S 154.84°E
	LmH	C	16 25.5	H = 15 05 32.3 h = 19 km MAG=5.4
				D = 127.0° Az = 331.5° (USCGS)

A Study of Relative Frequencies of Body-Wave Onsets in Seismic Registrations of the Station Moxa

by

PETER BORMANN[1])

1. Aim of the Researches

The clearness of seismic onsets is above all dependent on the difference of the periods and amplitudes of the signals to those of the noise, i. e. of signal noise ratio (SNR). Supposing that on an average equally distinct onsets are interpreted with the same frequency in registrations of the same type (short-period, broad-band, long-period respectively), then significant frequency differences are a hint as to the different spectres of periods and amplitudes of the corresponding wave onsets. The spectres of periods and amplitudes of the recorded seismic waves are dependent upon the character of the focal mechanism, on the released seismic energy, of the tectonics and the medium in the focal region, of the focal depth h, of the kind of the wave spreading in space, of the epicentral distance D, of the influences along the wave path (attenuation, dispersion, diffraction, reflection, mode conversation), of the station underground and the response characteristics of the seismographs. Only part of the factors of influence are well-known and can be taken into consideration in appropriate form in statistic analysis.

First we are interested in the frequency of the registration of all significant body wave phases at the station Moxa dependent on the focal region, epicentral distance, focal depth and the type of the seismograph. Frequency distributions attained in this way offers the possibility of evaluating weighting factors for the interpretation of seismic phases as well as of criteria for the identification and location of seismic events. In connection with a catalogue of the types of seismic records they also represent an essential help for bulletin work, particu-

[1]) Mitteilung aus dem Zentralinstitut Physik der Erde Nr. 137

larly in case that this work is to be transferred to technical forces without any loss of size or information content. Apart from this the results of the frequency analyses give valuable clues for a purposive study of details of the quantitative differences in the spectres of periods and amplitudes, which cause those differences in distinctness and thus frequency. Their finding, classification and statistic securing are conditions for an exploration of their causes. The knowledge of corresponding standard and their range of variation is of significance for a routine computer interpretation of the seismogrammes, since an unambiguous interpretation of the phases is not always possible in the light of the onset times of the seismic waves.

2. Starting Material and Preliminary Results of Frequency Analyses

Starting material for frequency analysis was the detail interpretations of the registrations performed by the author for the seismological bulletins of the station Moxa in 1965, 1966, and 1967. All evaluations of earthquakes with MAG $\geqq 5{,}5$ were taken into consideration. The SNR, as a rule, is then big enough to evaluate several body wave onsets in all three standard registrations of the type A, B, and C (vid. page 13). Some of the most interesting preliminary results of frequency analyses are summed up in the tables 1—4, from which interpretation aids and location criteria can be deduced for the registrations of the station Moxa. A graphic description of the frequencies of interpretation of seismic body wave onsets in registrations of earthquakes in various regions is given in the fig. 1—13.

Table 1

Relative frequencies of interpretation of the most significant deep earthquake phases in teleseismic registrations of earthquakes in different focal depth intervals

Phase	h < 70 km	70 km ≦ h ≦ 300 km	h > 300 km
pP, pPKP	28%	60%	67%
sP, sPKP	7%	18%	12%
pPP	3%	18%	12%
sS	<1%	13%	14%

Table 2
Significant differences in the distribution of focal depths of earthquakes registered from different regions equally distant from Moxa (error probability $\pi \leqq 10\%$)

Focal depths [km]	Frequency [%]	Region	D [°]
$70 \leqq h \leqq 300$	78 0	Hindukush all other regions	41…46
$h > 300$	57 0	N-Argentina all other regions	97…105
	22 0	Banda Sea all other regions	110…115
	66 0	Fiji Islands all other regions	144…156

Table 3

Significant differences in the frequency of the interpretation of equal phases in registrations of earthquakes from various regions equally distant from Moxa ($\pi \leqq 10\%$)

Phase	Frequency [%]	Region	D [°]
P	100	Philippine Is.	86…102
	60	Vulcano Is., Bonin Is., Mariana Is.	90…105
PP	46/33	Hokkaido/Aleutian Is.	75…79
	14	Kurile Is.	75…79
	100	Banda Sea	110…115
	22	South Sandwich Is.	110…115
S	39	W-Aleutian Is.	75…78
	68	E-Aleutian Is.	76…79
SKS	60	South America	92…110
	20	SE-Indonesia	100…110
PS	53	South America	92…110
	17	SE-Indonesia	100…110
SS	20	W-Aleutian Is.	75…78
	68	E-Aleutian Is.	76…79

Table 4

Frequency ratios of some phase pairs in registrations of earthquakes from various regions, equally distant from Moxa

Phase ratio	Frequency ratio	Region	D [°]
PP/P	0,16 0,50	Kurile Is. Hokkaido	75···79 77···79
	0,68 1,5	Philippine Is. Vulcano Is., Bonin Is., Mariana Is.	86···102 90···105
PP/S	0,26 0,60	Kurile Is. Hokkaido	75···79 77···79
	0,35 0,79	E-Aleutian Is. W-Aleutian Is.	76···79 75···78
	1,0 2,0	Philippine Is. Vulcano Is., Bonin Is., Mariana Is.	86···102 90···105
	1,5 3,0	South Sandwich Is. Banda Sea	110···115 110···115
PP/SS	0,75 1,80	South Sandwich Is. Banda Sea	110···115 110···115
PP/PS	0,69 1,64	South Sandwich Is. Banda Sea	110···115 110···115
PS/S	0,18 0,48	E-Aleutian Is. W-Aleutian Is.	76···79 75···78
	0,24 1,0	Philippine Is. Vulcano Is., Bonin Is., Mariana Is.	86···102 90···105
PS/SS	0,18 0,90	E-Aleutian Is. W-Aleutian Is.	76···79 75···78

Türkei, Kaukasus, östl. Mittelmeer
$20° \leq D < 26°$ $n = 10$

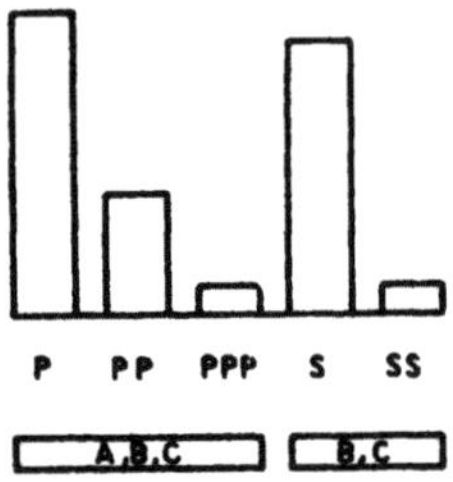

Rotes Meer, Golf v. Aden, Afrika
$36{,}5° < D < 69°$ $n = 16$

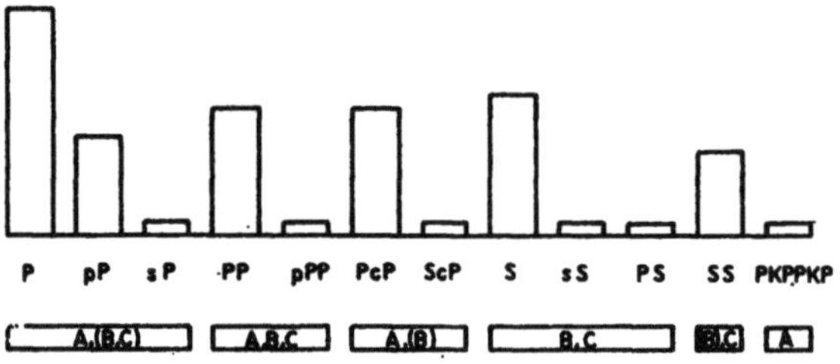

Fig. 1. Frequencies of the interpretation of body wave phases related to the total number in the earthquakes studied (1st column) with a hint to the type of the seismograph, by which these phases are registrated more or less distinctly (symbols in parentheses).
White field: $h < 70$ km, △: $70 \text{ km} \leq h \leq 300$ km, ◇: $h > 300$ km

Persischer Golf, S-Iran, Afghanistan, W-Pakistan, Hindukusch, Kaschmir, Pamir, Tientschan, Sinkiang Prov., mittelasiatische Unionsrepubliken der SU

$35^\circ < D < 50^\circ \quad n = 40$

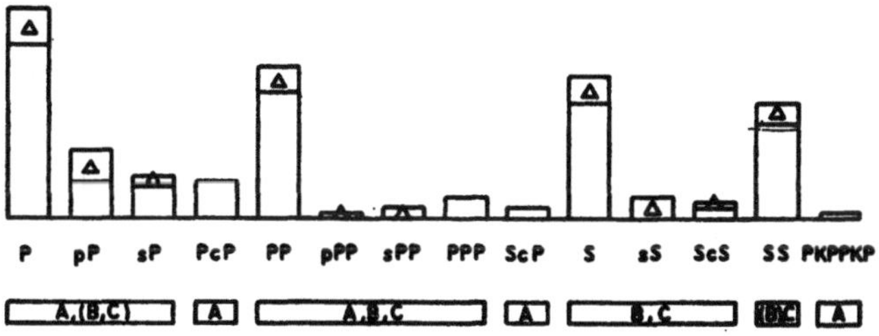

Himalaja, Nepal, Burma, Tibet, E-China, Mongolei, Pribaikal

$50^\circ < D < 74^\circ \quad n = 31$

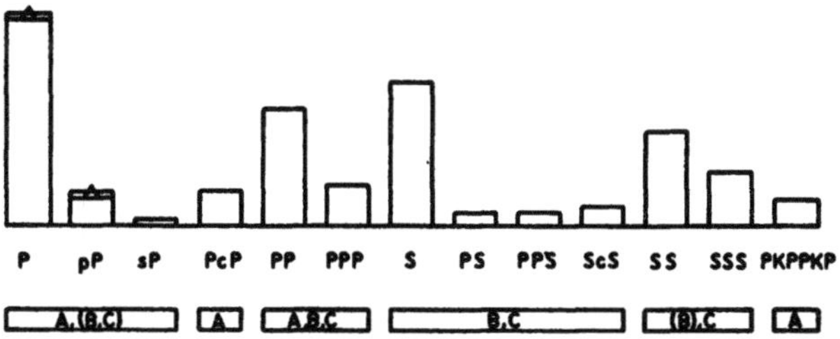

Fig. 2. Comments vid. fig. 1

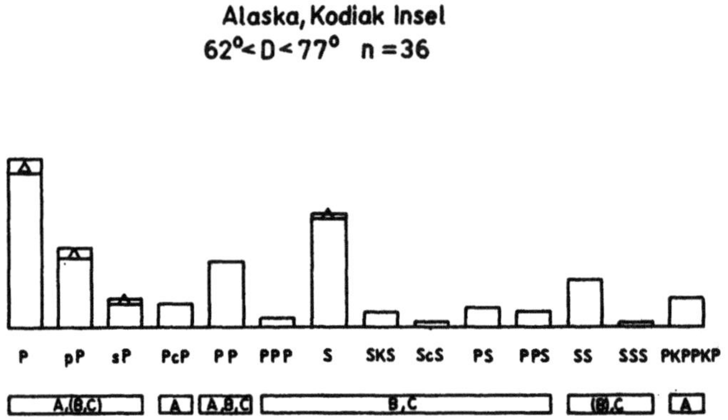

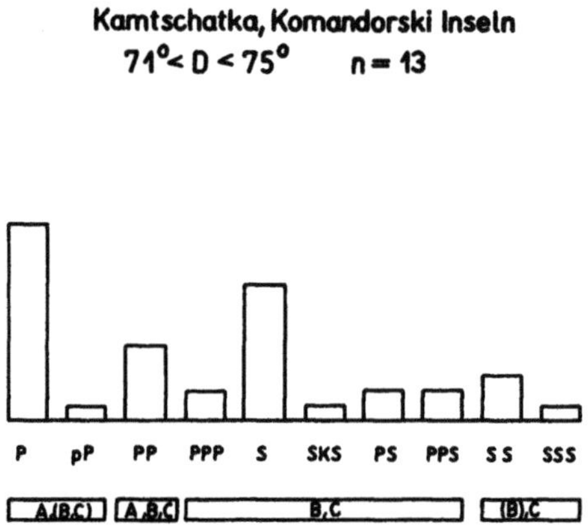

Fig. 3. Comments vid. fig. 1

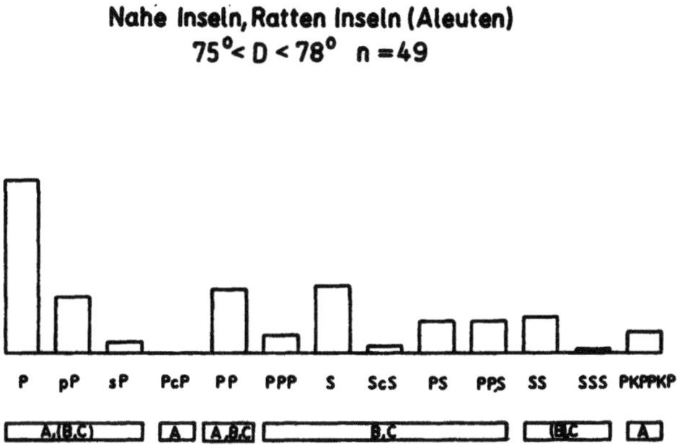

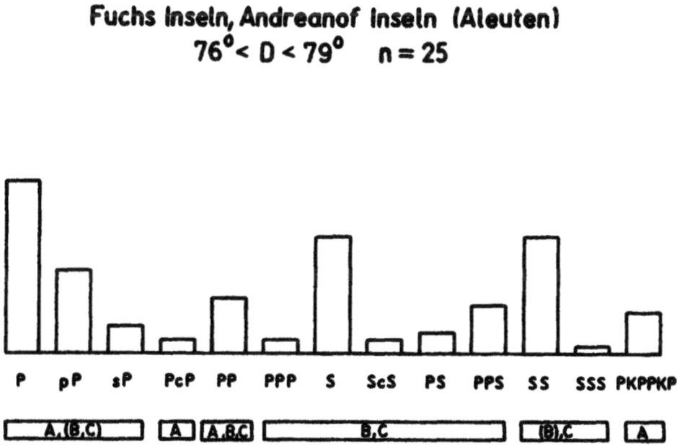

Fig. 4. Comments vid. fig. 1

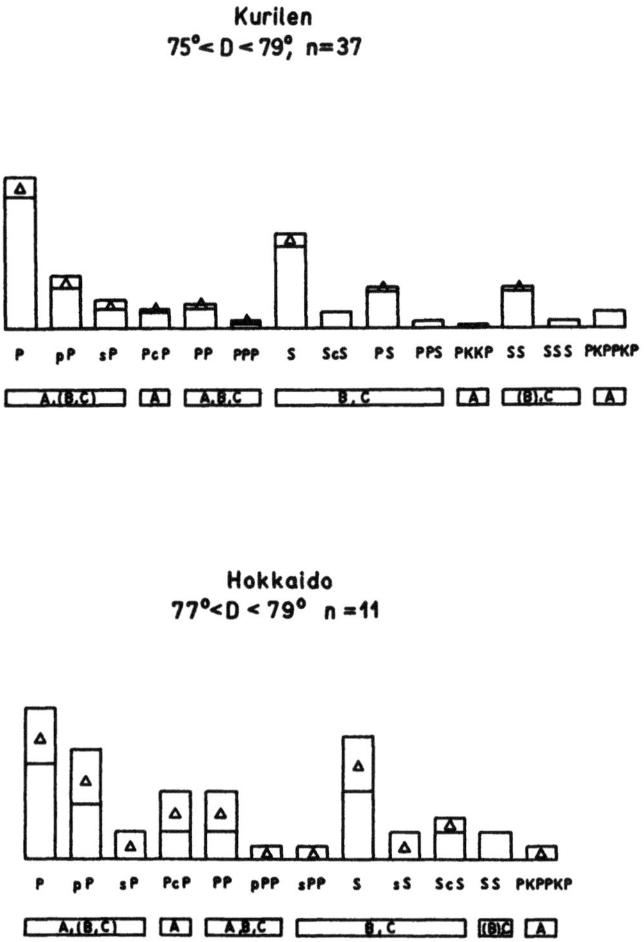

Fig. 5. Comments vid. fig. 1

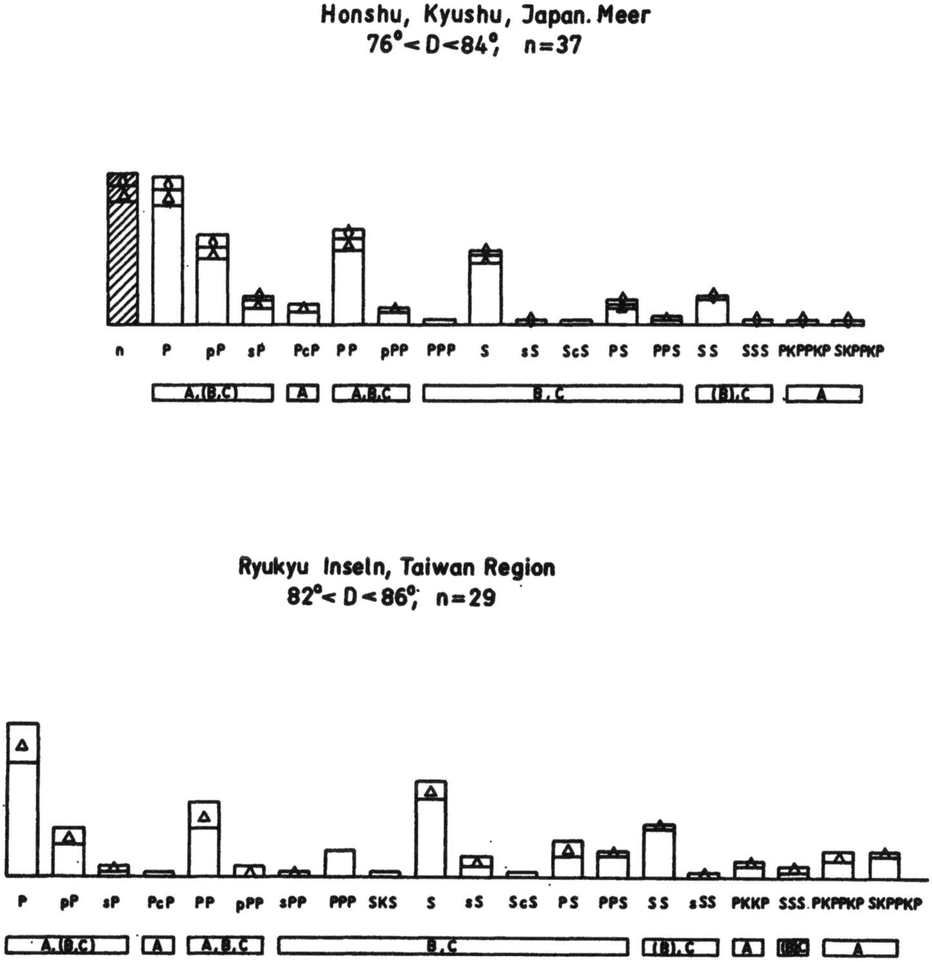

Fig. 6. Comments vid. fig. 1

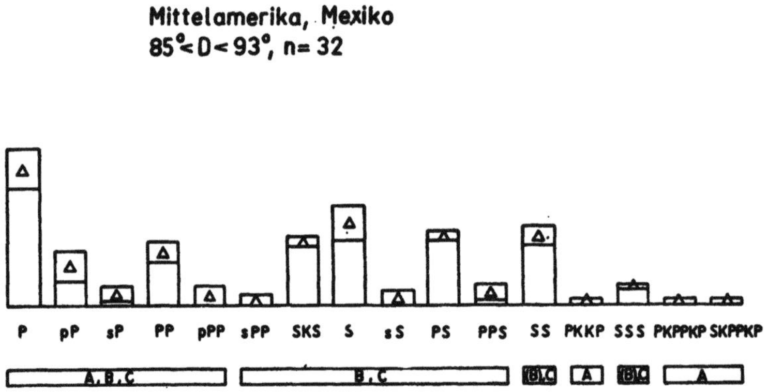

Fig. 7. Comments vid. fig. 1

Philippinen
86°< D < 102° n = 25

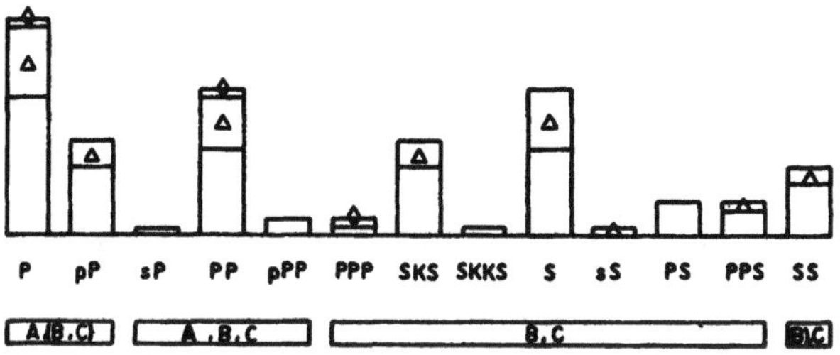

Vulkan Inseln, Bonin Inseln, Marianen
90°< D < 105° n=20

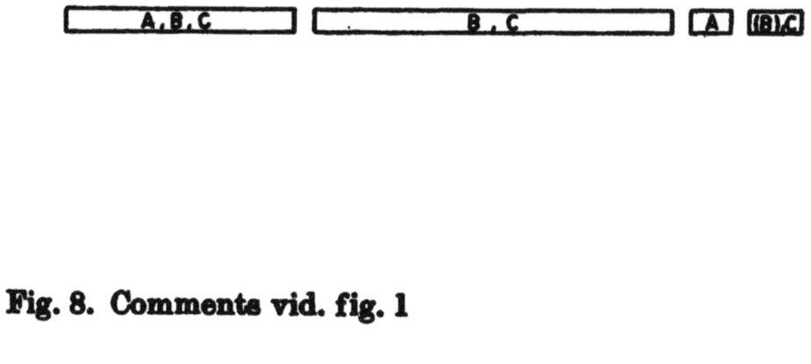

Fig. 8. Comments vid. fig. 1

Zentral – u. Nordchile, Argentinien, Peru u. Grenzgebiete
92° < D < 110°, n = 43

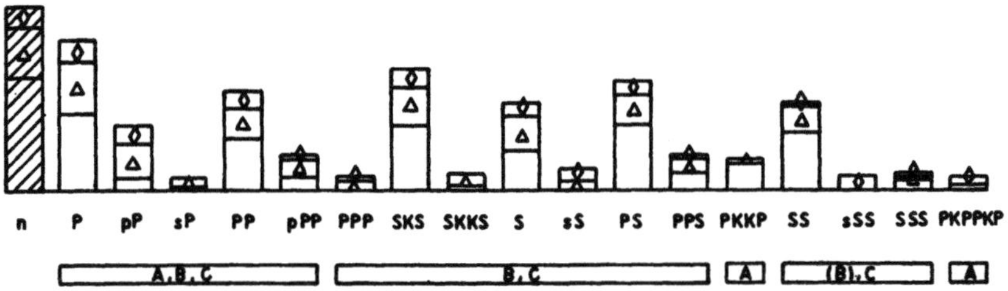

Halmahera, Celebes, Molukken, Java
100° < D < 110° n = 30

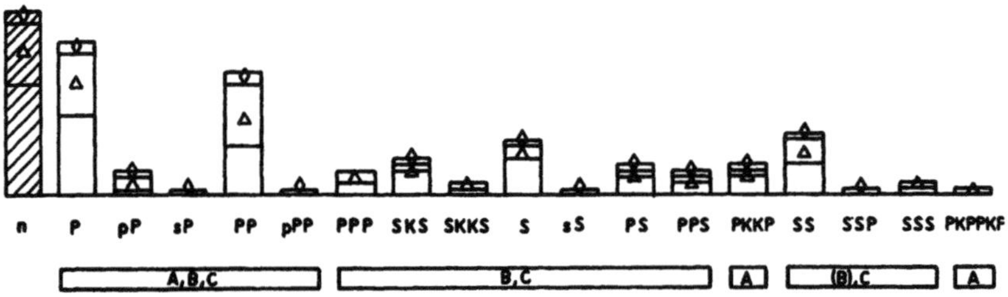

Fig. 9. Comments vid. fig. 1

Süd Sandwich Inseln, Süd Georgien
110° ≦ D < 115°, n = 20

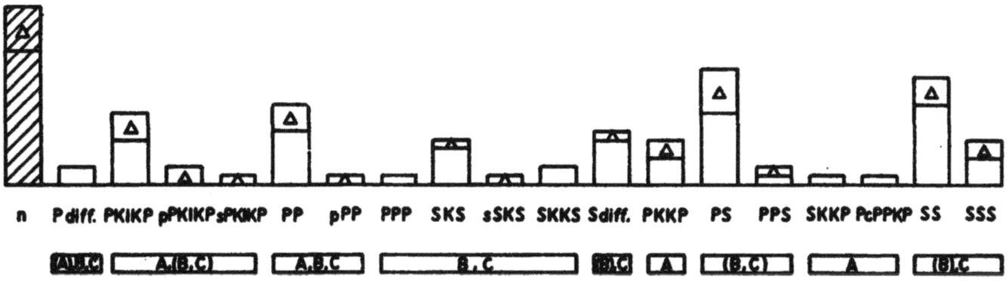

Banda See
110° ≦ D < 115°, n = 18

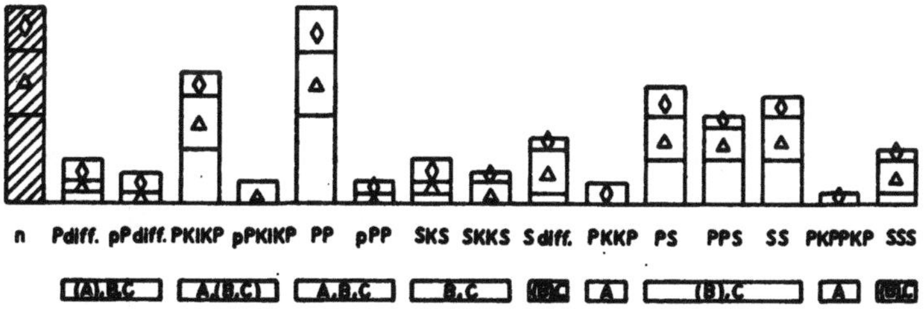

Fig. 10. Comments vid. fig. 1

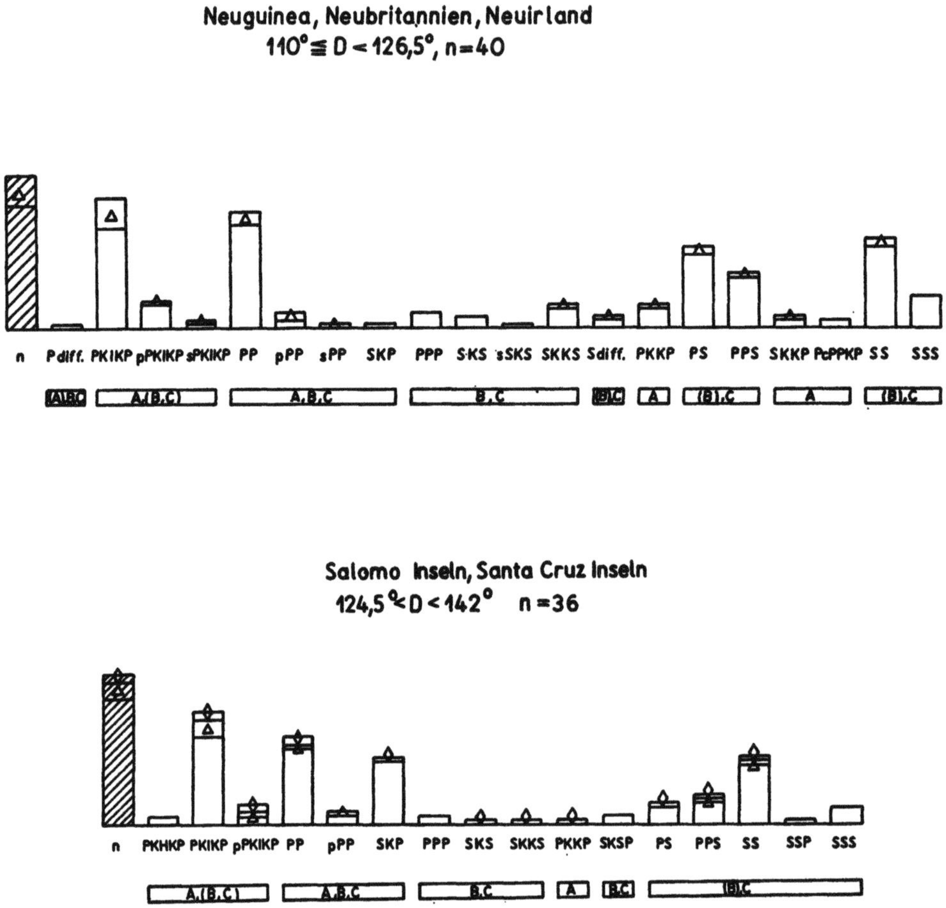

Fig. 11. Comments vid. fig. 1

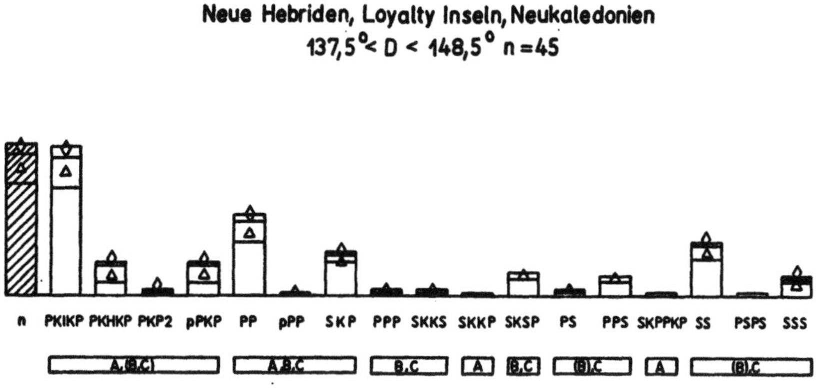

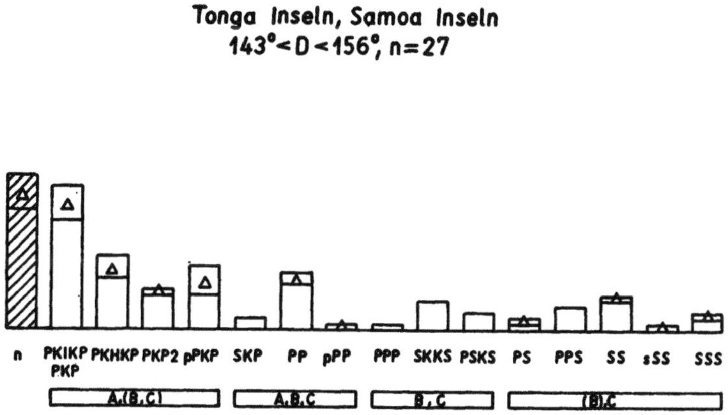

Fig. 12. Comments vid. fig. 1

Fiji Inseln
144°< D < 156°, n=55

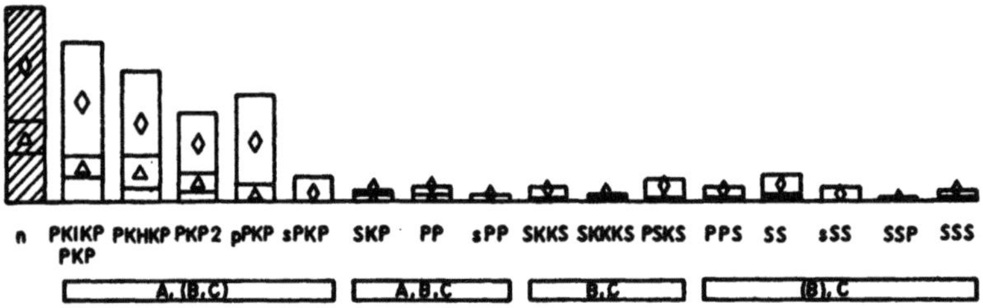

Kermadec Inseln, Neuseeland, Macquarie Inseln, Balleny Inseln
152°< D < 166° n = 24

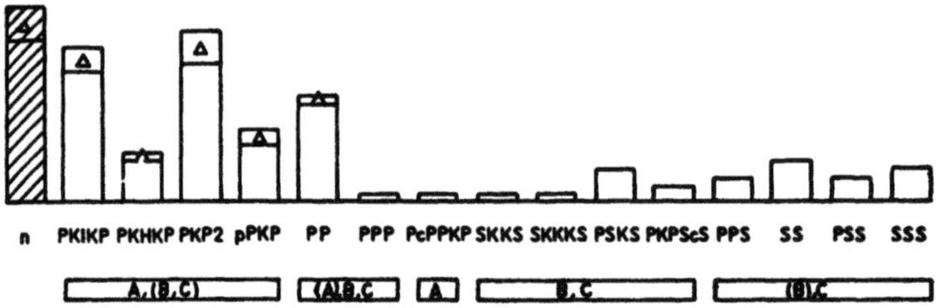

Fig. 13. Comments vid. fig. 1

Bei Fragen zur Produktsicherheit wenden Sie sich bitte an:
If you have any questions regarding product safety,
please contact:

Walter de Gruyter GmbH
Genthiner Straße 13
10785 Berlin
productsafety@degruyterbrill.com